T0155739

NEUROMETHODS

Series Editor
Wolfgang Walz
University of Saskatchewan
Saskatoon, SK, Canada

For further volumes:
http://www.springer.com/series/7657

Electrophysiological Analysis of Synaptic Transmission

Nicholas Graziane

Neuroscience Department, University of Pittsburgh, Pittsburgh, PA, USA

Yan Dong

Neuroscience Department, University of Pittsburgh, Pittsburgh, PA, USA

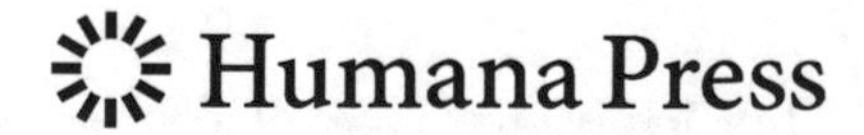

Nicholas Graziane
Neuroscience Department
University of Pittsburgh
Pittsburgh, PA, USA

Yan Dong
Neuroscience Department
University of Pittsburgh
Pittsburgh, PA, USA

ISSN 0893-2336 ISSN 1940-6045 (electronic)
Neuromethods
ISBN 978-1-4939-8009-3 ISBN 978-1-4939-3274-0 (eBook)
DOI 10.1007/978-1-4939-3274-0

Springer New York Heidelberg Dordrecht London

Softcover re-print of the Hardcover 1st edition 2016

Printed on acid-free paper

Humana Press is a brand of Springer
Springer Science+Business Media LLC New York is part of Springer Science+Business Media (www.springer.com)

Series Preface

Experimental life sciences have two basic foundations: concepts and tools. The *Neuromethods* series focuses on the tools and techniques unique to the investigation of the nervous system and excitable cells. It will not, however, shortchange the concept side of things as care has been taken to integrate these tools within the context of the concepts and questions under investigation. In this way, the series is unique in that it not only collects protocols but also includes theoretical background information and critiques which led to the methods and their development. Thus it gives the reader a better understanding of the origin of the techniques and their potential future development. The *Neuromethods* publishing program strikes a balance between recent and exciting developments like those concerning new animal models of disease, imaging, in vivo methods, and more established techniques, including, for example, immunocytochemistry and electrophysiological technologies. New trainees in neurosciences still need a sound footing in these older methods in order to apply a critical approach to their results.

Under the guidance of its founders, Alan Boulton and Glen Baker, the *Neuromethods* series has been a success since its first volume published through Humana Press in 1985. The series continues to flourish through many changes over the years. It is now published under the umbrella of Springer Protocols. While methods involving brain research have changed a lot since the series started, the publishing environment and technology have changed even more radically. Neuromethods has the distinct layout and style of the Springer Protocols program, designed specifically for readability and ease of reference in a laboratory setting.

The careful application of methods is potentially the most important step in the process of scientific inquiry. In the past, new methodologies led the way in developing new disciplines in the biological and medical sciences. For example, Physiology emerged out of Anatomy in the nineteenth century by harnessing new methods based on the newly discovered phenomenon of electricity. Nowadays, the relationships between disciplines and methods are more complex. Methods are now widely shared between disciplines and research areas. New developments in electronic publishing make it possible for scientists that encounter new methods to quickly find sources of information electronically. The design of individual volumes and chapters in this series takes this new access technology into account. Springer Protocols makes it possible to download single protocols separately. In addition, Springer makes its print-on-demand technology available globally. A print copy can therefore be acquired quickly and for a competitive price anywhere in the world.

Saskatoon, Canada ***Wolfgang Walz***

Overview

The central nervous system comprising the brain and spinal cord controls vertebrate responses to external stimuli. For example, our ancestors needed the central nervous system to visually locate predators and respond by making quick motor movements to a safe place. Simply reading this book requires information to be processed in the brain and stored so that it can be implemented at a future time in a research setting. This constant processing and responding to our external environment is mostly controlled by electrical signals, which form a quick and efficient means of communication between distinct regions of the central nervous system.

In modern neuroscience, electrophysiology often refers to the study or research approach implemented to investigate the electrical properties of brain cells and tissues. It is sensitive enough to measure basal voltage/potential or electrical currents from a single ion channel or a whole cell, while also possessing the ability to study electrical processing within an entire brain region.

The brain functions by integrating and computing electrical events from individual brain cells (including nervous cells and glial cells) and communication between these cells. These electrical events sometimes happen slowly, but most of the time they occur very fast, within the submillisecond to millisecond range. Electrophysiology is one of the finest approaches capable of detecting and analyzing these electrical events during the endeavor of understanding the function of neurons, neural systems, and the brain.

Specifically, for studying synaptic transmission, electrophysiology offers great advantages that other techniques may not provide. First, and most important in our opinion, electrophysiology detects functional readouts of synaptic transmission. Under the physiological and pathophysiological conditions, the number of synapse-like structures is often much larger than the number of synapses that are actually functional. By definition, a synapse is the connection between cells, functioning to transfer signals from one cell to another. Thus, although abundant synapse-like structures are observed, it cannot be concluded that transmission between these two cells is effective. In a typical electrophysiological measurement, only functional synapses, namely the synapses that mediate synaptic currents, are detected and recorded. To some extent, electrophysiology measures the electrical consequences of activating synapses. This property of the electrophysiological approach is significant for several reasons. First, it can detect synaptic efficacy. Under different physiological conditions or over different developmental phases, central synapses exhibit different efficacies, which are regulated by key mechanisms underlying synaptic plasticity and maturation. Many of these regulations occur at the molecular level, e.g., phosphorylation of synaptic proteins, without changing the shape/structure of synapses. Therefore, these changes can be readily and reliably detected electrophysiologically. An extreme example is that some synapses are generated but remain dormant; they do not conduct reliable synaptic activity and thus are non- or semi-functional. During development or upon experience-dependent regulations, these synapses may evolve into fully functional synapses or, under other circumstances, completely lose their function and are primed for pruning/removal. These dynamic processes can also be detected electrophysiologically. We will discuss this and other related procedures thoroughly in this book.

Second, electrophysiology measures the real timing of synaptic transmission. Timing is everything, and this is also true for synaptic transmission. From start to end (neurotransmitter release stimulated by a presynaptic action potential to postsynaptic receptor response to neurotransmitter release), synaptic transmission can be accomplished as fast as within a few milliseconds. In most central neurons, depolarization induced from a single excitatory synapse is usually not sufficient to trigger action potentials in the postsynaptic neuron. In order to trigger an action potential in a postsynaptic cell, a few synapses must be activated within the same time window in order to generate summed depolarization that is large enough to push the postsynaptic cell beyond the threshold of an action potential. With certain manipulations, electrophysiological measurements can distinguish the contribution of each synapse to the summed depolarization, and thus determine how effective/efficient the isolated set of synapses contributes to the temporal summation.

Third, electrophysiology sensitively measures the functional changes at synapses. Most synapse neuroscientists hold the hypothesis that external experiences change the function of synapses, thus reshaping future behaviors. Much of this hypothesis has been formulated based on electrophysiological studies of long-term potentiation (LTP) and long-term depression (LTD), two forms of long-lasting synaptic modifications critically contributing to learning and memory. Electrophysiology is powerful in cellular models of LTP and LTD in which the efficacy or strength of synaptic transmission is continuously monitored for several tens of minutes and sometimes hours. As such, an increase or decrease of synaptic efficacy can be detected during and after experimental manipulations mimicking external stimulation.

Fourth, combined with other techniques, electrophysiology can address very sophisticated molecular and cellular questions of synaptic transmission. For example, using current technology, neurotransmitter receptors can be easily labeled fluorescently such that their location and movement can be monitored in real time using imaging methods. The correlative synaptic recruitment/internalization of these receptors and increases/decreases in synaptic strength have been used as strong evidence showing the postsynaptic mechanism of synaptic plasticity.

Like all other experimental approaches, electrophysiology has its limitations in examining synaptic transmission. Probably the most glaring one is that it cannot unequivocally distinguish pre- vs. postsynaptic alterations upon changes in synaptic efficacy. Electrophysiological data alone can often be interpreted either way, although sometimes one way appears to be more parsimonious than the other. This limitation is indeed one of the major reasons for the decade-long debate of the expression mechanisms underlying LTP.

Another limitation is that spatial and temporal effects are often involved in synaptic electrophysiological experiments. If these factors are not sufficiently considered, false conclusions can be drawn. To help the new electrophysiologist avoid such mistakes, we make sure that the potential experimental and interpretational caveats are thoroughly discussed in each electrophysiological approach we introduce.

There are many outstanding books and published manuscripts, which are cited throughout this book, explaining the many facets of electrophysiology used to better assist scholars lacking accessibility to an experienced electrophysiologist. Our goal in writing this book is to create a guide, which introduces and highlights important topics in the field while at the same time attempting to extend these topics to practical electrophysiological approaches through the perspective of the two authors. We expect that this book holds enough technical information for graduate students or junior postdoctoral fellows to get started in their journey of synapse neuroscience.

The two authors, Dr. Nicholas Graziane and Dr. Yan Dong, each have over 10 years of experience in the field of electrophysiology. Dr. Graziane started his research career in 2003 and he continues to perform electrophysiologically relevant research as a senior postdoctoral scholar in Dr. Yan Dong's laboratory. Dr. Dong had been running an active electrophysiology laboratory and teaching electrophysiology for ~15 years. The partnership of these two authors was specifically designed to provide a student-friendly electrophysiology guide that explains complicated concepts in a straightforward manner. More importantly, the authors address the most common errors that occur during electrophysiological measurements and the most straightforward approaches that can be implemented to correct these errors. It is the hope of the authors that their efforts will guide and enrich the knowledge of an aspiring electrophysiologist with a passion for neuroscience.

Pittsburgh, PA, USA

Nicholas Graziane, Ph.D.
Yan Dong, Ph.D.

Contents

The original version of the authors on the cover has been revised. An erratum can be found at DOI 10.1007/978-1-4939-3274-0_2_23

Part I

Basic Concepts

Chapter 1

Extracellular and Intracellular Recordings

Nicholas Graziane and Yan Dong

Abstract

In 1766, Luigi Galvani discovered that electrical activity drove nerve function. His discovery pioneered contemporary electrophysiology, which now consists of extracellular and intracellular approaches used to study electrical signal transfer among neurons. Each approach has advantages/disadvantages and useful applications that once understood can greatly benefit an experimenter looking to design the appropriate experiments. The purpose of this chapter is to provide the reader with core understanding for each approach focusing on in vitro applications (in vivo electrophysiology is covered in Chap. 22).

Key words Multi-cell recording, Cell-attached patch, Inside-out patch, Outside-out patch, Sharp-electrode recordings, Whole-cell patch

1 Introduction

In 1766, Luigi Galvani discovered that electrical activity drove nerve function. His discovery pioneered contemporary electrophysiology, which now consists of extracellular and intracellular approaches used to study electrical signal transfer among neurons. Each approach has advantages/disadvantages and useful applications that once understood can greatly benefit an experimenter looking to design the appropriate experiments. The purpose of this chapter is to provide the reader with core understanding for each approach focusing on in vitro applications (in vivo electrophysiology is covered in Chap. 22).

This chapter covers multi-cell extracellular recordings followed by cell-attached patches. It concludes by describing intracellular recording techniques, including whole-cell, outside-out, inside-out, and sharp-electrode recordings. Discussions about some related techniques, such as cell-attached and whole-cell, include detailed explanations for how to achieve an appropriate patch. We advise readers interested in intracellular recordings to look over cell-attached patches since they are the gateway to many intracellular configurations. Other electrophysiology approaches discussed

Nicholas Graziane and Yan Dong, *Electrophysiological Analysis of Synaptic Transmission*, Neuromethods, vol. 112,
DOI 10.1007/978-1-4939-3274-0_1,

include references that the reader may feel useful while attempting to deepen his or her understanding about the technique.

2 Multi-cell (Multi-unit) Recording (In Vitro)

Multi-cell recording is used to study activity in a neuronal population as well as the interactions between neurons. This technique is commonly used by systems neuroscientists in order to study neuronal ensembles in a given neurocircuit. Multi-cell recording is normally performed using microelectrode arrays (MEAs). Standard MEAs contain multiple electrodes typically made of titanium nitride and are arranged in an 8×8 or 6×10 configuration on a glass substrate (Fig. 1). Using cell-cultures or tissue slices, the experimenter can record neuronal spikes (generated by action potentials), evoked field potentials, or spontaneous field potentials in a given neuronal population (for a detailed description of array types see [1]).

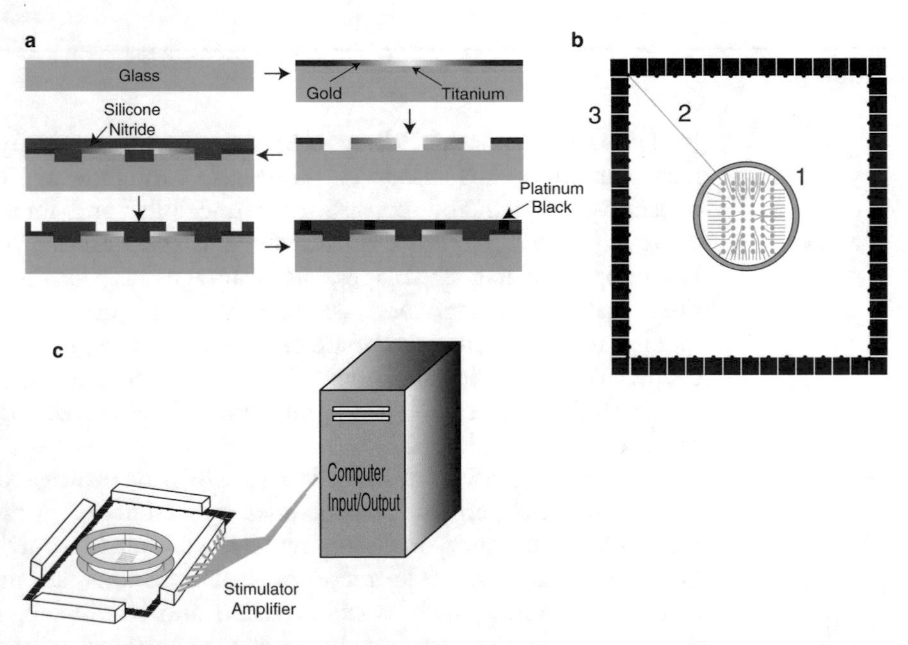

Fig. 1 (**a**) Steps for MEA fabrication. First, gold and titanium are layered on a glass slide. Then silicon nitride is added followed by platinum black, which forms the recording locations for neuronal activity. (**b**) Diagram of a microelectrode array (bird's eye view). The recoding/stimulating electrodes are situated inside the recording chamber (*1*), which retains the bath solution. The cultured cells or tissue slice is then situated on top of the electrodes inside the cylinder. The electrodes project (*2*) toward the contact pads (*3*). (**c**) Diagram illustrating an MEA setup. Data acquisition cables transmit information from the contact pads to a computer allowing for a digital readout of neuronal activity

2.1 Advantages/Disadvantages

There are several advantages to multi-cell recordings. Data can be gathered from multiple neuronal locations simultaneously, which is best for testing neuronal population activity and their potential synchronous activities within a given brain region. The experimenter can easily change recording and stimulating sites within the neuronal preparation, which can be useful for within slice controls [1].

The disadvantages to using multi-cell recordings are that the electrodes are fixed, preventing movement throughout the neuronal preparation. When using slice preparations the electrodes are not inserted inside the tissue creating smaller amplitude recordings. Stable baselines may be difficult to achieve in acute brain slice preparations due to fluid level changes. In addition, multi-cell recordings may be difficult to interpret since the multi-cell population may be physiologically heterogeneous.

2.2 Applications

Multi-cell recordings have been used to study how different pharmacological treatments affect spiking within different neuronal populations in a brain slice. This has useful benefits for the pharmaceutical industry during drug screening [2].

Each electrode in a multi-electrode array can be used to elicit an electrical stimulus evoking field potentials in a specific region. For example, researchers have used hippocampal brain slices arranged on a multi-electrode array to stimulate CA3 while recording from a neuronal population at CA1 [3].

Microelectrode arrays are useful for studying the spontaneous activity throughout a neuronal population. Drugs applied to cell cultures or brain slices can generate spontaneous potentials that can be recorded from the multiple electrode arrays situated under varying cell populations. This has been used to study hippocampal slice oscillations [4].

3 Cell-Attached Patch

A cell-attached patch is performed by sealing the micropipette to the cell membrane (Fig. 2b). The technique is easy to perform as long as the experimenter is familiar with the following two points: (1) apply positive pressure before inserting the micropipette into the bath (typically using a syringe, which is connected to the electrode holder's port via tubing) (Fig. 3a). Positive pressure prevents contaminants from coming in contact with the micropipette tip. In addition, positive pressure cleans the targeted neuron allowing for a tight seal. (2) Once the electrode tip is near the neuron (<a few microns away), a small dimple on the neuronal surface can be seen (Fig. 3b). The dimple is formed from the positive pressure. The experimenter should immediately remove the positive pressure and apply suction (negative pressure) to the recording electrode in

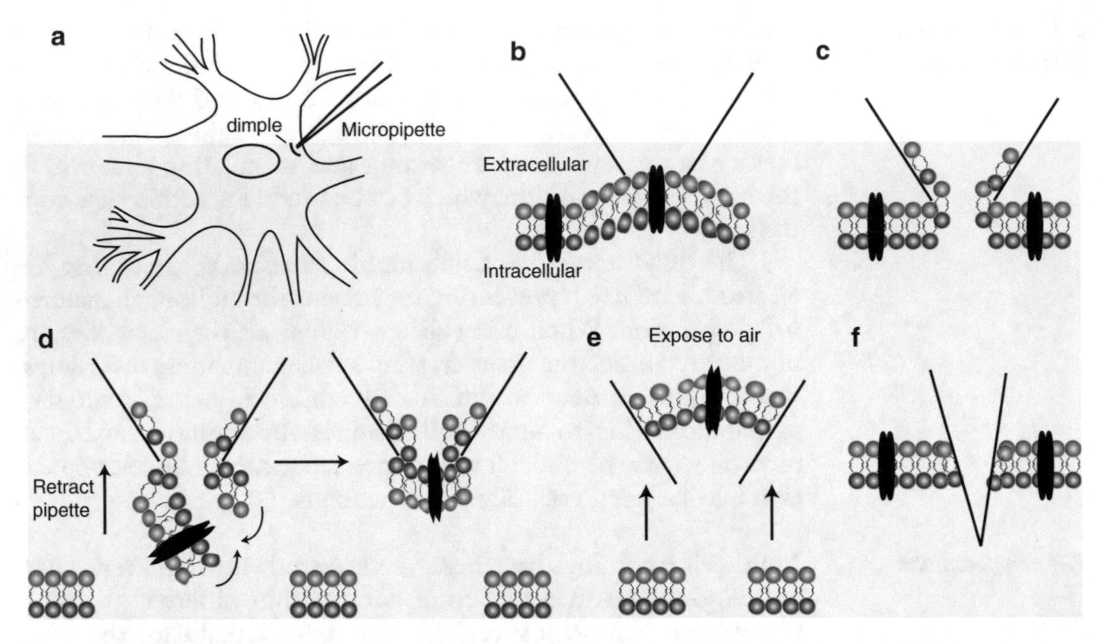

Fig. 2 Patch clamp configurations for electrophysiology recordings. (**a**) A somatically patched cell using a micropipette. (**b**) A cell-attached patch. The mild suction applied through the micropipette enables a GigaOhm seal, but also deforms the membrane. (**c**) A whole-cell patch. After obtaining a cell-attached patch, further negative pressure is applied via the micropipette to rupture the membrane allowing access to the intracellular compartment. (**d**) An outside-out patch. After obtaining a cell-attached patch, the micropipette is slowly pulled away causing the membrane to break. Naturally, hydrophilic and hydrophobic forces cause the membrane to reform creating a microcell with the extracellular side of the membrane facing away from the micropipette towards the external solution. (**e**) An inside-out patch. After forming an outside-out patch, the membrane is exposed to air by lifting the micropipette out of the bath. Exposing the membrane to air causes the intracellular side of the membrane to face toward the bath solution

order to form a GigaOhm seal. A GigaOhm seal can be identified by passing a voltage step (5 mV) step through the recording electrode (Fig. 3c). Following a GigaOhm seal, the negative pressure should be removed otherwise a whole-cell configuration is achieved. A whole-cell patch can sometimes be unintentionally achieved even when the negative pressure is removed. Obtaining a cell-attached patch is the first step in many intracellular recordings discussed below (excluding sharp-electrode recording).

3.1 Advantages/ Disadvantages

There are several advantages to using a cell-attached patch. The micropipette does not penetrate the neuron, thus maintaining the cytosolic integrity. Because of this, ionic channels can be studied in their normal environment. The measurement resolution is high due to a reduced current noise level (0.1 pA for a 1 kHz bandwidth) produced by the high resistance GigaOhm seal [5]. In addition, ionic gradients are contained at the membrane patch (a fraction of the whole cell surface) limiting potential harm to the entire neuron.

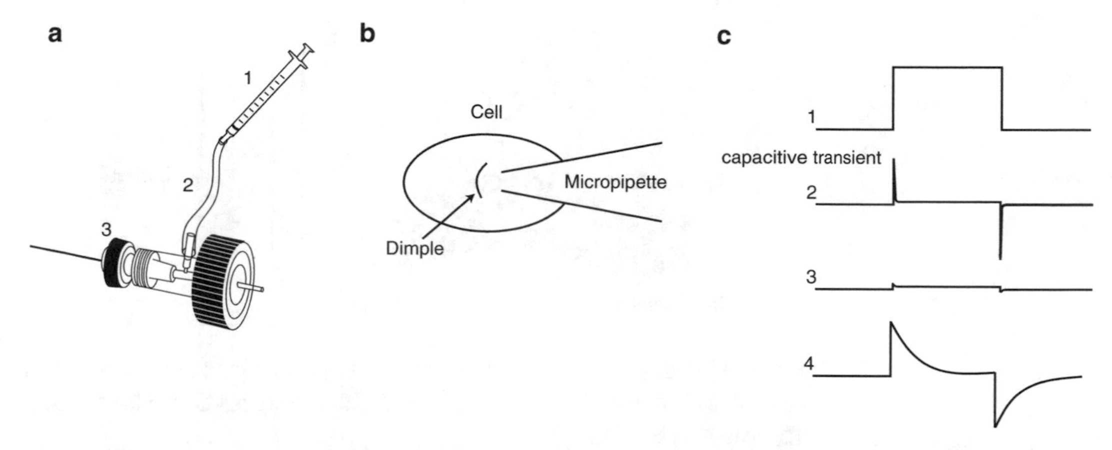

Fig. 3 (**a**) A syringe (*1*) is connected to an electrode holder (*3*) via a piece of Tygon tubing (*2*). (**b**) Positive pressure is applied to the micropipette via the syringe. As the micropipette approaches the cell soma, the positive pressure pushes on the membrane causing a dimple to form. (**c**) After releasing the positive pressure, the membrane begins to form a seal around the micropipette, which is evident from the increase in the pipette resistance (*2*). The voltage-step induced square wave (*1*) disappears as the pipette forms a GigaOhm seal with the membrane leaving only capacitive transients (i.e., as the recording electrode forms a tight seal, the resistance increases causing the square wave amplitude to reduce to zero). These capacitive transients are caused by the capacitive properties of the micropipette and can be corrected producing minimal transients (*3*). By applying more negative pressure, the membrane ruptures leading to a whole-cell configuration, which produces the characteristic capacitive transients shown (*4*)

The disadvantage is that intracellular access is absent. Therefore the resting membrane potential is unknown and intracellular regulation is prevented. Forming a high-resistance seal deforms the neuronal membrane (Fig. 2b) [6, 7]. This can lead to alterations in channel function such as local leaks or increased receptor mobility from damaged cytoskeletal components [8]. It can also lead to (1) a cytoplasmic bridge being formed between the membrane in the pipette tip and the cell-surface membrane or (2) poorly defined voltage gradients in the membrane forming the high resistance seal (rim effects) [8]. The cytoplasmic bridge and rim effects are identified by heterogeneous current amplitudes or rounded rising phases.

3.2 Applications

This patch clamp technique has at least three useful applications. One, it is useful for single channel recordings. Two, it is useful for studying neuronal action potential firing in neurons with pacemaker activity (i.e., dopamine neurons, Purkinje cells, etc.). Three, channel activity in a membrane section can be studied during a physiological event.

For single channel recordings, both ligand-gated and voltage-gated channels can be recorded. For recording ligand-gated channels, the internal solution contains agonists for the receptor of interest. For example, acetylcholine can be added to the internal solution in order to activate nicotinic acetylcholine receptors (Fig. 4a).

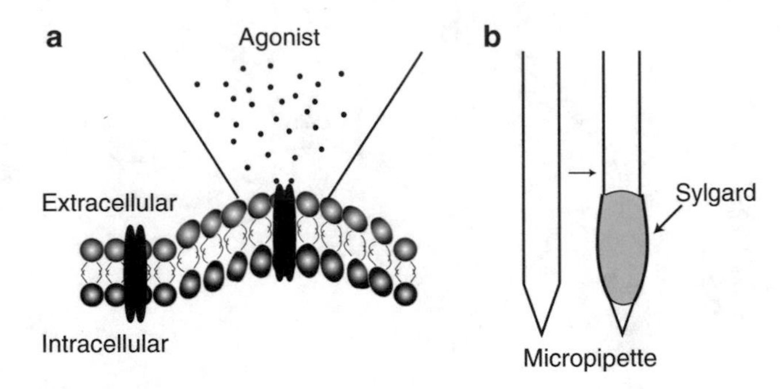

Fig. 4 (**a**) Single-channel recordings in a cell-attached patch with the internal solution spiked with the receptor's agonist. (**b**) A micropipette before and after application of Sylgard

For voltage-gated channels, depolarizing or hyperpolarizing steps can be introduced through the recording electrode activating the channel of interest. Since more than one channel can be present in a patch, it is likely that two or more receptors are activated at the same time. Knowing the receptor's conductance can help identify how many receptors are activated. Since single channel current is small (very often <5 pA), it is important to have minimal baseline noise. Baseline noise can be reduced by decreasing the micropipette capacitance. The micropipette capacitance is dependent upon the wall thickness, the immersion depth, or the material composition [9]. To reduce the capacitance, and therefore reduce baseline noise, Sylgard can be used. Sylgard is a silicone elastomer (Dow Corning), which acts as an insulator. It is applied to the tip of the recording electrode as shown in Fig. 4b.

For studying pacemaker activity, the recording electrode is filled with the external solution. This is important so that the receptors located directly under the recording electrode and the rest of the neuron are bathed in the same solution; thus keeping the resting membrane potential consistent throughout the entire neuron. Current clamp can be used in the cell-attached patch configuration in order to measure potentials directly across the membrane. These potentials require Na^+ and K^+ receptor activation. In voltage-clamp (−70 mV) the Na^+ and K^+ channel activation can be directly seen from inward currents produced by Na^+ channels and outward currents produced by K^+ channels [10].

Lastly, channel activity during a physiological event can be studied. For example, Mazzanti and DeFelice looked at voltage-gated channel activity during an action potential in a spontaneously beating 7-day-old chick ventricle using a two-electrode patch. One patch in the whole-cell configuration (discussed below), while the other patch was in the cell-attached configuration. By performing this experiment, Mazzanti and DeFelice were able to relate the opening/closing of a channel to the time course of an action potential spike [11].

4 Intracellular Recording

There are four configurations commonly used for intracellular recordings: whole-cell, outside-out, inside-out, and sharp electrode. Compared to extracellular recordings, intracellular recordings allow for the direct control of the neuron or receptor channels. Intracellular recordings can provide information on ionic reversal potentials, resting membrane potentials, single-channel conductance, second messenger roles in receptor function, and synaptic plasticity in neurons.

However, unlike extracellular recordings, intracellular recordings are invasive to the neuron. Therefore, maintaining high quality stable recordings for a long period of time (>1–2 h) are less likely (but see Chap. 7—Perforated Patch Clamp Techniques).

4.1 Whole-Cell

There are two methods for obtaining a whole-cell patch configuration. The first is referred to as the perforated patch, which uses substances that form channel pores in the cellular membrane. This patch technique was implemented in order to minimize bioactive molecules from being washed-out of the intracellular compartment by the internal solution (for an in-depth description of the perforated patch technique refer to Chap. 7). The second method for obtaining a whole-cell patch configuration is to use negative pressure applied to a recording electrode after achieving a cell-attached patch. The negative pressure ruptures the lipid membrane allowing the recording electrode access to the cytosol (Fig. 2c). This access is clearly seen by applying a positive voltage step, which produces a capacitive transient (Fig. 3c). For a positive voltage step the initial capacitance transient is upward followed by a downward transient at the end of the voltage step. The upward current represents the current produced by the electrode to charge the lipid bilayer (the lipid bilayer is a capacitor—see Chap. 2). Capacitive transients are not formed from ionic currents. They are purely a result of the electrode removing negative ions (normally chloride if you are using a Ag/AgCl wire and applying a positive voltage step) from the neuron's intracellular side.

During any whole-cell recording the experimenter should check the access resistance of the neuron. Any access resistance change can lead to inaccurate current or voltage measurements. Typically, a recording should be discarded if a change by more than 20 % occurs. This is because the recorded channel amplitude/kinetics can be altered. However, it is advisable to empirically test how changes in access resistance can affect the channel's amplitude/kinetics as the accepted electrophysiological range may be too lenient. In voltage-clamp recordings, a hyperpolarizing step (—0.5 to 5.0 mV for 50 ms) can be applied. The access resistance can then be calculated from the capacitive transient produced (Fig. 3d). Using Ohm's

$$V = IR$$

where (V) = voltage, (I) = current, and (R) = resistance, the access resistance can be calculated by the voltage of the hyperpolarizing step plus the holding voltage divided by the peak amplitude of the capacitive transient.

To check the access resistance in current clamp mode, simply check the bridge balance (the bridge balance corrects for the pipette resistance. See Chap. 3) before and after the recording. Again, any recordings that have a change in the access resistance >20 % should be discarded.

4.1.1 Advantages/Disadvantages

The major advantage of the whole-cell configuration is that the electrode can directly sample electrical signals from the whole neuron with minimal signal loss [12]. This is due to the low resistance access to the cell provided by the low resistance tip of the micropipette. Therefore, synaptic currents generated on proximal/distal dendrites can be recorded with resolutions of quantal size (see Chap. 8).

Another advantage of the whole-cell configuration is that the intracellular compartment can be perfused with any desired pipette solution. This allows the experimenter to manipulate ionic reversal potentials as well as resting membrane potentials. In addition, the neuron can be perfused with dyes (biotin) [13], second messengers [14], 'caged' second messengers [15], and antibodies to intracellular proteins [16]. These ions, second messengers, antibodies, and/or proteins need time to equilibrate as the internal solution perfuses the neuron. Work has shown that Na^{+} ions equilibrate in whole-cell mode with a time constant of 5 s in chromaffin cells [17]. Larger molecules (500–1000 Da) take several minutes to equilibrate [18], and when using antibodies, it is advised to wait 5–10 min for the antibody to perfuse the neuron and bind to its target molecule. Of course, equilibration is dependent upon the recording patch pipette's tip size as well as the access resistance.

The disadvantage of a whole-cell configuration is that the internal solution perfuses the neuron, completely substituting for the intracellular solution (the intracellular solution is negligible compared to the much larger internal solution volume ~10 pl vs. 10 μl, respectively). Therefore, endogenous cellular proteins (i.e., second messengers) are lost into the recording electrode. This can cause ionic currents to rundown (a gradual decrease in peak amplitude). For example, Ca^{2+} current rundown occurs after a recording period of about 10 min [19, 20] (see Chap. 17).

4.1.2 Applications

Whole-cell mode has many useful applications. One, channels with small conductance can be studied since the channel number is the greatest with this configuration. This is due to access to channels throughout the soma, distal dendrites, and axon hillock. Two, channels with fast kinetics can be studied. When performing channel kinetic experiments, it is advised to keep the peak amplitude ≤200 pA when voltage clamping the cell at −70 mV. Larger currents could result in space clamp problems, thus affecting channel decay time (i.e., the software cannot hold the cell at −70 mV). Three, channels of interest can be isolated. For example, Ca^{2+} currents are usually contaminated with K^+ currents. In whole-cell mode, Cs^+ can be applied to the internal solution, thus blocking K^+ channels. Four, the cell can be voltage clamped allowing for isolation of currents from channels of interest.

4.2 Outside-Out (Cell-Free Patch)

From a whole-cell configuration the recording electrode is slowly pulled away (a few millimeters away from the neuron) forming a thin fiber that eventually breaks. Once the fiber breaks, the cell membrane reseals at the tip of the recording electrode resulting in a micro-cell (less frequently inside-out configurations can be directly obtained following this process) (Fig. 2d). This micro-cell configuration allows the experimenter to study the influence that extracellular factors have on single channels, a few channels, or small population of channels. The channel number depends on the vesicle size and the channel density.

4.2.1 Advantages/Disadvantages

An apparent advantage of an outside-out patch is that it allows the experimenter to easily alter the bath composition during a recording. This process can be done very quickly within a timescale similar to neurotransmitter release at a synapse (1 ms) [21, 22]. This fast solution change contrasts whole-cell mode, which can require tens of milliseconds. Some experimenters also use this configuration to exclude the involvement of intracellular components. Formation of an outside-out patch maintains a small piece of membrane on the tip of the electrode, with most cell content being removed. Thus, ion channels or receptors can be examined in a relatively isolated manner. However, it has been a concern that some signaling proteins are firmly attached to ion channels and receptors and the isolated membrane in the outside-out configuration cannot entirely remove them.

There are a few disadvantages to an outside-out patch. One, major structural rearrangements can occur altering channel properties (i.e., kinetics, see [23, 24]. Two, cytosolic factors are washed out, and three, the cytoskeletal structure is disrupted.

4.2.2 Applications

The outside-out patch is used to study externally located receptors. Such studies include identifying receptor conductance states [25] and gating properties [26]. The outside-out patch can also be used to identify ionic channels activated by specific neurotransmitters or hormones [27–30].

4.3 Inside-Out (Cell-Free Patch)

From a cell-attached patch (see cell-attached above for description on how to obtain a cell-attached patch), the recording electrode (filled with external solution) is pulled away. Doing this pulls off a membrane patch leaving a vesicle (Fig. 2e). The most common way to remove the vesicle is to expose it to the bath–air interface. This is done by lifting the recording electrode out of the bath solution. Other methods for vesicle removal include exposing it to a low Ca^{2+} solution or by making contact with a cured piece of Sylgard or a droplet of paraffin [31].

4.3.1 Advantages/Disadvantages

The advantage of using an inside-out patch configuration is that the extracellular environment can be controlled and the intracellular or cytosolic environment can be manipulated by solution exchange.

The disadvantage of using an inside-out patch configuration is that it disrupts the cytoskeletal structure [32]. In addition, perfusing the bath with internal solution exposes the entire preparation (i.e., brain slice or cell culture) to solutions that contain high K^+ concentrations. K^+ at high concentrations can depolarize neurons causing apoptosis or trouble getting a seal with the recording electrode. Because of this, the cell-culture or brain slice must be discarded after one attempt from an inside-out recording. Therefore, this technique should be used sparingly if the number of cultures or tissue slices are minimal or if multiple inside-out patches form the same neuron are necessary [33].

4.3.2 Applications

With the intracellular membrane exposed, the experimenter can test single channel activity regulated by intracellular factors, which is useful for studying a mechanism involving intracellular components. For example, studies have used inside-out patches to test how changes in intracellular Ca^{2+} concentration affect Ca^{2+}-dependent K^+ channel activity [34]. In addition, channels regulated by protein kinases and ATP have been studied using inside-out configuration [35, 36].

4.4 Sharp Electrode Recording

Sharp electrode recordings get their name from the recording pipette that is fabricated for cell impalement. The recording pipette has a tip approximately 0.01–0.1 μm in size. This leads to a high resistance tip that can be used to impale the cellular membrane

(Fig. 2f). Impalement can be accomplished by briefly oscillating the capacity compensation circuit (typically referred to as a zap on amplifier) or by applying short 20 ms intense current injections [37]. Impaling the membrane inflicts significant damage to the cell so it is advisable to allow for a 15-min recovery before taking measurements [38].

4.4.1 Advantages/Disadvantages

The advantage to sharpelectrode recording is that a solution exchange between the pipette and the neuron is limited because of the high resistance pipette tip (25–125 MΩ). Since sharp electrode recordings lack solution exchange, perfusion or current run down is potentially avoidable. This is in direct contrast to whole-cell recordings where the internal solution perfuses the neuron leading to second messenger/intracellular protein loss. Sharp electrode recordings can also be used to study electrically coupled cells since the electrical coupling is minimally disrupted [38]. Despite the limited solution exchange, cellular markers can still be incorporated intracellularly using iontophoreses. This can be accomplished by passing negative current pulses (0.2–0.6 nA, 250 ms, 1 Hz for 5–30 min) [39]. Another advantage to sharp-electrode recording is that both current clamp and voltage clamp measurements can be performed. However, current clamping is more commonly as voltage clamping using sharp-electrodes is typically poor due to the small tip/high electrode resistance.

The disadvantage to sharp electrode recording is that the noise level is high because obtaining a GigaOhm seal prior to cell membrane puncture is not guaranteed. Sharp electrodes also have high resistances, which can impede signal measurements. To reduce the resistance, a high ionic strength internal solution is used (typically 0.5–3.0 M KCl). Since the internal solution and surrounding environment (i.e., bath or cytoplasmic) have differences in ionic concentrations, electrochemical gradients are introduced leading to net ion flux. This net ion flux forms liquid junction potentials that need to be corrected for during the recording. The high resistance pipette tips also cause junction potentials at the cell–electrode interface [40], which are difficult to predict. The recording electrodes used can be inadequate (i.e., fragile, susceptible to tip blockage). Additional leak current between the sharp electrode and membrane is created because the sharp electrode penetrates the membrane.

4.4.2 Applications

Sharp electrodes were developed in 1949 by Ling and Gerard and used predominately in cellular recordings until whole-cell mode was developed [41]. However, sharp electrodes are still used today in order to measure membrane potential, resistance, time constants, synaptic potentials, and action potentials.

5 Summary

There are many useful in vitro electrophysiological techniques that can be implemented to investigate a particular question of interest. For questions related to neuronal activity in a given ensemble, multi-cell recordings are highly useful. If single channel conductance or channel kinetics are of interest, cell-attached patches can be used. Finally, there are many different forms of intracellular recording patch-clamp techniques (i.e., whole-cell, outside-out, inside-out, sharp electrode) that can be used to study intracellular and extracellular functioning of surface receptors. Despite the advantages of each technique, experimenters should be aware of potential disadvantages. In understanding the shortcomings of a particular patch-clamp technique, the experimenter can design appropriate experimental investigations, thus limiting potential confounds in the future.

References

1. Taketani M, Baudry M (2006) Advances in network electrophysiology: using multi-electrode arrays. Springer, USA
2. Stett A, Egert U, Guenther E, Hofmann F, Meyer T, Nisch W, Haemmerle H (2003) Biological application of microelectrode arrays in drug discovery and basic research. Anal Bioanal Chem 377(3):486–495
3. Parameshwaran D, Bhalla US (2012) Summation in the hippocampal CA3-CA1 network remains robustly linear following inhibitory modulation and plasticity, but undergoes scaling and offset transformations. Front Comput Neurosci 6:71
4. Plenz D, Stewart CV, Shew W, Yang H, Klaus A, Bellay T (2011). Multi-electrode array recordings of neuronal avalanches in organotypic cultures. J Vis Exp(54):2949
5. Kettenmann H, Grantyn R (1992) Practical electrophysiological methods: a guide for in vitro studies in vertebrate neurobiology. Wiley, New York
6. Hamill OP, Marty A, Neher E, Sakmann B, Sigworth FJ (1981) Improved patch-clamp techniques for high-resolution current recording from cells and cell-free membrane patches. Pflugers Arch 391(2):85–100
7. Sakmann B, Neher E (1983) Geometric parameters of pipettes and membrane patches. In: Sakmann B, Neher E (eds) Single-channel recording. Springer, USA, pp 37–51
8. Sakmann B, Neher E (1984) Patch clamp techniques for studying ionic channels in excitable membranes. Annu Rev Physiol 46:455–472
9. Zhao Y, Inayat S, Dikin DA, Singer JH, Ruoff RS, Troy JB (2008) Patch clamp technique: review of the current state of the art and potential contributions from nanoengineering. Proc Inst Mech Eng N J Nanoeng Nanosyst 222(1):1–11
10. Hodgkin AL, Huxley AF (1952) Currents carried by sodium and potassium ions through the membrane of the giant axon of Loligo. J Physiol 116(4):449–472
11. Mazzanti M, DeFelice LJ (1990) Ca channel gating during cardiac action potentials. Biophys J 58(4):1059–1065
12. Staley KJ, Otis TS, Mody I (1992) Membrane properties of dentate gyrus granule cells: comparison of sharp microelectrode and whole-cell recordings. J Neurophysiol 67(5):1346–1358
13. Margolis EB, Mitchell JM, Ishikawa J, Hjelmstad GO, Fields HL (2008) Midbrain dopamine neurons: projection target determines action potential duration and dopamine D(2) receptor inhibition. J Neurosci 28(36): 8908–8913
14. Okada Y, Fujiyama R, Miyamoto T, Sato T (2000) Comparison of a Ca(2+)-gated conductance and a second-messenger-gated conductance in rat olfactory neurons. J Exp Biol 203(Pt 3):567–573
15. Gjerstad J, Valen EC, Trotier D, Doving K (2003) Photolysis of caged inositol 1,4,5-trisphosphate induces action potentials in frog vomeronasal microvillar receptor neurones. Neuroscience 119(1):193–200
16. Graziane NM, Yuen EY, Yan Z (2009) Dopamine D4 receptors regulate GABAA

receptor trafficking via an actin/cofilin/ myosin-dependent mechanism. J Biol Chem 284(13):8329–8336
17. Fenwick EM, Marty A, Neher E (1982) A patch-clamp study of bovine chromaffin cells and of their sensitivity to acetylcholine. J Physiol 331:577–597
18. Pusch M, Neher E (1988) Rates of diffusional exchange between small cells and a measuring patch pipette. Pflugers Arch 411(2):204–211
19. Byerly L, Moody WJ (1984) Intracellular calcium ions and calcium currents in perfused neurones of the snail, Lymnaea stagnalis. J Physiol 352:637–652
20. Kostyuk PG (1980) Calcium ionic channels in electrically excitable membrane. Neuroscience 5(6):945–959
21. Liu Y, Dilger JP (1991) Opening rate of acetylcholine receptor channels. Biophys J 60(2):424–432
22. Maconochie DJ, Knight DE (1989) A method for making solution changes in the submillisecond range at the tip of a patch pipette. Pflugers Arch 414(5):589–596
23. Fernandez JM, Fox AP, Krasne S (1984) Membrane patches and whole-cell membranes: a comparison of electrical properties in rat clonal pituitary (GH3) cells. J Physiol 356(1):565–585
24. Trautmann A, Siegelbaum S (1983) The influence of membrane patch isolation on single acetylcholine-channel current in rat myotubes. In: Sakmann B, Neher E (eds) Single-channel recording. Springer, USA, pp 473–480
25. Hamill OP, Sakmann B (1981) Multiple conductance states of single acetylcholine receptor channels in embryonic muscle cells. Nature 294(5840):462–464
26. Sine SM, Steinbach JH (1984) Activation of a nicotinic acetylcholine receptor. Biophys J 45(1):175–185
27. Cull-Candy SG, Ogden DC (1985) Ion channels activated by L-glutamate and GABA in cultured cerebellar neurons of the rat. Proc R Soc Lond B Biol Sci 224(1236):367–373
28. Gardner P, Ogden DC, Colquhoun D (1984) Conductances of single ion channels opened by nicotinic agonists are indistinguishable. Nature 309(5964):160–162
29. Hamill OP, Bormann J, Sakmann B (1983) Activation of multiple-conductance state chloride channels in spinal neurones by glycine and GABA. Nature 305(5937):805–808
30. Nowak L, Bregestovski P, Ascher P, Herbet A, Prochiantz A (1984) Magnesium gates glutamate-activated channels in mouse central neurones. Nature 307(5950):462–465
31. Ogden D, Stanfield P (1994) Microelectrode techniques: the Plymouth workshop handbook. Patch clamp techniques for single channel and whole-cell recording. Company of Biologists, Cambridge, UK
32. Molleman A (2003) Basic theoretical principles. Patch clamping. Wiley, Chichester, pp 5–42
33. Windhorst U, Johansson H (1999) Modern techniques in neuroscience research: 33 tables. Springer, Berlin
34. Marty A (1981) Ca-dependent K channels with large unitary conductance in chromaffin cell membranes. Nature 291(5815):497–500
35. Noma A (1983) ATP-regulated K+ channels in cardiac muscle. Nature 305(5930):147–148
36. Shuster MJ, Camardo JS, Siegelbaum SA, Kandel ER (1985) Cyclic AMP-dependent protein kinase closes the serotonin-sensitive K+ channels of Aplysia sensory neurones in cell free membrane patches. Nature 313(6001):392–395
37. Ogden D (1994) Microelectrode techniques: the Plymouth Workshop handbook. Company of Biologists Limited, Cambridge, UK
38. Langton PD (2012) Essential guide to reading biomedical papers: recognising and interpreting best practice. Wiley, Chichester, West Sussex
39. Xu Z-Q (2011) Electrophysiology. In: Merighi A (ed) Neuropeptides, vol 789. Humana, New York, pp 181–189
40. Brette R, Destexhe A (2012) Handbook of neural activity measurement. Cambridge University Press, Cambridge
41. Ling G, Gerard RW (1949) The normal membrane potential of frog sartorius fibers. J Cell Comp Physiol 34(3):383–396

Chapter 2

Electrical Theory

Nicholas Graziane and Yan Dong

Abstract

The electrical properties of a cell are maintained by ions moving into and out of the cell. This ionic movement produces electrical potentials, which regulate cellular excitability. The purpose of electrophysiology is to measure cellular excitability by looking at ionic flow and potentials across the cell membrane. This chapter discusses the interpretation of electrophysiological measurements taking into account two forms of in vitro electrophysiology: current clamp and voltage clamp. The chapter begins by looking at field potentials, which are measured extracellularly in the current clamp configuration. Special attention should be paid to the direction of potentials (sinks and sources), which are dependent on positioning of the recording electrode as well as the type of ions moving into or out of a cell. We then discuss field potentials at an axon, a synapse, and the types of fields typically observed. We finish the chapter discussing interpretations of voltage clamp recordings in which currents can be measured.

Key words Nernst equation, Goldman-Hodgkin-Katz equation, Field potentials, Postsynaptic currents, Current rectification, Biological capacitors

1 Introduction

The electrical properties of a cell are maintained by ions moving into and out of the cell. This ionic movement produces electrical potentials, which regulate cellular excitability. The purpose of electrophysiology is to measure cellular excitability by looking at ionic flow and potentials across the cell membrane. This chapter discusses the interpretation of electrophysiological measurements taking into account two forms of in vitro electrophysiology: current clamp and voltage clamp. The chapter begins by looking at field potentials, which are measured extracellularly in the current clamp configuration. Special attention should be paid to the direction of potentials (sinks and sources), which are dependent on positioning of the recording electrode as well as the type of ions moving into or out of a cell. We then discuss field potentials at an axon, a synapse, and the types of fields typically observed. We finish the chapter discussing interpretations of voltage clamp recordings in which currents can be measured.

Nicholas Graziane and Yan Dong, *Electrophysiological Analysis of Synaptic Transmission*, Neuromethods, vol. 112,
DOI 10.1007/978-1-4939-3274-0_2,

2 Field Potentials (In Vitro)

The extracellular fluid consists of charged ions (e.g., Na^+, K^+, Ca^{2+}, Mg^{2+}, Cl^-, HCO^-) and is therefore a conductive medium. A conductive medium is any medium that can transmit electrical energy. In order to transmit electrical energy, a difference in electrical potential between adjoining regions is needed. The adjoining regions in this case are the extracellular and intracellular space, which are separated by a lipid membrane (Fig. 1a). Since neurons have a negative resting membrane potential (e.g., −70 mV; the intracellular side of the membrane is more negative than the extracellular side), negatively charged ions, like Cl^-, accumulate along the intracellular side of the membrane. Meanwhile, positively charged ions (e.g., Na^+, Ca^{2+}, K^+) accumulate along the lipid membrane on the extracellular side. What is produced is an electrical field across the lipid membrane. This field is created as positive ions tend to flow toward the negatively charged ions in the cytosol (with the electric field), while the negatively charged ions tend to flow toward the positively charged ions in the bath (against the electric field). This push or pull of a charge due to its interaction with another charge is known as force and this force is what a field potential measures. Therefore, a field potential is the force exerted on an ion in a conductive medium measured in the form of potential difference.

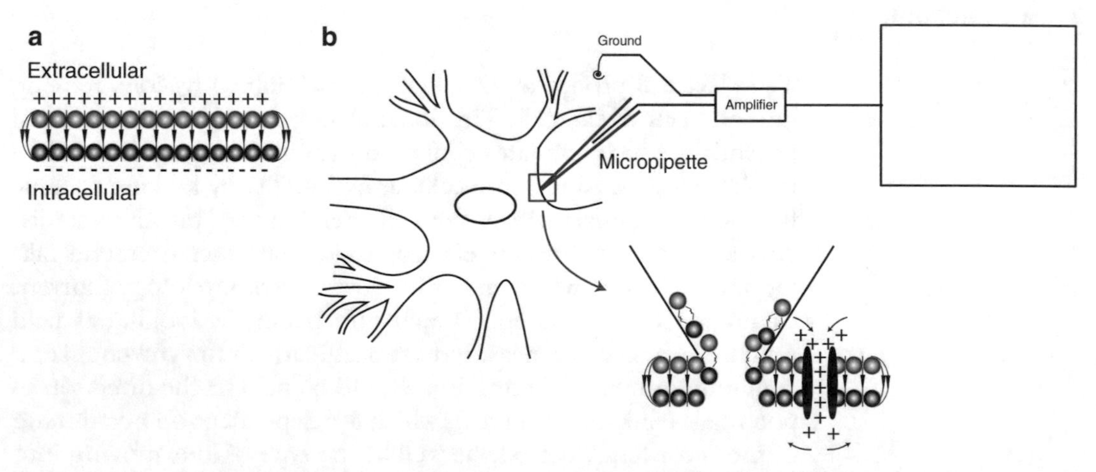

Fig. 1 (**a**) The separation of charge across the lipid membrane and the electric fields produced (*arrows*). The positively charged ions accumulate at the extracellular side of the membrane, while the negatively charged ions accumulate at the intracellular side. (**b**) A somatic whole cell recording showing positively charged ions passing through a transmembrane channel down the electrochemical gradient. This flow of ions is detected via a chloride silver wire connected to an amplifier, which transmits the signal to a digital output. The signal produced is the difference in potential between the recording electrode and the ground electrode, which is submerged in the bath solution

3 Nernst Equation

The force exerted on an ion is subject to change when there is a change in the ionic equilibrium potential or the membrane potential. The ionic equilibrium potential can be calculated using the Nernst equation, which defines the equilibrium potential of an ion in terms of its intracellular and extracellular concentrations ([C]in and [C]out, respectively):

$$= \frac{RT}{zF} \ln \frac{[C]\text{out}}{[C]\text{in}}$$

where R is the gas constant (8.314 J/K mol), T is the temperature in Kelvin (K) (K = °C + 273.15), z is the ionic charge, and F is Faraday's constant (96,480 Coulombs/mol). At T = 20 °C and z = +1 the ionic equilibrium potential can be calculated using the following formula:

$$= 58\text{mV} \log \frac{[C]\text{out}}{[C]\text{in}}$$

For warm blooded animals when T = 37 °C and z = +1 the equation becomes to following:

$$= 62 \text{ mV} \log \frac{[C]\text{out}}{[C]\text{in}}$$

When z is negative as is the case for Cl^- and T = 37 °C the equation becomes

$$= -62\text{mV} \log \frac{[C]\text{out}}{[C]\text{in}}$$

or

$$= 62\text{mV} \log \frac{[C]\text{in}}{[C]\text{out}}$$

When z represents a divalent cation (i.e., Ca^{2+}) and T = 37 °C the equation then becomes

$$= 31\text{mV} \log \frac{[C]\text{out}}{[C]\text{in}}$$

Table 1 lists ionic concentrations and equilibrium potentials for K^+, Na^+, Ca^{2+}, and Cl^- [1]. Typically, the concentration distribution of these major ions follows these general rules: $[K^+]\text{in} > [K^+]\text{out}$, $[Na^+]\text{in} < [Na^+]\text{out}$, $[Ca^{2+}]\text{in} < [Ca^{2+}]\text{out}$, and $[Cl^-]\text{in} < [Cl^-]\text{out}$ [1].

Table 1
Ionic concentrations and equilibrium potentials

Mammalian cell	Inside (mM)	Outside (mM)	Equilibrium potential $= \frac{RT}{zF} \ln \frac{[C]out}{[C]in}$ $T = 37$ °C
K^+	140	5	−89.7 mV
Na^+	5–15	145	+61.1 mV to +90.7 mV
Ca^{2+}	1–2	2.2–5	+136 to +145 mV
Cl^-	4	110	−89 mV

The equilibrium potential values discussed indicate that if the membrane potential is equal to a particular ion's equilibrium potential, the ionic movement will remain stagnant. However, as the membrane potential moves further away from the ionic equilibrium potential, the driving force for that ion to flow down its electrochemical gradient increases (Fig. 13.1A).

4 Goldman–Hodgkin–Katz Equation

The Nernst equation discussed above can be used to calculate the equilibrium potential of a single ionic species. The membrane potential equals the equilibrium potential of that ion if it is the only ion present in solution. However, if two or more ions coexist than the membrane potential (V_m) can be calculated using a simple linear equation:

$$= \left[\frac{g1}{g1 + g2 + \cdots} \right] E_{m1} + \left[\frac{g2}{g1 + g2 + \cdots} \right] E_{m2} + \cdots$$

where g refers to the conductance and Em refers to the equilibrium potential of each ionic species. Using this equation is convenient and is theoretically solid approach. However, this equation does not apply to ions with low concentrations. When the concentrations of ions become important, the Goldman–Hodgkin–Katz equation can be used:

$$= \frac{RT}{F} \ln \left\{ \frac{px[C]out}{px[C]in} + \frac{px[C]out}{px[C]in} + \frac{px[C]in}{px[C]out} + \ldots \right.$$

where p is the membrane permeability for a given ion (x). Note that in the third additive group of the equation the intracellular concentration moves to the numerator and the extracellular concentration moves to the denominator. This occurs when anions are included in the equation (e.g., Cl^-).

5 Ionic Gradients Maintained

The potential difference can only be measured when there is a flow of ions between the intracellular and extracellular space. The ionic flow is made possible by transmembrane channel activation. Once the activated channels open, ions flow down their electrochemical gradient, creating an electrical potential difference between the recording and ground electrodes (Fig. 1b).

Generally, there are three types of receptors/channels that are responsible for creating electrical potential differences. They are excitatory receptors, inhibitory receptors, and voltage-gated channels. As ions flow into or out of these receptors/channels, a current is generated, thus creating a potential change of the membrane (Voltage (V) = current (I) × resistance (R)). Depending on the position of the recording electrode, it detects an active current sink, an active current source, a passive current source, or a passive current sink (Fig. 2a).

When excitatory receptors become activated at synapses, Na^+ or Ca^{2+} flows from the extracellular space into the intracellular compartment. This flow of ions produces a potential known as an excitatory postsynaptic potential (EPSP). At the site where positive ions enter the neuron an active current sink is formed. This occurs as the positively charged ions rush intracellularly leaving behind a net negative charge (Fig. 2a). As positive ions flow inside the cell, there must be an equivalent flow of positively charged ions flowing outside the cell, known as a passive current source. The passive current source is generated at a distance away from the active current sink. As positively charged ions rush into the cell at the active current sink, a flow of positive charged ions flows out of the cell creating the passive current source.

To help grasp this concept, picture a cylinder lying horizontal with holes located equidistant apart on the top surface. Now picture the cylinder completely filled with water. As additional water is poured through one hole (active current sink), excess water inside the cylinder flows out (passive current source) from the adjacent holes. This flow of water out of the cylinder decreases as the distance increases from the active current sink. This example is similar to how current flows into and out of the neuron. See the electric field in Fig. 2b. The arrows indicate the direction that the positive ions flow (positive charge always flows with the electric field), meaning that negative ions must flow opposite (negative charge always flows against the electric field). The force is strongest at the site of receptor activation. Therefore, the electric field is the strongest and is illustrated by small distance between lines. The force decreases as the distance increases from the active current sink. This is illustrated by the increase in the distance between electric field lines further from the active current sink.

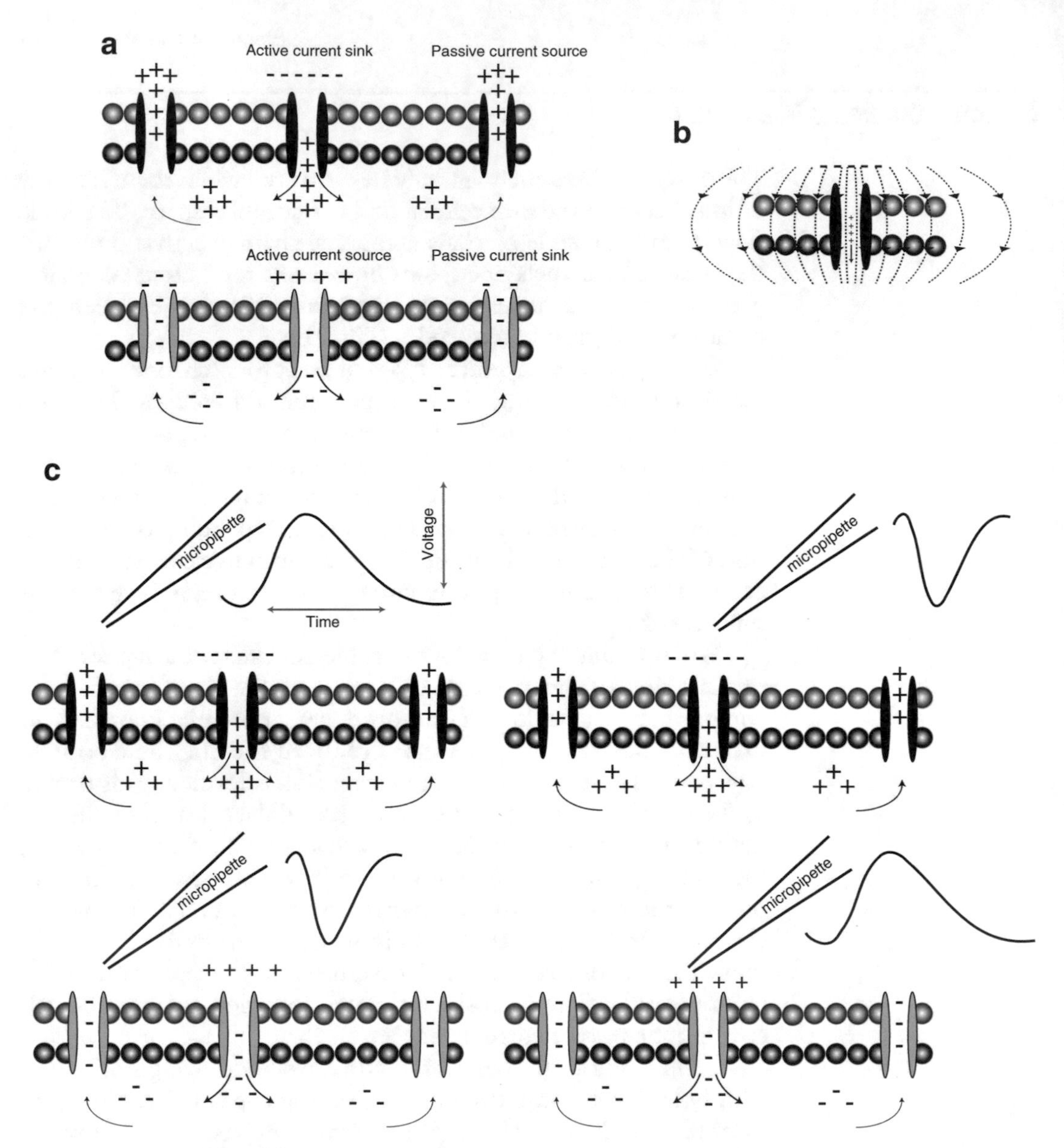

Fig. 2 (**a**) (*Top*) The flow of positively charged ions down the electrical gradient following cationic channel activation. As positively charged ions move through the channel intracellularly, a net negative charge is left extracellularly (active current sink). As positively charged ions move from the extracellular compartment to the cytosol, an equivalent flow of positively charged ions moves in the opposite direction creating a passive current source adjacent to the active current sink. (*Bottom*) The flow of negatively charged ions against the electrical gradient following anionic channel activation. As negatively charged ions move through the channel intracellularly, a net positive charge is left extracellularly (active current source). As negatively charged ions move from the extracellular compartment to the cytosol, an equivalent flow of negatively charged ions moves in the opposite direction creating a passive current sink adjacent to the active current source. (**b**) An electric field generated following activation of a cationic channel and subsequent ionic movement. The *arrows* indicate the flow of positively charged ions flowing with the electric field. The force is the strongest at the site of receptor activation and decreases as the distance from the active current sink increases. (**c**) The potentials that are generated in the signal detected are dependent upon the position of the micropipette in relation to the sinks and sources. (*Top*) Shows a positive potential detected at the passive current source (*left*), while a negative potential is measured at the active current sink (right). (*Bottom*) As anions move from the extracellular space into the cytosol, a negative potential is measured at the passive current sink (*left*), while a positive potential is measured at the active current source (*right*)

Where the recording electrode is placed dictates the field potential direction (positive or negative) as well as the field potential magnitude. If the recording electrode is positioned at the active current sink, a negative potential is recorded. However, if the electrode is located at the passive current source, a positive potential is recorded (Fig. 2c). The magnitude of the field potential is dependent on the distance from the recording electrode (i.e., the field potential decays as the square of the distance).

In addition to active current sinks and passive current sources, there are active current sources and passive current sinks. When inhibitory receptors at synapses become activated there is an influx of negatively charged ions (Cl^- or HCO_3^-) into the neuron. This flow of ions is known as an inhibitory postsynaptic potential (IPSP). As negatively charged ions flow intracellularly, there is a net positive potential extracellularly at this site (active current source). Since negatively charged ions are flowing into the cell, they also must be flowing out of the cell adjacent to the active current source (passive current sink) (Fig. 2c). If the recording electrode is positioned at the active current source, a positive potential is recorded. However, if the electrode is located at the passive current sink, a negative potential is recorded (Fig. 2c).

IPSPs are also detected when K^+ channels open causing K^+ to flow from inside the cell to outside. When K^+ channels open, an active current source is detected nearby the activated K^+ channel site while a passive current sink is located adjacent to this site.

Finally, there is an important point to discuss. Conductance (the ability to conduct electrical charge) is dependent upon temperature. Increasing temperature always increases the solution's conductivity by 1.5–5 % per degree Celsius. Therefore, when performing an in vitro field potential experiment, it is wise to keep your recordings at a consistent temperature. Otherwise, discrepancies in your results can mount.

6 Field Potentials Along an Axon

When an action potential is generated in the soma and moves along the axon, sodium channels open. Subsequently, an influx of positively charged ions enters the cell leaving behind an excess negative charge (active current sink). The current that is produced flows into the intracellular fluid and exits the membrane at more distal locations (passive current source) (Fig. 3a). As the action potential propagates down the axon towards the dendrites, the sinks and sources move to accompany the current flow. So if an action potential was elicited at point A and the recording electrode was positioned at point B, a positive potential would be detected since the electrode would be located at the passive current source. If the action potential moves right below the recording electrode, the

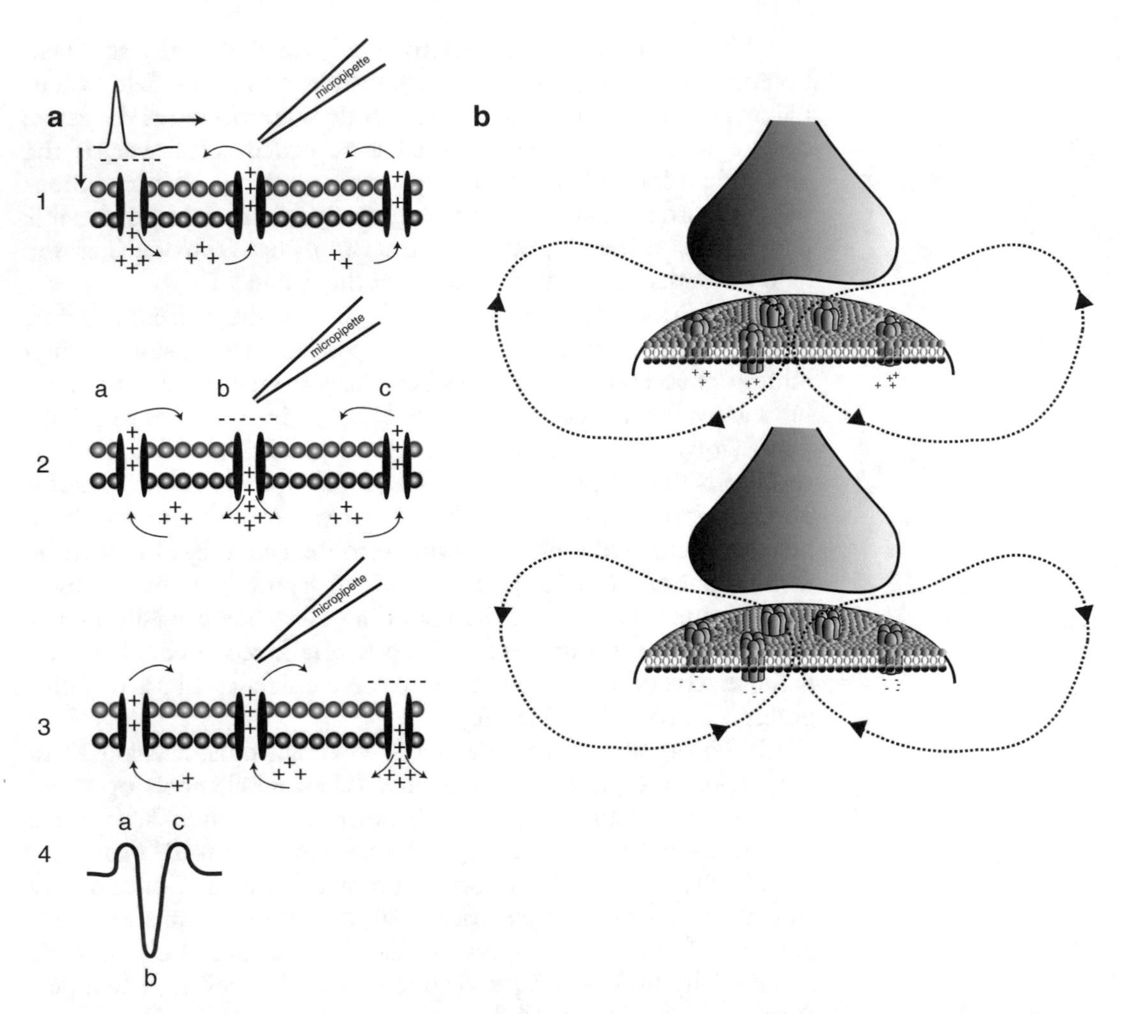

Fig. 3 (**a**) An action potential is generated and propagates along an axon while extracellular potentials are measured by a micropipette. The micropipette is stationary and detects potential differences (relative to ground) as the action potential moves along the axon. (*1*) An action potential elicits sodium channel activation upstream of the micropipette creating a passive current source at the recording site. (*2*) The action potential propagates down the axon opening sodium channels adjacent to the recording site creating an active current sink. (*3*) The action potential continues to propagate down the axon creating a passive current source at the recording site. This sequence of events creates a positive–negative–positive detected signal *n*. The negative potential produces the largest amplitude because the electrical gradient is strongest at the site of the active current sink. Where the passive current sources are measured a weaker electrical gradient exists. (**b**) *Top*. Positive ions flow intracellularly through activated excitatory receptors creating an active current sink at the synapse and passive current sources at adjacent regions. *Bottom*. Negative ions flow toward the cytosol through activated inhibitory receptors creating an active current source at the synapse and passive current sinks at adjacent regions. *Lines* represent the electrical field generated at each synapse

recording electrode would detect a negative potential (active current sink). Finally, the recording electrode would detect a positive potential as the action potential moves further along the axon past the recording electrode (passive current source). The result is a positive–negative–positive potential. The generation and propagation of action potentials produces a field potential known as a population spike. A population spike refers to the summed synchronous action potentials of a pool of neurons. The relevance of population spikes is discussed in the analysis section below.

7 Field Potentials at Synapses

Synaptic potentials are generated as postsynaptic receptors become activated by presynaptic neurotransmitter release. As receptors are opened, current can flow into the neuron creating a potential. At excitatory synapses an EPSP appears when positive ions flow intracellularly (active current sink) and exit the membrane at more distal locations (passive current source) (Fig. 3b). At inhibitory synapses, an IPSP appears when negative ions flow intracellularly (active current source) and exit the membrane at more distal locations (passive current sink) (Fig. 3b). The representative traces recorded for EPSPs and IPSPs at active and passive current sinks and sources are illustrated in Fig. 3b.

8 Types of Fields

The central nervous system generates at least three distinct field potential patterns. The field patterns are open fields, closed fields, and open-closed fields [1, 2].

Open fields are generated in neurons that have a long apical dendrite, which extends away from the soma. The neurons with this phenotype are arranged side-by-side in a columnar fashion (Fig. 4a). This neuronal arrangement is seen in the hippocampus, cerebellum, and cerebral cortex. When measuring field potentials in an open field, a negative extracellular potential is measured at an active or passive current sink. In contrast, positive extracellular potentials are measured at an active or passive current source. For example, if a backpropagating action potential is generated at the soma, the soma has a negative extracellular field potential. The dendrites, on the other hand, have a positive extracellular field potential. This generates a dipole (equal and oppositely charged poles separated by a distance) (Fig. 4a). Between the charged poles is a zero potential line (a region in space where there is no potential generated) that extends ad infinitum horizontally (Fig. 4a). However, the amplitude of field potentials decays as the square of the distance [2, 3]. Because of this, the zero potential as well as the negative and positive isopotentials (a region in space where every point is at the same potential) can only be detected if they are nearby the potential difference created by the current flow.

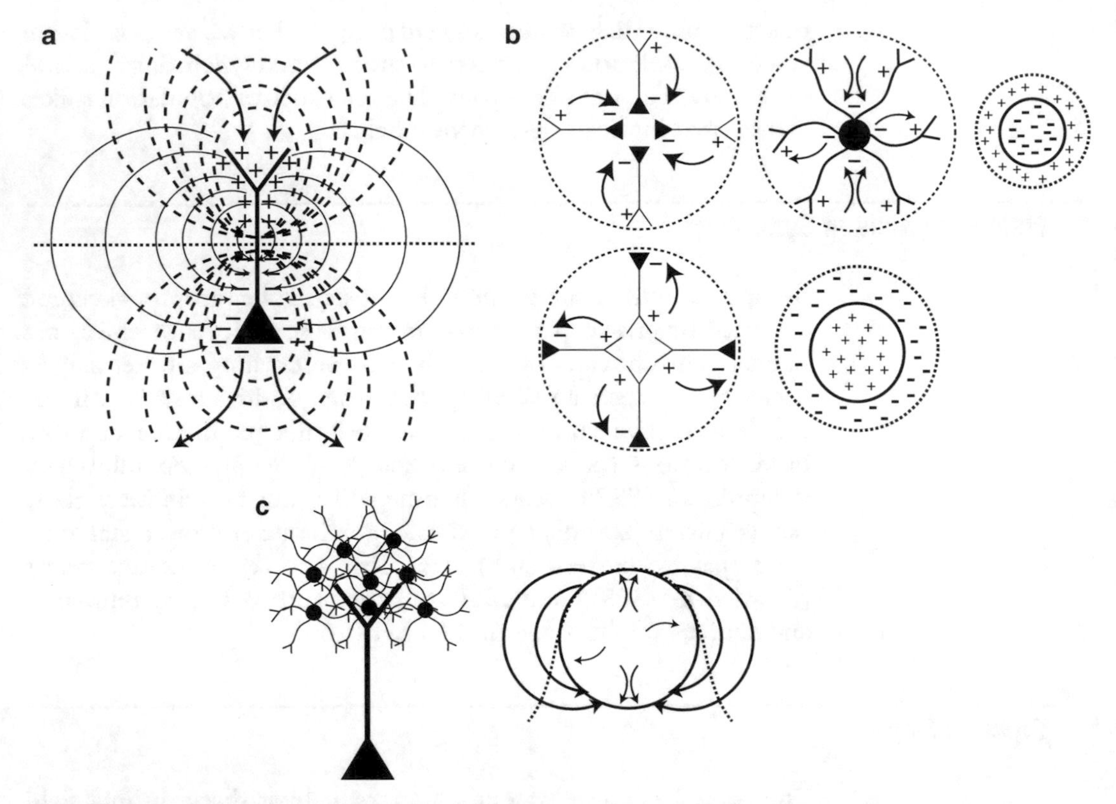

Fig. 4 Types of field potentials. (**a**) An open field generated from a neuron with an apical dendrite extending vertically away from the soma. The illustration shows a backpropagating action potential initiated at the soma with the corresponding extracellular potentials produced. Above the zero potential line (*dotted horizontal line*), positive isopotentials are generated at the passive current source (*curved dashed lines*), while below the zero potential line, negative isopotentials are generated at the active current sink. *Arrows* indicate the direction of the electric field produced by the backpropagating action potential. The separation of charges in an open field produces extracellular dipole currents (*solid curved lines*) flowing from the source to the sink. (**b**) (*Top*) A closed field generated by (1) a group of neurons situated with centralized somas and dendrites positioned radially (*left*) or (2) a single neuron with a central soma and radially extending dendrites (*middle*). Figure shows a backpropagating action potential with an active current sink at the soma and a passive current source at the dendrites producing closed potentials (*right*) and a zero potential line (*dotted line*). (*Below*) A closed field produced backpropagating action potentials in a group of neurons with somas positioned peripherally with dendrites extending centrally (*left*) generating closed field potentials (*right*). In all examples of a closed field, the extracellular currents flow radially from sources to sinks. (**c**) An open-closed field generated by an apical dendrite extending vertically away from the soma into a group of neurons situated with centralized somas and dendrites positioned radially. The isopotential line is illustrated as the *dashed black line*

Closed fields are generated from neurons that extend dendrites radially from a central soma (i.e., stellate-shaped spinal motor neurons). They are also generated from neurons that extend dendrites centrally from somas positioned peripherally (Fig. 4b). The isopotential is spherical with a diameter dependent upon dendritic electrical excitability (Fig. 4b). Recording field potentials from neurons arranged as shown in Fig. 4b results in active or passive current sinks where a negative potential is generated and active or passive current sources generated where a positive potential exists. Unlike open fields, where potentials extend ad infinitum horizontally, closed fields produce potentials that extend spherically in a closed circle. In a closed field, where the dendrites extend radially from the central soma, action potential generation results in negative–positive spikes at all distances from the soma [4].

Open-closed fields are the third type of field potential. They are generated when both radial dendrites and columnar dendrites coexist in a neuron pool. As seen from Fig. 4c, positive potentials can be recorded above the isopotential line and negative potentials can be recorded from below the isopotential line.

9 Postsynaptic Currents

So far we have discussed electrical potential differences between the recording electrode and the reference electrode. Now we are going to discuss current measurements, which directly measure the amount of ions moving through a channel. Current is measured in ampere (amp) and 1 amp equals 1 coulomb (unit of electrical charge)/second. Therefore, when we measure currents, we are measuring how many ions pass through a given point/second. The major advantage is that electrophysiologists are able to measure how many ions flow through a specific channel.

10 Inward vs. Outward Current (Whole-Cell Configuration)

Inward current describes any flow of ions that makes the cell more depolarized (Fig. 5a). For example, positive ions flowing into the cell depolarizes the cell causing an inward current (Fig. 5a). Similarly, negative ions flowing out of the cell also depolarize the cell causing an inward current (Fig. 5a).

Outward current describes any flow of ions that makes the cell more hyperpolarized. This occurs if positive ions flow out of the cell (typical of K+ channels, although there are some exceptions—inward rectifier K+ channels) or if negative ions flow into the cell (typical of GABAR and glycine receptors which are permeable to Cl–).

Electrophysiologists should be aware that the flow of ions depends on the ionic concentrations of the internal and external environment. Physiologically, Cl– flows into the cell meaning an outward current. However, many publications show a Cl– channel with inward current [5]. In these experimental settings, Cl– flows out of the cell since the internal solution contains a high Cl– concentration, shifting the physiological Cl– reversal potential from –70 mV to ~0 mV (Fig. 5a). The reason this is done experimentally is because the neuron can be voltage clamped at a more physiologically relevant potential (i.e., –70 mV) while recording Cl– currents.

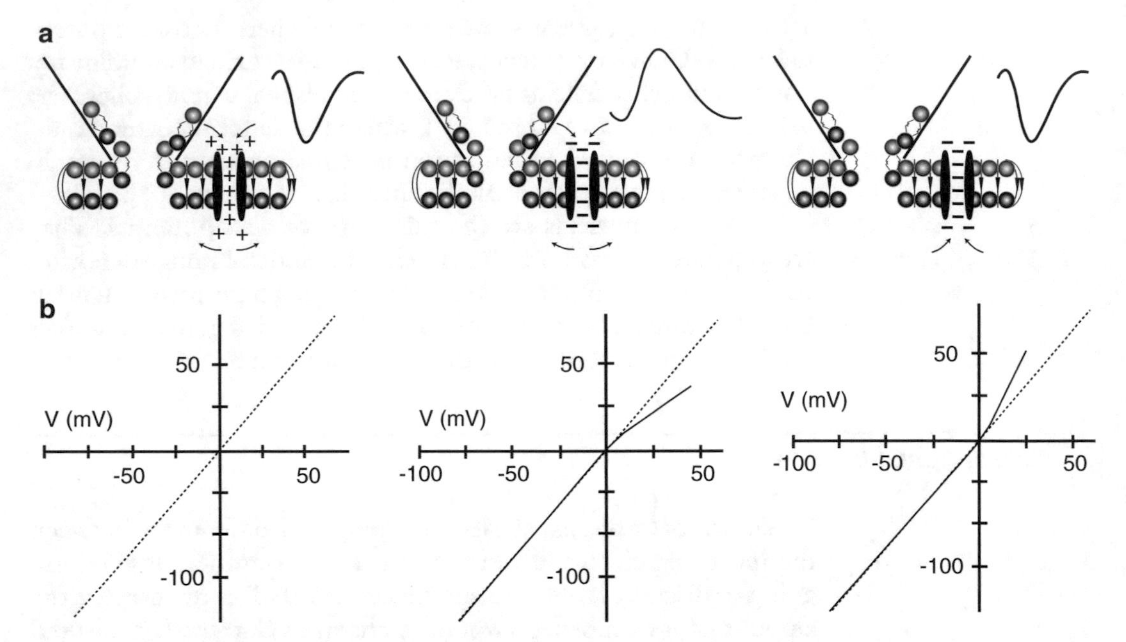

Fig. 5 Inward vs. outward current and current rectification. (**a**) Inward currents recorded in the whole cell configuration following the flow of positive ions from the extracellular to the intracellular side of the membrane (*left*). Outward currents recorded in the whole cell configuration following the flow of negative ions from the extracellular to the intracellular side of the membrane (*middle*). Inward currents recorded in the whole cell configuration following the flow of negative ions from the intracellular to the extracellular side of the membrane (*right*). (**b**) Illustration of the characteristic rectified currents seen in neurobiology including no rectification (*left*), inward rectification (*middle*), and outward rectification (*right*)

11 Current Rectification

Rectification describes voltage-dependent changes in channel conductance. Electrophysiologists have investigated both inward and outward rectifier channels. Inward rectifier channels pass more inward current than outward current, which is characteristic of calcium permeable AMPA receptors (GluR2-lacking) or inward-rectifier potassium channels. Calcium permeable AMPARs are inward rectifiers because at membrane potentials >0 mV the channel pore is blocked by polyamines, thus minimizing the flow of outward current (Fig. 5b). Similarly, inward-rectifying potassium channels are blocked by polyamines at depolarizing membrane potentials (Fig. 5b). Conventionally, K^+ flows out of the neuron, but inward-rectifier K^+ channels are open at hyperpolarizing potentials more negative than potassium's reversal potential. Therefore, K^+ flows into the cell to bring the cell back to its resting potential. However, as stated before, at more positive potentials inward-rectifier K^+ channels are inhibited by polyamines, reducing the flow of K^+ out of the cell.

Outward rectifier channels on the other hand pass more outward than inward current, which is characteristic of TREK-1 K^+ channels (Fig. 5b) Therefore, upon the same driving force, the outward flow of K+ is greater than the inward flow with the opposite direction (Fig. 5b).

12 Biological Capacitors

Capacitors have the ability to store charges (Q) when a voltage, ΔV, is applied (Fig. 6a). The ability for a capacitor to store charge is measured as capacitance (C; measured in farads, F) so that:

$$C = Q / DV$$

In electrophysiology, capacitance can be introduced by the electrode (immersed part), the biological membrane, and the stray capacitance (from the recording electrode to grounded surfaces or from the recording electrode holder).

The electrode's capacitance is formed across the glass walls. Therefore, the glass used to construct recording electrodes can greatly affect the capacitance (see Chap. 4). However, typically when immersed in the bath the capacitance of the recording electrode is 1 pF/mm of immersion depth [6].

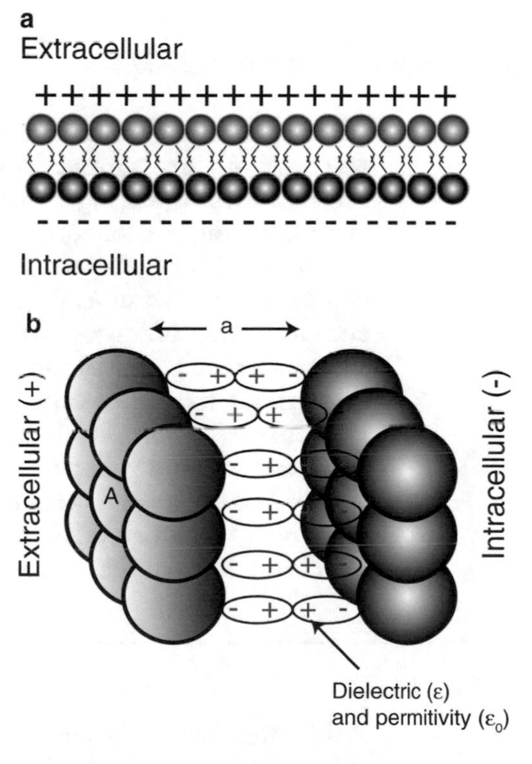

Fig. 6 Biological membranes as capacitors. (**a**) The properties of a lipid membrane enable charges to be separated along the extracellular and intracellular surface. Positive charges accumulate extracellularly, while negative charges accumulate at the intracellular surface. (**b**) The capacitive properties of the lipid membrane are illustrated. Contributing to the capacitance is the distance separating the extracellular and intracellular lipids (*a*), the surface area (*A*), and the dielectric and permittivity constants (ε and ε_0, respectively). The dielectrics are created as polar molecules situated in the lipid bilayer are oriented according to the extracellular and intracellular charges

Biological membranes are analogous to parallel-plate capacitors due to similar geometry (Fig. 2b). Their capacitance is directly proportional to their surface area (A) (the larger the cell, the greater the capacitance), to the dielectric constant of the medium separating the membrane (ε) (figure showing this), to the permittivity constant (ε_0) (referring to the ability of a substance to store electrical energy in an electric field), and it is inversely proportional to the distance (a) separating the plates.

$$C = A\varepsilon\varepsilon_0 / a$$

Biological membranes are typically less than 10 nm thick with an approximate capacitance of 1 $\mu F/cm^2$.

Stray capacitance is the capacitance that exists between conductive elements in a circuit. Typically, this capacitance is very low (a few picofarads) [6]. However, when stray capacitance couples to high impedance points, such as the micropipette input, circuit operation can be severely affected.

13 Electrophysiology Clamping Techniques

Current clamp: Current clamp maintains or clamps the current such that no net current flows through the membrane. This is accomplished by passing a time-varying or known constant current and measuring the changes in membrane potential.

Voltage clamp: In voltage clamp mode, the cell's potential is maintained at a command voltage by passing positive or negative current from the recording electrode into the cell. This way the membrane potential can be maintained at a consistent voltage throughout the entire experiment. In voltage clamp, the experimenter is essentially measuring the current that is needed to maintain the command voltage.

Dynamic clamp: Dynamic clamp refers to injecting conductance into a recorded cell in order to reproduce electrical effects of ion channels when they become activated [7].

14 Summary

Field potentials are maintained by ionic movement between intracellular and extracellular compartments. Depending on the movements of cations and anions, and the location of recording pipettes (e.g., axon, synapse, and soma), field potential sinks and sources can be measured. These potentials depend upon the driving force pushing ions down the electrochemical gradients, which can be calculated via the Nernst equation listed above. In addition to measuring cellular potentials, currents produced by ionic flow can also be measured. Much like field potentials, cation or anion movement dictates the direction of the current trace measured during experimentation. Since electrophysiological measurements take into account electrical activity, biological as well as experimenter factors, such as capacitance, can be introduced affecting the electrical circuit. Attention must be paid to neutralizing capacitance if possible in order to prevent unwanted artifacts from being introduced into the electrophysiological measurements.

References

1. Johnston D, Wu SMS (1995) Foundations of cellular neurophysiology. MIT Press, Cambridge, MA
2. Hubbard JI, Llinás RR, Quastel DMJ (1969) Electrophysiological analysis of synaptic transmission. Arnold Edward, London
3. Einevoll GT, Kayser C, Logothetis NK, Panzeri S (2013) Modelling and analysis of local field potentials for studying the function of cortical circuits. Nat Rev Neurosci 14(11): 770–785
4. Boulton AA, Baker GB, Vanderwolf CH (1990) Neurophysiological techniques: basic methods and concepts. Humana, Clifton, NJ
5. Graziane NM, Polter AM, Briand LA, Pierce RC, Kauer JA (2013) Kappa opioid receptors regulate stress-induced cocaine seeking and synaptic plasticity. Neuron 77(5):942–954
6. Axon Instruments I (1993) The Axon guide for electrophysiology & biophysics laboratory techniques. Axon Instruments, Foster City
7. Destexhe A, Bal T (2009) Dynamic-clamp: from principles to applications. Springer, New York

Chapter 3

Amplifiers

Nicholas Graziane and Yan Dong

Abstract

Synaptic transmission often occurs at the current level of pico-amperes. Because of the microscopic currents generated, they must be collected and amplified before they can be examined and analyzed. This job is done by the micro-amplifier. The amplifier directly connects to the electrode acquiring electrical signals from the targeted cell or brain regions. Thanks to advances in semiconducting technology, most amplifiers are light-weight, but extremely efficient in achieving stable, low-noise recordings. Combined with user-friendly software, operating the amplifier requires little effort. We have witnessed several occasions that without any prior experience, an undergraduate student successfully recorded synaptic transmission and collected useful data in a well-tuned setup. Nonetheless, these user-friendly features of the hardware and software are also risky because investigators need to know far less about the underlying mechanisms of their operation before they can start collecting data. This can lead to erroneous results. Here we summarize several basic properties of the preamplifier, with the hope that by knowing these properties, some obvious mistakes can be avoided.

Key words Data acquisition frequency, Gain, Compensation, Leak subtraction, Filtering, Troubleshooting noise

1 Introduction

Synaptic transmission often occurs at the current level of pico-amperes. Because of the microscopic currents generated, they must be collected and amplified before they can be examined and analyzed. This job is done by the micro-amplifier. The amplifier directly connects to the electrode acquiring electrical signals from the targeted cell or brain regions. Thanks to advances in semiconducting technology, most amplifiers are light-weight, but extremely efficient in achieving stable, low-noise recordings. Combined with user-friendly software, operating the amplifier requires little effort. We have witnessed several occasions that without any prior experience, an undergraduate student successfully recorded synaptic transmission and collected useful data in a well-tuned setup. Nonetheless, these user-friendly features of the hardware and software are also risky because investigators need to know far less about

Nicholas Graziane and Yan Dong, *Electrophysiological Analysis of Synaptic Transmission*, Neuromethods, vol. 112,
DOI 10.1007/978-1-4939-3274-0_3, © Springer Science+Business Media New York 2016

the underlying mechanisms of their operation before they can start collecting data. This can lead to erroneous results. Here we summarize several basic properties of the preamplifier, with the hope that by knowing these properties, some obvious mistakes can be avoided.

2 Data Acquisition Frequency

Electrical signals, often called traces in everyday recordings, are continuous. Most electrophysiological information contained in a trace is two-dimensional. It consists of signal amplitude values (e.g., ohm for voltage and ampere for current) and the elapsed time. At any given time point there is an amplitude value, which defines the properties of the recorded electrical signals. To acquire and store these signals in the computer, the amplifier and its connected digitizer must sample these amplitude-time values at an appropriate frequency, which is often referred to as the acquisition frequency. Setting the acquisition frequency determines how often the amplitude of the trace is sampled. In the example shown in Fig. 1a, a hypothetical trace (green) can be sampled four times (left) or 17 times (right) during the whole course. Thus, the acquired data are a set of non-continuous points. The experimenter

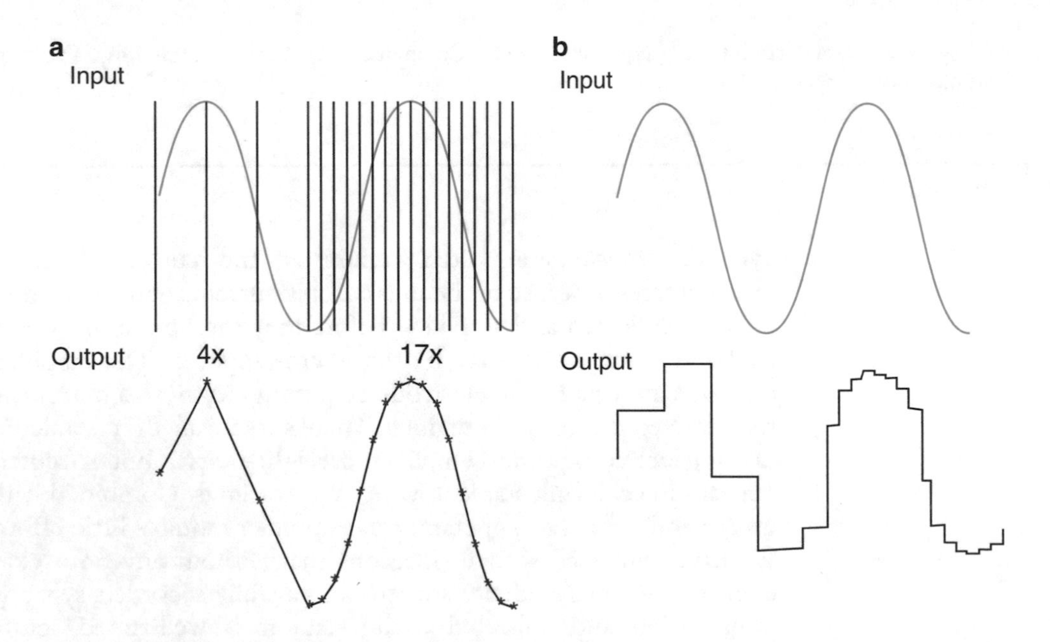

Fig. 1 Acquisition frequency determines the shape of the recorded trace. (**a**) A hypothetical trace whose input is sampled either 4 times or 17 times. The output shows that with a higher sampling rate, the shape of the output trace approaches the shape of the input trace. (**b**) A low sampling rate (*left*) can inaccurately represent an input trace. In contrast, increasing the sampling rate provides a more accurate representation of the input trace (*right*)

or the computer later on can connect all the sampled values with the hope that the reconstructed curve is close to the original trace.

As we can see in Fig. 1a, if the acquisition frequency is high, more data pairs are collected, and the reconstructed curve (dashed green line) is closer to the original trace. This sounds so straightforward that most beginners do not pay attention to the acquisition frequency before collecting data, which can potentially lead to erroneous data acquisition. The error can occur if acquisition frequency is set too low. As shown in Fig. 1b, the two example excitatory postsynaptic currents (EPSCs) are identical; they are recorded from the same set of synapses, with the same activation and inactivation kinetics and peak amplitude. In the real data collection process, the initiation of EPSCs often slightly varies, which is known as jitter. In other words, even from the same synapses and under the same experimental conditions, the onset of synaptic events can be different. If the sampling rate (acquisition frequency) is low, it is possible that different sets of values are obtained from EPSCs with the same peak amplitude (Fig. 1b). Consequently, different peak amplitudes are obtained from the reconstructed traces (Fig. 1b lower). In many cases, the acquisition frequency is high enough not to miss the peak, but if it is not sufficiently high, uneven or unsmooth reconstructed traces are observed. This is relatively trivial, but still indicative of a lack of understanding about the amplifier and experimental conditions.

Conversely, it is also not true that the higher the acquisition frequency, the better the recording. When doubling the acquisition frequency, the stored data size is doubled accordingly. This slows down the data acquisition and subsequent data analysis processes. However, nowadays computers are fast enough and the hard drive is large enough to make this less worrisome.

So, what is the best acquisition frequency we should set in our experimentation? This question was carefully considered by Harry Nyquist and Claude Shannon, who contributed different aspects to the Nyquist-Shannon sampling theorem. This theorem basically states that the sampling frequency should be set to at least twice the highest frequency within a given signal.

Although there is no common rule for what is the best sampling frequency, experienced electrophysiologists often regard that 5–10 times the data bandwidth (a range of frequencies that make up the recorded signal) is sufficient. This sampling frequency is selected because it prevents aliasing, which is the distortion of a signal frequency (Fig. 2). By sampling at 5–10 times the data bandwidth, artifacts within the noise can be reduced.

3 Gain

Signals from synaptic transmission are often very small, typically around several picoampere or millivolts. When collecting these tiny electrical signals, the amplifier can increase the signal by setting the

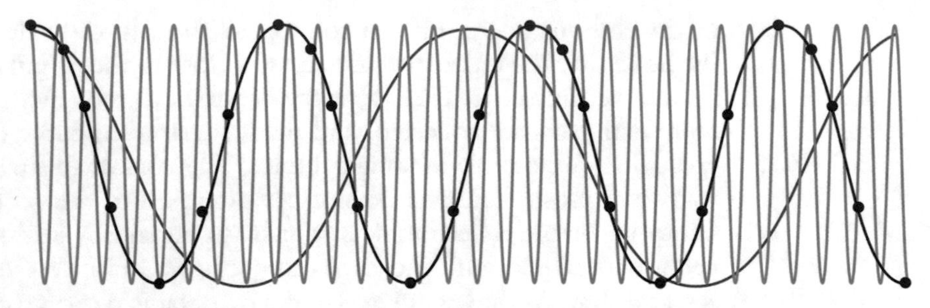

Fig. 2 Aliasing of frequency signals. The picture illustrates a frequency of interest (*black trace*) that is sampled over time (*black dots*). However, within the signal there lies higher (*gray trace*) and lower (*red trace*) frequency bands that are also sampled within the acquisition rate. This is called aliasing as the signal of interest is distorted by other signals sampled at the same time. To prevent aliasing, anti-aliasing filters are used, which remove higher and lower frequency signals leaving only the desired signals in the passband. In the illustrated figure, the *red trace* may be of interest. If this is the case, low-pass filters are used to remove high frequency signals. If the *gray trace* is of interest, the acquisition rate is increased

parameter "gain." The term gain that is used in electrophysiologists' everyday life is actually the gain factor, referring the ratio of output (e.g., voltage of current) to input. Thus, to set an appropriate gain factor, the investigator should have a clear mind about how much the signal she/he is going to measure and how much of the signal they would like to see.

When reconstructing the collected signals, the gain factor must be incorporated in order to generate the actual values of the original signal. A common mistake would occur when the gain factor was not offset during data analysis resulting in the current amplitude values being several folds higher than the actual values. Recently, powerful software can often trace the gain factor and offset the compensation "automatically."

During data acquisition, the electrical signals measured by the electrode are first amplified within the headstage (i.e., the headstage gain factor). The gain factor of the headstage is usually set by the manufacturer, thus predetermining the headstage's ability to acquire electrical signals. Setup of the headstage gain is often achieved by a resistor. The resistor is critically involved in preventing the acquired current from surpassing the measurable range. Therefore, the resistance is set depending on the size of electrical current. Assuming the input electrical current is large the resistance must also be large. In most voltammetric measurements of neurotransmitter release, electrical currents are large with respect to whole-cell recordings (microamperes vs. pico/nanoampere, respectively). Because of this, normal headstages for whole-cell patch clamp recordings are no longer feasible as the gain of these headstages is set to measure smaller currents. Normally headstages with small gain are available from the manufacturers. Thus, changing the headstage is sufficient to switch a patch amplifier to a voltammetric amplifier.

4 Capacitance Compensation

When current is applied to a biological membrane via channels or by the electrode that current charges the membrane before changing the membrane voltage; a characteristic caused by the biological membrane or the micropipette acting as a capacitor (Figs. 3a and 4a).

$$V(t) = V_{\text{inf}}\left(1 - e^{-t/\tau}\right)$$

As seen from the figure the voltage is not reached immediately, but instead it is approached with the time constant τ, defined by:

$$\tau = RC$$

Therefore, increases in capacitance or resistance (see Sect. 7 below for more information on resistance) increases the time it takes for changes in membrane voltage to be reached. For high-frequency recordings, a large tau is undesirable as signals of interest go undetected during the charging phase of the capacitor.

Similarly to passing a current, applying a voltage step charges and discharges the capacitor (seen as a current spike) (Fig. 3b). Unfortunately, these spikes known as capacitive transients can saturate the headstage circuit or later circuits, distorting signals of interest.

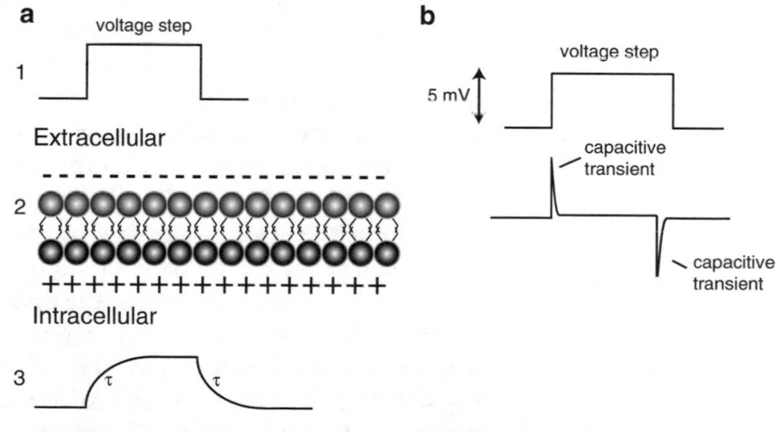

Fig. 3 The cellular membrane acts as a capacitor. (**a**) When a positive charge is applied to the intracellular compartment (*1*), the cellular membrane becomes charged as positive ions accumulate intracellularly and negative ions extracellularly (*2*). Charging of the membrane is illustrated in *3*. The time it takes for the membrane to reach a steady-state is determined by tau (τ). (**b**) A positive voltage step is applied to a cellular membrane via a micropipette in whole-cell mode. The positive voltage step elicits capacitive transients as the cellular membrane charges (*bottom*)

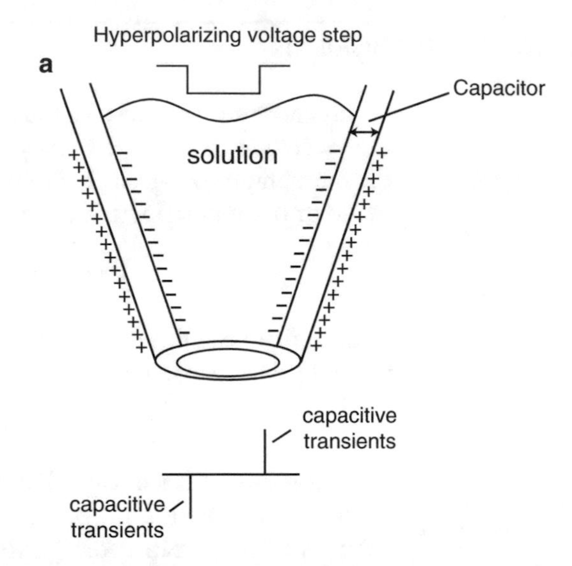

Fig. 4 A glass micropipette acts as a capacitor. (**a**) A hyperpolarizing step is passed through the micropipette generating charge separation across the glass as negative charges accumulate along the inside of the micropipette and positive charges accumulate along the outside of the micropipette. In a cell attached patch, the capacitance of the micropipette can be seen as capacitive transients (*bottom*)

Since capacitors can alter electrophysiological recordings, there are useful strategies to minimize their affects. They include manual or electrical techniques.

An experimenter can manually reduce micropipette capacitance by applying Sylgard to the micropipette tip, which thickens the recording electrode wall (remember capacitance is inversely proportional to distance or separation of a parallel plate). Minimizing the amount of bath solution covering the micropipette can also reduce capacitance. To accomplish this, the depth that the micropipette is submerged in the bath can be reduced or the micropipette's surface tension can be reduced. Decreasing the surface tension prevents the bath solution from coating the micropipette. To reduce surface tension, the micropipette can be coated in a hydrophobic compound such as mineral oil (or silane) by submerging it immediately before use. However, it is important to only submerge a filled micropipette as the aqueous solution in the pipette prevents the mineral oil from entering the tip.

To prevent stray capacitance, the holder or micropipette should not be placed too close to any grounded surfaces (e.g., objective, ground wires). The choice of pipette glass can also reduce micropipette capacitance. Glass with a low dielectric constant and a low dissipation factor are desirable. Borosilicate is the most common glass used for pipettes (dielectric constant = 4.5–6, dissipation factor = 0.002–0.005), while quartz is considered the gold standard,

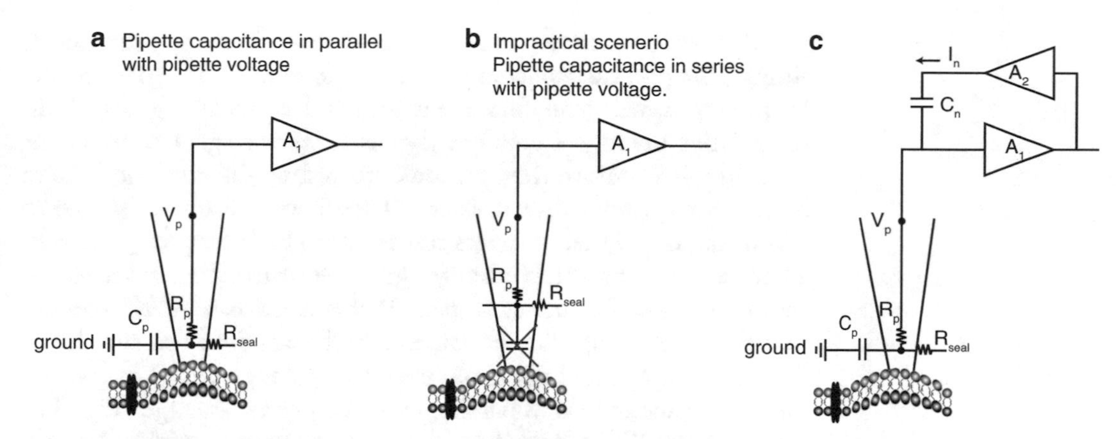

Fig. 5 Electrical diagram explaining how pipette capacitance distorts rapid ionic currents. (**a**) Electrical circuit diagram illustrating the pipette capacitance (C_p) is situated in parallel to the pipette voltage (V_p). R_p refers to pipette resistance, Reseal refers to the seal resistance and A1 refers to amplifier. (**b**) Electrical circuit diagram showing that the pipette capacitance has to be situated in series with the pipette voltage in order to act as a high pass filter. In electrophysiology, this is never the case as the pipette capacitance sits in parallel to the pipette voltage. (**c**) Electrical circuit diagram showing how the pipette capacitance is corrected for in an electrical circuit. Here, a current is passed through the capacitor connected to the headstage input equaling the pipette current

but with drawbacks (see Chap. 5) (dielectric constant = 3.86, dissipation factor = 10^{-5} to 10^{-4}) [1].

In addition to these techniques, electrical manipulations can be performed. For example, capacitance can be compensated for by injecting current that is equivalent to the stray capacitance current that is lost to ground (Fig. 5c) (see Sects. 5 and 6 below).

5 Pipette Capacitance Compensation

Pipette capacitance is introduced from a charge separation across the electrode and the bath. This charge separation can occur during a voltage step as current passes through the recording electrode (Fig. 4a). The result is a capacitive transient easily noticeable in cell-attached patch (Fig. 4a). Using pipette capacitance compensation can decrease root mean square noise (r.m.s.) (baseline) noise, it can quickly bring the electrode to the desired potential preventing rapid-onset ionic currents from being distorted, and it can remove capacitive transients.

How does capacitance increase r.m.s. noise? The association is given by

$$si = 2\pi \text{fc} \cdot Cs \cdot V$$

where "*s*" is the r.m.s. noise for current (i) and voltage (V), fc is the bandwidth, and C is the total capacitance from input to earth [2]. Clearly then, any increases in capacitance can lead to increases in r.m.s. noise.

Pipette capacitance distorts rapid-ionic currents; to put it simple, the pipette capacitance acts as a low-pass filter (passes low-frequency signals and filters out higher frequency signals). This means that any high frequency signals are sent to ground and consequently lost before they can be detected by the electrode. Here is why; capacitors act as resistors. They have greater resistance to low frequency signals and less resistance to high frequency signals. That means that high frequency signals pass through a capacitor far easier than low frequency signals. If this is the case, it is expected that the pipette capacitance acts as a high-pass filter (passing high-frequency signals and filters out lower frequency signals). However, pipette capacitance is in parallel with the pipette voltage (Fig. 5a), not in series. Because of this, high-frequency signals take the path of least resistance through the capacitor to ground. For pipette capacitance to act as a high-pass filter, it would have to be situated in series with the pipette voltage (Fig. 5b), but this is not the case in electrophysiology. Therefore pipette capacitance distorts rapid-ionic currents by acting as a low-pass filter with a time constant, $\tau_P = R_P C_P$ where R_P and C_P refer to the pipette resistance and capacitance, respectively. From the equation we can see that by reducing the pipette resistance we can compensate for this time constant. However, this is typically not feasible for electrophysiology experiments so capacitance compensation is used.

So how does pipette compensation work? In cell-attached mode the capacitive transients are directly related to the pipette capacitance. To remove these transients and offset the pipette capacitance a current is passed through the capacitor connected to the headstage input to exactly equal the pipette current (Fig. 5c) ($I_c = C\mathrm{d}V/\mathrm{d}t$). This correction compensates for the net charge referred to as the magnitude and the time constant referred to as tau. Both the net charge and the time constant have a fast and slow component, which is corrected by most amplifiers [3].The fast component is related to the initial charge of the capacitor while the slow component occurs as the charge moves to the pipette tip.

6 Whole-Cell Capacitance Compensation

After breaking into the cell, two more components are added to the circuit. They are the membrane resistance (R_m) and the membrane capacitance (C_m) (Fig. 6a). The membrane resistance is regulated by membrane channels. Low membrane resistance means many ionic channels are opened and therefore they are passing current (it can also mean that there is a leaky whole-cell patch!). Membrane capacitance is given by

$$C = \varepsilon_0 A / d$$

where ε_0 describes lipid membrane properties, d is the membrane thickness, and A is the cell surface area. Calculating the capacitance

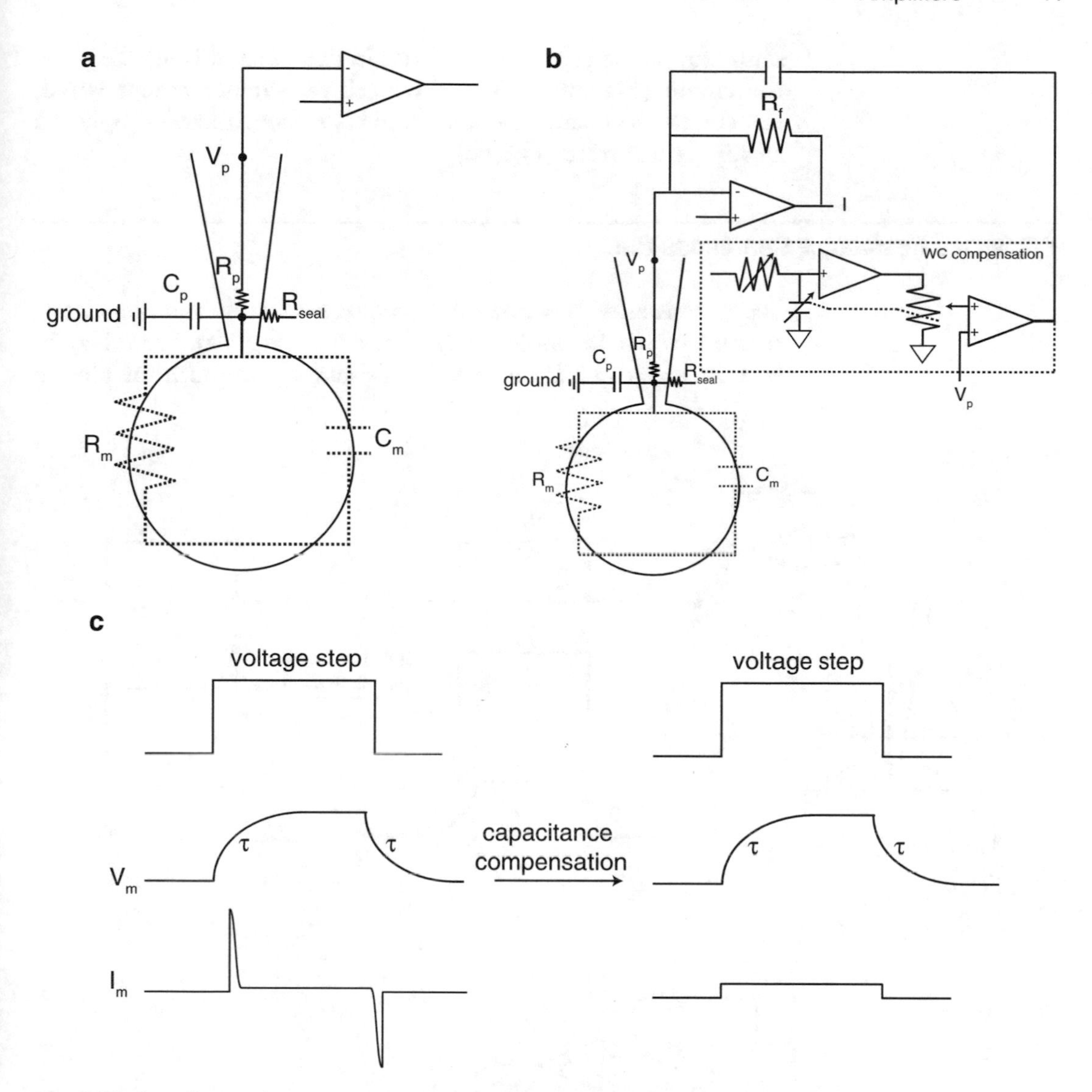

Fig. 6 Whole-cell capacitance compensation. (**a**) Whole-cell configuration introduces membrane resistance (R_m) and capacitance (C_m) to the pipette capacitance (C_p), pipette resistance (R_p), and seal resistance (R_{seal}). (**b**) A whole-cell compensation circuit showing how current is injected into the circuit in order to compensate for C_m. R_f refers to the feedback resistor. (**c**) With whole-cell capacitance compensation, during a voltage step, the membrane voltage (V_m) rises exponentially, but the membrane current (I_m) rises slowly (without transients)

(typically calculated by software after whole-cell access) can provide useful information regarding the cell size, which can help identify neurons of interest. However, like pipette capacitance, whole-cell capacitance can have negative effects on recording (see Sect. 7). Like pipette capacitance, whole-cell capacitance can lead to distortion of rapid-ionic currents since the capacitor sits in parallel with the pipette voltage (see Sect. 5 for explanation above). Because of this, the whole-cell capacitance is compensated for in much the same way as the pipette capacitance in that current is injected into the circuit (Fig. 6b). With whole-cell capacitance compensated, the current to charge the membrane is removed leaving only the steady state current (Fig. 6c).

7 Series Resistance Compensation

Series resistance describes the pipette resistance and the access resistance once in whole-cell configuration (Fig. 7a). From Fig. 7a you can see why it is called series resistance since both the pipette

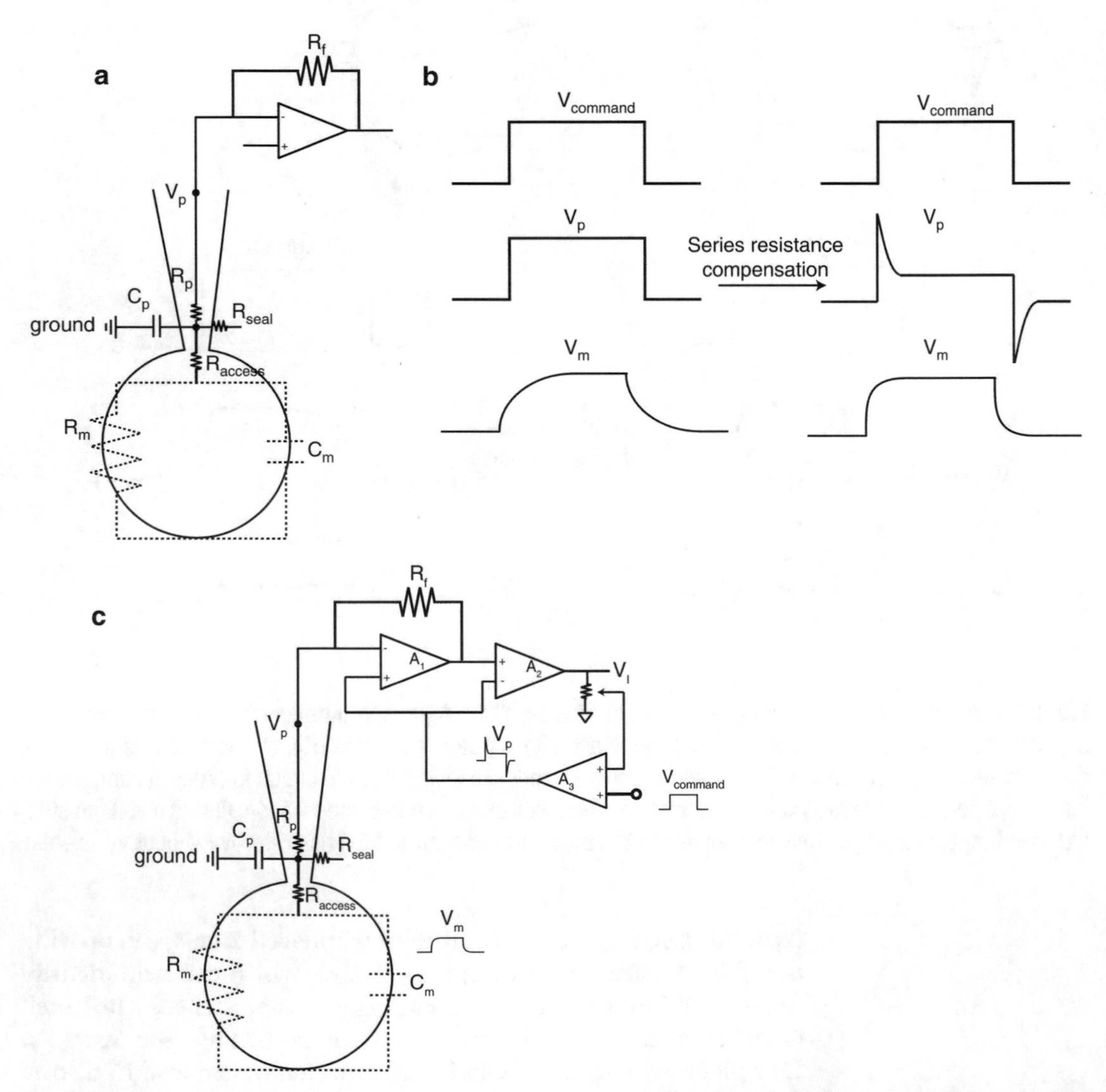

Fig. 7 Series resistance compensation explained. (**a**) Electrical circuit diagram showing the series resistance, which includes the pipette resistance and the access resistance once in whole-cell configuration. (**b**) By compensating for the series resistance, current is injected (**c**) An electrical circuit diagram including the series resistance compensation circuit

and access resistance are in series (not in parallel) with the pipette voltage. During an electrophysiological recording, the pipette resistance will not change. This means that any changes in the series resistance are due to changes in cellular access. Many electrophysiologists monitor the series resistance during a recording and discard recordings with >20 % changes in series resistance over experiments. This is because changes in cellular access in whole-cell mode can have drastic consequences on measured currents and potentials.

If the series resistance is compensated, monitoring changes in the access resistance during a recording is not practical. However, series resistance compensation may still be necessary depending on the experimental design. The necessity is because series resistance adds steady-state errors and dynamic errors that may need to be reduced. Steady-state errors arise because the membrane voltage (V_m) is given by

$$V_m = V_P - I_m R_{series}$$

which means that V_m does not equal V_P, but rather V_m relies also on the membrane current (I_m) and the series resistance (R_{series}). By compensating for the series resistance, V_m approaches V_P reducing the probability of steady-state errors.

Dynamic errors occur during rapid voltage changes like those performed in voltage steps. The membrane voltage changes with lag time (τ) determined by

$$\tau = R_{series} C_m$$

where C_m refers to the membrane capacitance. This equation should look familiar since it closely resembles the time constant for the pipette lag time, $\tau_P = R_P C_P$, which if left uncompensated can lead to dynamic errors as well. So this means that it is necessary to apply series resistance compensation in order to have V_m approach V_P (Fig. 7b) or to be able to record rapid-ionic currents during a voltage step.

The concept behind series resistance compensation is similar to pipette and whole-cell capacitance compensation discussed above in that current is injected into the circuit. Series resistance compensation removes some of the load from the feedback resistor (R_f) when R_f supplies the current to charge C_m (Fig. 7c). This allows V_m to approach V_P, thus reducing the lag time (τ).

An important note is that in order to perform series resistance compensation accurately, you must perform the pipette and whole-cell capacitance compensations. This is because the membrane capacitance is directly related to the lag time associated with dynamic errors ($\tau = R_{series} C_m$) and the pipette capacitance acts as a low-pass filter also contributing to rapid-onset current distortion.

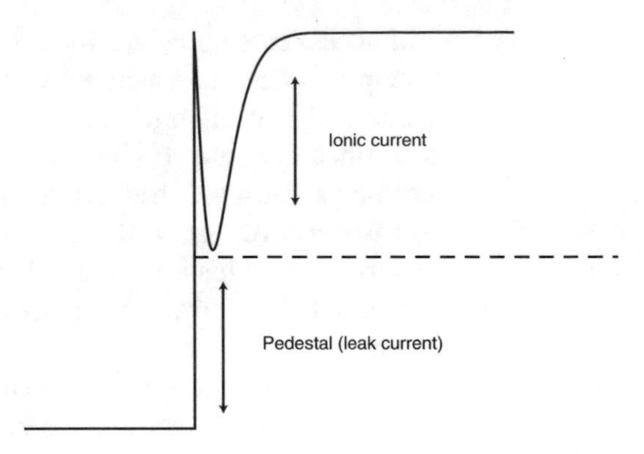

Fig. 8 An illustration of leak current, which is seen as a pedestal rising prior to the ionic current

8 Leak Subtraction

Leak conductance occurs at the pipette–membrane interface or from ungated "leak" channels expressed in the membrane. Uncorrected leak conductance results in "pedestals" in the trace (Fig. 8). To subtract leak currents, two approaches can be used. The first approach takes an average of sweeps during a time when no channel activity has occurred ($n = 10$ makes any root mean square noise negligible; $\sqrt{1+1/n}$) and subtracts that average form each sweep to be analyzed in the data set. The second technique, described by Bezanilla and Armstrong [4], is called the P/4 subtraction and can be used for macroscopic current and single-channel current recordings. The idea is to use a scaled-down version of the test pulse typically one fourth of the value [4]. So, for example, if the control pulse were to go from −40 mV to 0 mV the scaled down pulses would be from −60 mV to −50 mV. The experimental pulses can then be subtracted from the P/4 scaled pulses. Alternatively, some software includes a leak subtraction; simply clicking the button results in automatic leak subtraction, which is determined by the software (e.g., Multiclamp 700B).

9 Junction Potential Compensation

Junction potentials are created when interfaces with different ionic compositions come into contact (a junction). In electrophysiology, a liquid–liquid junction potential occurs at the pipette tip because internal solution (liquid) comes into contact with the external solution (liquid) of different ionic concentration and mobility causing electrochemical gradients to form. Consequently, ions move between the two solutions creating a potential. The ionic movement is dependent upon the mobility of each ion. So if slower moving ions are negatively charged (e.g., aspartate) and faster moving ions are positively charged, one side of the junction will be negatively charged while the other side of the junction will be positively charged, thus creating a potential difference.

Liquid–liquid junction potentials need to be accounted for in electrophysiology experiments because errors can be introduced when the pipette current is zeroed in the bath. When the pipette current is zeroed, it sets the pipette potential with respect to the external solution or bath potential (determined by the ground). In other words, an opposing voltage is applied to offset the potential difference between the internal and external solutions. However, the liquid–liquid junction potential will be present at the pipette tip altering the potential detected by the electrode wire (Fig. 9). Because of this, the pipette offset is not truly at zero, but instead incorporates the liquid–liquid junction potential. Correcting the junction potential error is fairly simple. Fill the pipette and the bath with internal solution and zero the pipette offset. Then change the bath to the external solution and measure the change in the zero current potential. Input the value into the equation describing the

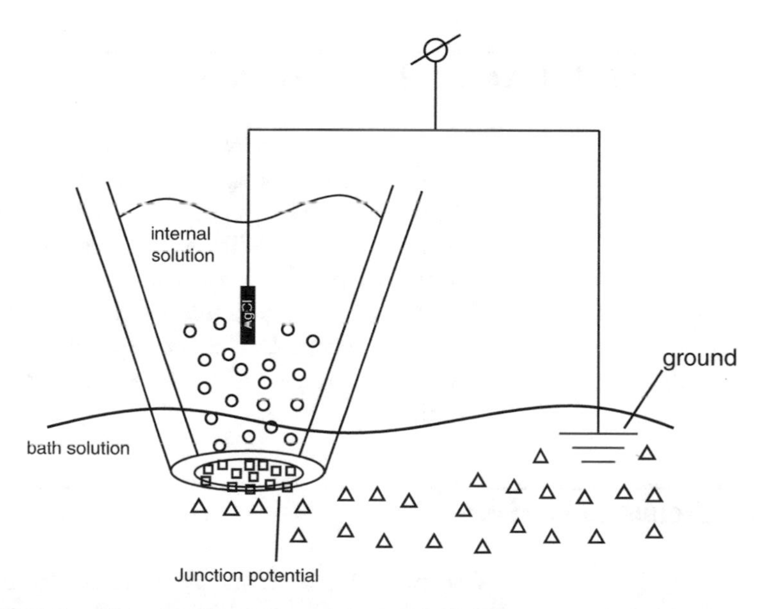

Fig. 9 Junction potentials arise at the interface between internal and external solutions. The illustration shows electrical gradients as shapes. *Circles* represent electrical gradients contained in the internal solution, which are detected by the AgCl electrode. *Triangles* represent electrical gradients contained in the external solution, which are detected by the ground wire placed in the bath. The pipette offset is subtracting the difference between the electrical gradients in the internal and external solutions (i.e., the subtracting the difference between the *circle* and the *squares*). However, the potentials generated at the internal-to-external solution interface (*squares*) go undetected, thus altering the true potential

patch-clamp mode used (V_M = membrane potential, V_A = patch-clamp amplifier potential, V_{LJ}, for liquid junction potential, V_C = cell potential).

Inside-out	$V_M = -V_A + V_{LJ}$
Cell-attached	$V_M = V_C - V_A + V_{LJ}$
Outside-out	$V_M = V_A + V_{LJ}$
Whole-cell	$V_M = V_A + V_{LJ}$

The above equation for whole-cell patch is a simplification since a few problems arise as the pipette tip comes into contact with the cytosol. The whole-cell equation is acceptable for small cells (10–20 μm in diameter). This is because small mobile ions in the cytosol equilibrate across the pipette tip within a few seconds. However, this is not always the case in larger cells. Another potential error can occur if the cytosol contains immobile anions, which cannot be accounted for when calculating the junction potential. This causes the cell interior to be negative in respect to the pipette. These immobile anions eventually diffuse out of the cell (e.g., 15 min for some cells) correcting the error [5, 6].

10 Bath-Related Errors (Bath Error Potentials)

Bath-related errors arise when the bath solution is not relative to ground. This error can be caused by the ground electrode (typically an Ag/AgCl pellet) when the resistance of the ground electrode is too high. With a high resistance, the error voltage can be as much as 10 mV or more leading to potential shifts in *I–V* curves. Therefore, the resistance of the ground electrode can be minimized by reducing the access resistance or by using a grounding pellet with a large surface area that is totally submerged in the bath solution.

11 Ground-Related Error

Ground-related errors are often easy to identify and easy to fix. For example, on the oscilloscope before the recording micropipette is placed in the bath, a flat line can be detected. If the ground is not in the bath solution or the ground is not connected to the headstage, large oscillatory waves can be detected (Fig. 10). Another error can occur if the ground is not coated with chloride. This can lead to current or voltage drift easily detected on the oscilloscope. A poorly coated reference pellet can also change the pipette offset correction, which could lead to inaccurate current or voltage measurements. Therefore, when current or voltage drift is detected, the ground pellet should be coated with chloride.

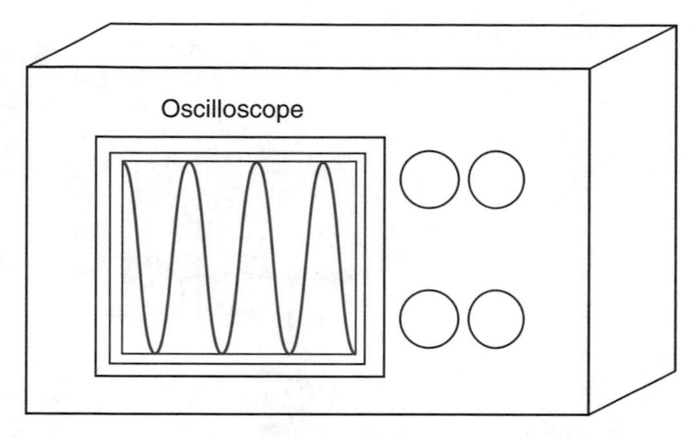

Fig. 10 An oscilloscope showing large sinusoidal waves (*red*) that occur when the ground wire is removed from the bath solution

12 Bridge Balance

In current clamp recordings rapid current injections are typically used. As current passes from the electrode through the pipette to the cell, it is hindered by the pipette and access resistance. This should sound familiar since we already discussed series resistance compensation. Bridge balance is very similar to series resistance compensation in that if left unaccounted for, errors can arise shown mathematically by

$$V_m = V_p - IR_{series}$$

$$\text{and} \quad \tau = R_{series} C_m$$

The bridge balance essentially removes the pipette resistance preventing the series resistance from distorting V_m (Fig. 11). Passing negative current pulses while adjusting the bridge balance can remove the steady-state pulses, thus balancing the bridge (Fig. 11). Passing positive current should be avoided because it can lead to the activation of voltage-gated channels. By balancing the bridge the instantaneous voltage step is removed leaving only V_m. Cellular access during current clamp recordings can be monitored by checking the bridge balance before and after a recording. An electrophysiological recording with changes in bridge oscillation >20 % over the experiment is often regarded as unstable recording and thus discarded.

All things considered, why is it necessary to go to all of this trouble to compensate for resistance? It is because of the voltage drops that occur across resistors. Basically, the voltage before the resistor is greater than the voltage after the resistor. This is because the resistor (pipette and access resistance) prevents the equal flow of current from the source to the cell membrane. Therefore, the resistance prevents V_m from equaling V_p.

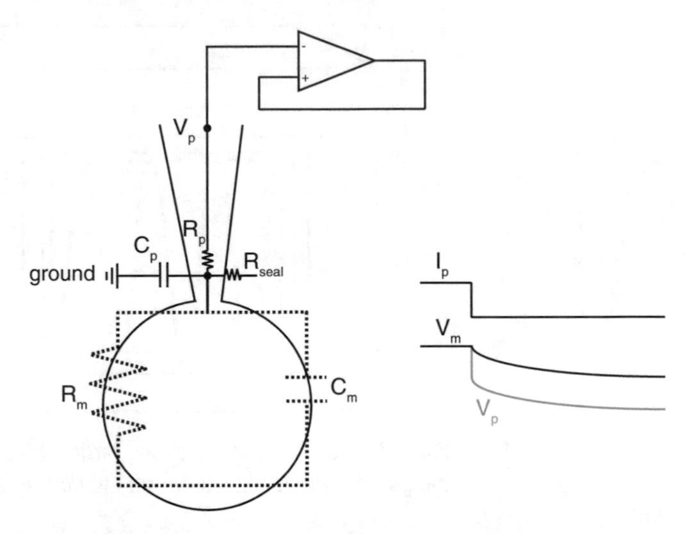

Fig. 11 An electrical circuit diagram showing how the pipette resistance (R_p) is included in the electrical circuit. By balancing the bridge, an instantaneous voltage drop is removed leaving only V_m

13 Electrode

Coating of the electrode wire: In many circumstances, beginners complain that the signals they record from are not stable; the signals drift or oscillate. A common reason is that the electrode is not well prepared. Current generated from the recorded cell must flow smoothly to the electrode wire in order for an authentic signal to be detected. Current, which is carried by ions, must cross the interface between the internal solution and the metal wire (most commonly silver wire). Crossing this interface can often cause signal loss or distortion. To prevent signal loss or distortion and ensure the smooth flow of current at the solution–wire interface, the silver wire is coated with a thin layer of silver chloride. When electrons flow from the wire through the silver–silver chloride interface, they (electrons) reduce silver + back to silver atom, and at the same time free the chloride ion, which goes into the solution. If the electrons flow in the opposite direction, the silver atom loses one electron resulting in silver+, which then forms silver chloride with the nearby chloride ion. The newly formed silver chloride is insoluble and therefore stays within the coating layer. If no such interface is provided, other known or unknown electrochemical processes may occur. This may not only cause signal loss or delay but also generate ions that change the recording solution. For example, naked copper wire was used in earlier electrophysiology studies. Through electrolysis, unwanted gases (e.g., O_2) and unwanted ions (e.g., $H+$) were generated around the wire reducing the solution–wire interface and changing recording conditions (e.g., pH).

Several common mistakes should be considered when using a silver chloride electrode. Probably the most common one is to use a silver chloride electrode in non-chloride-containing internal solutions. In many experimental designs in which chloride-mediated conductance is attempted to be minimized, gluconate, methosulfate, or other negatively charged organic molecules are used to replace chloride ions. If the internal solution does not contain chloride ions, the reversible silver-silver chloride transition cannot be achieved, and all the advantages we have talked about with this strategy no longer exist. To overcome this, many experienced investigators add a small amount of chloride ions (e.g., 5 mM cesium chloride) to set up the condition. Another common mistake is to regard the sliver chloride electrode as permanent. The electrode does wear out after repeated use, especially during repeated experiments in which current always flows in the same direction.

14 Electrode Circuit and Headstage

The exchange of ions occurring at the electrode interface is transmitted from the electrode to a gold pin, which is connected to the headstage. Excessive drift in the pipette offset (even with a chloride coated electrode) can occur if contact between the electrode and the gold pin is diminished. This happens when the gold pin becomes tarnished or when the electrode's surface area available to contact the gold pin is insufficient. Most often it is the latter of the two so we suggest one of two options; either spiral the electrode at one end or flatten one end with needle nose pliers (Fig. 12). In doing this, it ensures that the gold pin makes sufficient contact with the electrode, and thus forms a solid electrical connection.

Connected to the headstage is the gold pin and the reference/bath electrode. The reference electrode is commonly an AgCl pellet (coated with chloride) connected to a copper wire, which is then attached to the headstage via a gold pin. Pipette offset drift can be caused by a poor ground. It is important that the ground is coated well with silver chloride and submerged in the external solution throughout the recording. Another helpful tip is to make

Fig. 12 Illustrations of various options to ensure that the AgCl wire makes good contact with the gold pin, which connects the electrode to the headstage. An AgCl wire (*left*) can be spiraled (*middle*) at the end facing the gold pin or flattened (*right*)

sure that the ground has a large surface area allowing contact to as many available ions as possible. A ground with a large surface area cuts down on the time it takes for the pipette-offset drift to equilibrate.

15 Troubleshooting Noise

Thermal noise: Thermal noise is caused by thermally excited charge carriers moving randomly in a conductor.

Dielectric noise: Dielectric noise is produced from insulators that occupy the space between capacitors. These insulators dissipate electrical potential energy in the form of heat, thus making dielectric noise a form of thermal noise (e.g., glass is the insulator of a pipette capacitor. See Chap. 5).

Excess noise: Excess noise describes any noise that is present in the in the circuit excluding thermal noise, shot noise (arises when current crosses a potential barrier), and dielectric noise. Excess noise can be generated when DC current passes through a resistor (referred to as $1/f$ noise since this noise is observed in addition to the thermal noise of the resistor).

Amplifier noise: Amplifier noise refers to the intrinsic noise of the amplifier. This noise is generated by an input voltage noise in series with the negative input and an input current noise between the positive and negative inputs (Fig. 13). Nowadays, amplifiers manufactured for electrophysiology have very low intrinsic noise making amplifier noise negligible.

Electrode noise: Noise stemming from the electrode comes from the capacitance generated from the electrode holder and the patch pipette. As discussed previously in this chapter and in Chap. 5, the capacitance of the patch pipette can be reduced by using thick glass with a small dissipation factor, applying Sylgard to the patch

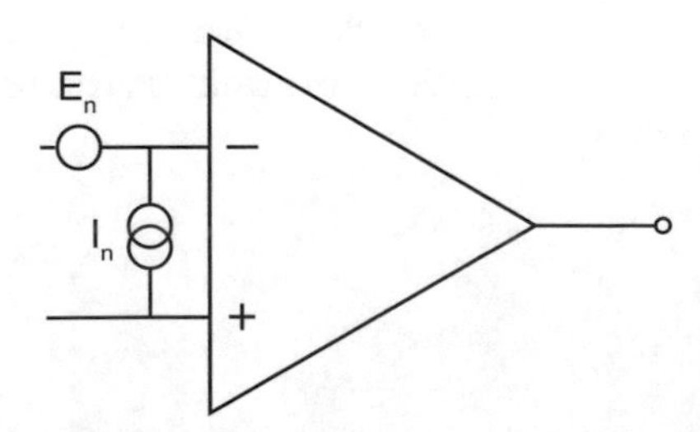

Fig. 13 An illustration of intrinsic amplifier noise and electrode noise. The amplifier noise is generated by the input current noise (I_n), which occurs between the positive and negative inputs of the amplifier. The electrode noise is introduced at the negative input, which is connected to the electrode holder and patch pipette (micropipette)

pipette tip, and minimizing the patch pipette depth in solution during recording. In cell-attached patch, the patch pipette resistance is considered to be negligible as long as the Ag/AgCl electrode protrudes as far as possible to the tip of the patch pipette (typically 4 mm from the tip). This is because the Ag/AgCl wire shorts out the resistance of the pipette-filling solution.

External noise sources: External noise comes from line-frequency pickup (harmonics and 60 Hz) generated from power supplies, fluorescent lights, laptops, monitors, cell phones, etc. One of the most often seen noises in electrophysiology is "60 cycle noise" produced from 60 Hz frequencies from power supplies (AC current in America oscillates at 60 Hz). Sixty cycle noise can be removed, like most external noise, by grounding, filtering, and shielding (aluminum foil wrapped around electrical wires). It is best to shield electrical wires, but refrain from using metal shielding for the pipette holder because the metal shielding increases noise. The best approach to determining the source of external noise is to systematically turn off/on the power of any piece of equipment nearby until the noise disappears. Once identified, shielding, grounding or removing (cell phone) the source should do the trick.

16 Filtering

Low-pass filters: Low-pass filtering eliminates signals and noise above the cutoff frequency passing only low-frequency signals. Say for example that the current you are trying to record is masked by noise in a 10 kHz bandwidth. By using a low-pass filter at 1 kHz, the current may become easily distinguishable (Fig. 14).

High-pass filters: High-pass filters (or AC coupling) remove signals and noise below the frequency range of the desired signals. In other words, the signals that pass through the high-pass filter will be larger than the filtered frequency.

Notch filters: Notch or band-reject filters, which are typically used in electromyogram (EMG) recordings, are used to eliminate line-frequency pickups such as 60 Hz (60 cycle) noise. Since line-frequency pickups are common in electrophysiology recordings, adding a notch filter seems like a reliable solution. However, notch filters should be used cautiously in time-domain analysis (time-domain analysis refers to signal analysis as it would appear on a conventional oscilloscope). This is because the signals of interest may encompass the notch frequency causing them to become distorted (Fig. 15). A commonly used filtering method for time-domain analysis is the Bessel filter because this filter adds less than 1 % overshoot to pulses. This minimal distortion is because in a Bessel filter, the phase change is linear with respect to frequency [7].

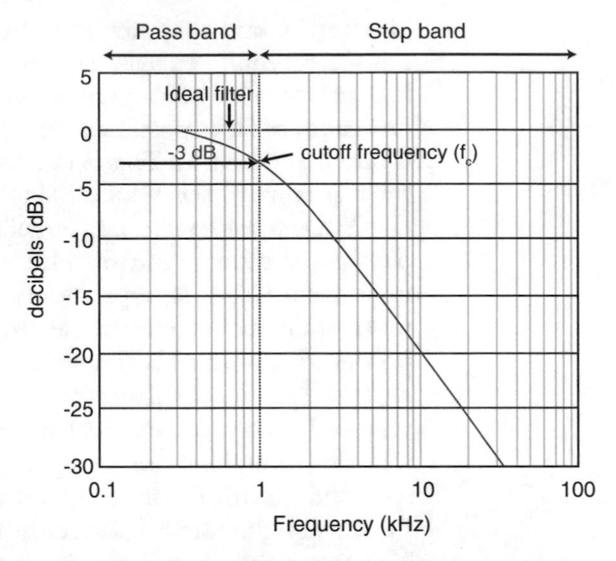

Fig. 14 A low pass Bessel filter passing 1 kHz frequency. At the cutoff frequency or −3 dB frequency, the energy passing through the system is reduced. A perfect filter would have a pass band to 1 kHz with a stop band >1 kHz, but as illustrated in the figure, the pass band is the frequency region below the cutoff frequency, while the stop band is defined as the frequency region above the cutoff frequency

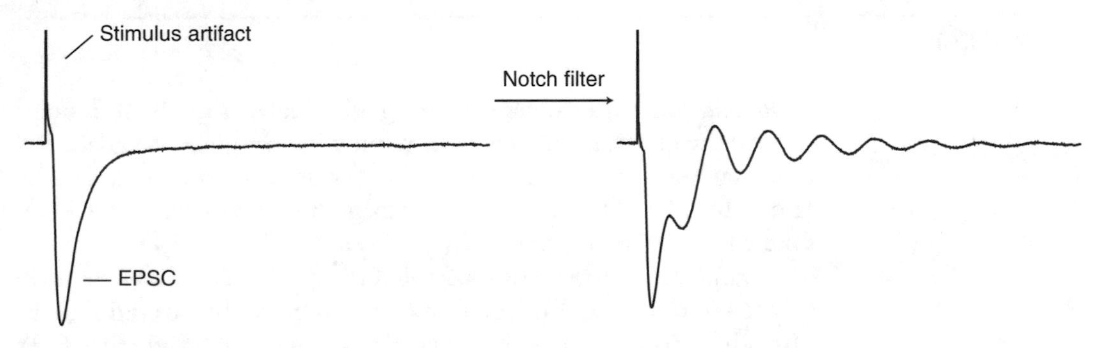

Fig. 15 Applying a notch filter to a current trace can distort the trace. In this illustration, the trace has a 60 Hz component that is filtered out by the Notch filter causing distortion

17 Summary

Currently, electrophysiologists have state-of-the-art equipment enabling corrections of unwanted electrical signals. Unfortunately, this comes at a cost as many parameters that adjust for capacitance, resistance, or bandwidths are done so without any knowledge as to why or how they work. By covering many different aspects of amplifier function in this chapter, we hope that beginning electrophysiologists have gained sufficient knowledge with respect to amplifier and software functions. With such knowledge, accurate measurements of cellular excitability can be collected, thus appropriately advancing our knowledge of cellular function.

References

1. Levis R, Rae J (2002) Technology of patch-clamp electrodes. In: Walz W, Boulton A, Baker G (eds) Patch-clamp analysis. Humana, Totowa, pp 1–34, 35
2. Ogden D (1994) Microelectrode techniques: the Plymouth Workshop handbook. Company of Biologists Limited, Cambridge
3. Simon SA, Nicolelis MAL (2001) Methods in chemosensory research. Taylor & Francis, Boca Raton
4. Armstrong CM, Bezanilla F (1977) Inactivation of the sodium channel. II. Gating current experiments. J Gen Physiol 70(5): 567–590
5. Barry P, Lynch J (1991) Liquid junction potentials and small cell effects in patch-clamp analysis. J Membr Biol 121(2):101–117
6. Neher E (1992) Correction for liquid junction potentials in patch clamp experiments. Methods Enzymol 207:123–131
7. Axon Instruments (1993) The Axon guide for electrophysiology & biophysics laboratory techniques. Axon Instruments, Foster City

Chapter 4

Salt Environment

Nicholas Graziane and Yan Dong

Abstract

In order to develop physiologically relevant experiments that test electrical activity in and between neurons, it is necessary to closely model the physiological milieu. This chapter discusses external and internal solution components for in vitro brain slice electrophysiology. These recipes have been developed throughout the years to model the physiological environment and to help produce viable, healthy neurons.

Key words Ionic concentrations, Brain slice preparation, Osmolality, Osmolarity

1 Introduction

In order to develop physiologically relevant experiments that test electrical activity in and between neurons, it is necessary to closely model the physiological milieu. This chapter discusses external and internal solution components for in vitro brain slice electrophysiology. These recipes have been developed throughout the years to model the physiological environment and to help produce viable, healthy neurons.

Below we include an introduction to the ingredients used in external and internal solutions along with reasons for each component. In addition, knowledgeable references are provided to assist the reader with troubleshooting strategies to use when encountering unhealthy neurons.

2 External Bath Solution

Typical components of cerebral spinal fluid are listed in Table 1. Mimicking these components exactly is experimentally impossible due to the molecules (e.g., proteins) that are present in vivo. However, typical extracellular recording solutions used today

Nicholas Graziane and Yan Dong, *Electrophysiological Analysis of Synaptic Transmission*, Neuromethods, vol. 112,
DOI 10.1007/978-1-4939-3274-0_4, © Springer Science+Business Media New York 2016

Table 1
Typical components of cerebral spinal fluid in humans and rats

Ion/molecule	Concentration (human)	Concentration (rat)
Osmolality	291.5 ± 4.0 mOsm/kg[a]	302 ± 4[b]
Na^+	147 mEq/kg[c]	148 ± 0.8 mEq/L[d]
K^+	2.86 mEq/kg[c]	3.16 ± 0.04 mEq/L[d]
Cl^-	113 mEq/kg[c]	117.9 ± 0.9 mEq/L[d]
Ca^{2+}	2.28 mEq/kg[c]	40.4 ± 4.1 μg/L[e]
Mg^{2+}	2.23 mEq/kg[c]	18.8 ± 0.6 μg/L[e]
HCO_3^-	23.3 mEq/kg[c]	24.3 ± 0.7 mEq/L[d]
Iron	61.01 ± 18.3 μg/L[f]	
Glucose	59.7 mg/dL[c]	65 mg/dL[b]
Amino acids	0.72 mEq/kg[c]	414.8 μmol/L[g]
Protein	39.2 mg/dL[c]	
Albumin	304 ± 126 mg/L[h]	10.9 ± 0.3 mg/100 mL[i]
Lactate	1.4–1.5 mmol/L[j]	2.8 ± 0.2 mmol/L[b]
Creatinine	54.4 nmol/mL[k]	
Phosphorus	1.2–2.0 mg/dL[c]	
Urea	4.7 mmol/L[c]	
pCO_2^-	50.5 mmHg[c]	
pH	7.3[c]	7.35[b]

mEq milliequivalent
[a][1]
[b][9]
[c][2]
[d][7]
[f][8]
[e][3]
[g][10]
[h][4]
[i][11]
[j][5]
[k][6]

(see Table 2) include many of the critical ionic components that are necessary for electrical energy transfer for an in vitro neuronal preparation.

2.1 Typical Ionic Concentrations

The typical external solution used for in vitro electrophysiological measurements consists of the components listed in Table 2. This artificial cerebral spinal fluid (aCSF) mimics the physiological ionic

Table 2
Typical aCSF components for electrophysiological recordings.

Compound	Concentration (mM)
$NaCl^-$	130
KCl^-	3
$MgCl_2^-$	5
$CaCl_2^-$	1
$NaHCO_3^-$	26
$NaH_2PO_4^-$	1.25
Glucose	10

concentrations thereby maintaining synaptic and neuronal function similar to what would be seen in vivo. To further mimic physiological conditions, the pH and osmolality of the aCSF should be adjusted accordingly to the values provided in Table 1. To maintain a physiological pH, aCSF should be continuously bubbled with 95 % oxygen/5 % carbon dioxide and contain sodium bicarbonate ($NaHCO_3^-$) in order to keep the pH at ~7.4. How? Carbon dioxide gas dissolved in water can react with H_2O to form carbonic acid causing the pH to drop (pH = 5.7). When carbon dioxide gas escapes it can react with sodium ions forming sodium carbonate making the solution alkaline. By adding the right amount of sodium bicarbonate to a bubbling solution containing 5 % CO_2, the pH will be buffered preventing carbonic acid or sodium carbonate formation. After adjusting the pH, the osmolality of the solution needs to be set to the physiological ranges. If the aCSF is heated during brain slice recovery (see below) or during electrophysiological recordings, it is best to check the osmola*L*ity and not the osmola*R*ity. This is because osmola*R*ity measures osmoles of solute per liter of solution (Osmoles/L). That means that any changes in the volume of solution caused by temperature and/or pressure can cause the osmola*R*ity to fluctuate. Because of this osmola*L*ity is preferred because it will remain constant regardless of changes in temperature or pressure. Osmola*L*ity is measured as moles of solute per kilogram of solvent (Osmol/kg) and thus is not dependent upon volume changes.

Often it is necessary to adjust the osmolality of the external solution to 280–300 Osmol/kg. If the osmolality is too high, water can be added. A quick tip to determine how much water to add: divide the desired value by the actual value, subtract 1, and multiply this value by the total volume of solution (in mL). If the osmolality is too low it can be raised by adding some more 10× stock aCSF solution or by adding glucose.

3 Selection of Bath Solution

3.1 Cutting

The most critical step for in vitro electrophysiological recordings is the preparation of brain slices. Healthy brain slice preparation takes into account multiple factors such as the solutions used and the speed with which the slices are cut. In addition, when using vibrating blades, the vibrations along the *z*-plane must remain minimal in order to reduce damage to the tissue.

The cutting solutions must be pH balanced and have the osmolality adjusted to the proper range (280–300 mOsm). aCSF can be used for tissue slicing, but the cell viability can decrease due to Na^+ rushing into the cell. As Na^+ rushes in, extracellular solution enters the cell leading to cell swelling and eventual death. Therefore, solutions that lack Na^+ such as sucrose or *N*-methyl-d-glutamine (NMDG) can be substituted for aCSF [12]. The temperature of the solutions is up to the experimenter since evidence for healthy brain slice preparations has come from varying temperatures (e.g., 0–4, 20–23, or 37 °C) [12–14]. Although, from personal experience, colder brain tissue is easier to section compared with warmer preparations.

Another contributing factor to healthy brain slices is the speed with which the blade passes through the brain tissue. Nowadays, most electrophysiology laboratories have a machine specifically manufactured for brain slice preparations. It is advised when using these brain slice machines to keep the speed at or below 0.12 mm/s and the frequency fixed at 85 Hz for unfixed, soft tissue. Higher speeds will cause the blade to push, instead of glide through the tissue creating uneven brain slices. It is also important to adjust the clearance angle of the blade to the desired position. For soft tissues, clearance angles of 15° or 18° are commonly used (Fig. 1a). However, this can be altered if cell quality is poor and, if necessary, should be optimized by the experimenter. In addition to the cutting speed and clearance angle of the blade, the vibration amplitude along the *z* axis for soft tissue is typically set at 1.0–1.5 mm (Fig. 1b). If cells are suddenly appearing unhealthy, the vibration amplitude should be checked for accuracy, which is service provided by some manufacturers (e.g., Leica Biosystems).

If all of the above approaches are used and cell viability is still poor, transcardial perfusions can be implemented and may troubleshoot the problem. Transcardial perfusion replaces the blood with the cutting solution thereby preserving the neurons during the dissection process. To perform transcardial perfusions, a needle is placed into the left ventricle of a deeply anesthetized animal and the descending aorta is cut. The cutting solution is pumped through the needle into the left ventricle. The time of perfusion depends on the animal size, but typically 50 mL is sufficient to remove blood

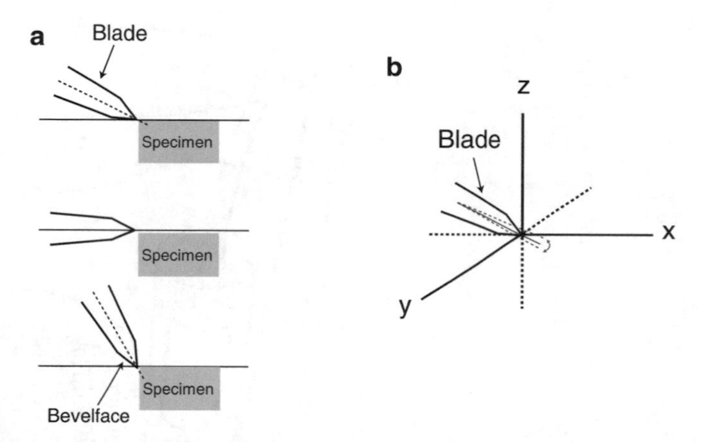

Fig. 1 Technical considerations for tissue sectioning. (**a**) The clearance angle of the blade is typically set to 15° or 18° making the bevelface parallel to the cutting motion (*top*). When the blade center line is parallel with the cutting motion (clearance angle not steep enough), unwanted pressure is applied to the specimen (*middle*). When the bevelface is above the cutting motion (clearance angle is too steep), the specimen can be deformed (*bottom*). (**b**) The blade's vibration along the *z*-axis is illustrated. Large vibrations along the *z*-axis can damage the slice causing unhealthy cells

and its associated proteins from the meninges in both mice and large rats. A successful cardiac perfusion is clearly seen as the brain appears white after it has been dissected from the animal.

3.2 Recovery

The incubation chamber (Fig. 2a) is typically filled with aCSF. After cutting, the brain slices are gently transferred to the incubation chamber, which is kept at 27–37 °C for 30 min to 1 h before being left at room temperature for the rest of the day (Table 2). As another alternative, NMDG can be used (Table 3), but for no more than 15 min. Slices then need to be transferred to aCSF (Table 2) or a sodium-based holding solution (Table 4) [12].

Of course there are variations to recovery solutions. Our laboratory uses the following ingredients for recovery ringer proven effective for nucleus accumbens slices. Recovery ringer (in mM): 119 NaCl, 2.5 KCl, $NaH_2PO_4{\cdot}H_2O$, 26.2 $NaHCO_3$, 1.3 $MgCl_2$, 2.5 $CaCl_2$, 11 d-glucose. Our slices are heated at 32 °C for 30 min and then moved to room temperature for the rest of the recording day.

There are two important points to mention before moving on to internal solutions. If an electrophysiologists desires to make a better solution for cutting or recovery, an investigation testing possible physiological alterations needs to performed. This is due to the fact that receptor or channel function may be altered depending

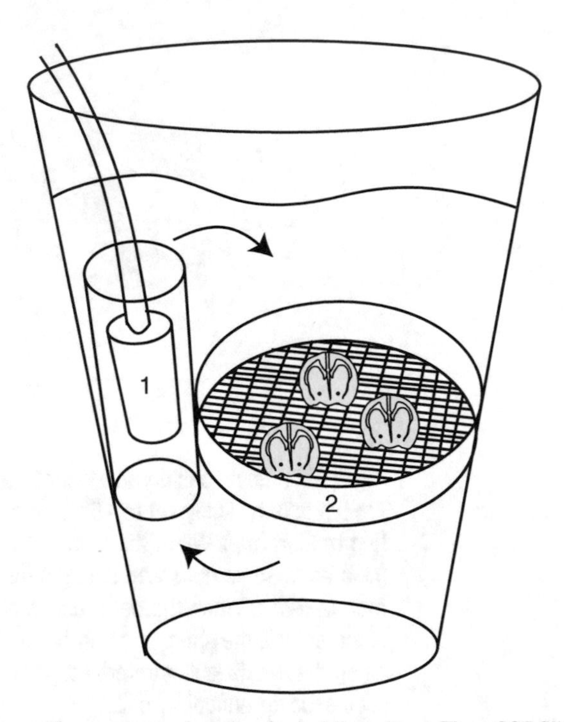

Fig. 2 Diagram of a recovery chamber for holding slices. The aCSF fills the chamber and is bubbled using an air stone (*1*), which is placed inside a hollow piece of tubing. The carbogen (O_2/CO_2, 95 %/5 %) circulates throughout the chamber (*arrows*). The bottom of a petri dish (*2*) is removed and replaced with gauze providing proper circulation of aCSF to both the top and bottom of the slices

Table 3
Cutting ringer (Na^{2+} removed) [15]

Compound	Concentration (mM)
N-Methyl-d-glutamine	135
KCl	1.0
KH_2PO_4	1.2
$CaCl_2$	0.5
$MgCl_2$	1.5
Glucose	11
Choline-HCO_3	20

pH to 7.3–7.4 using hydrochloric acid
Osmolality = 290

upon the ionic concentrations used. Second, careful consideration must be implemented when comparing results that have been formulated using different ionic concentrations/solutions. If the experimenter is trying to reproduce completed work, consistency in the solutions used is a must.

Table 4
Recovery ringer [12]

Compound	Concentration (mM)
NaCl	92
KCl	2.5
NaH_2PO_4	1.25
$NaHCO_3$	30
HEPES	20
Glucose	25
Thiourea	2
Na-ascorbate	5
Na-pyruvate	3
$CaCl_2 \cdot 4H_2O$	2
$MgSO_4 \cdot 7H_2O$	2

Osmolality 290

4 Internal Solution

Internal solution fills the micropipette, which is placed on, inside, or surrounding the cell membrane. Unlike external solution components, which vary only slightly, internal solution components can vary significantly depending on the currents or potentials being recorded. The tables below list typical ionic concentrations used by electrophysiologists for different recording types (readers can compare these tables with Table 5, which shows the cellular intracellular components) (Tables 6, 7, and 8).

Varying the ionic concentrations of internal solutions has its benefits. For example, it is often desired to record inhibitory postsynaptic currents (IPSCs), while keeping the cell at its physiological resting membrane potential (−70 mV). By raising the chloride concentration, the Cl^- reversal potential shifts from −70 to 0 mV. In doing this, Cl^- rushes out of the cell resulting in inward currents (when the cell is voltage-clamped at −70 mV). This approach is commonly practiced despite earlier investigations suggesting chloride containing salts (i.e., NaCl) (3–20 mM) increase the affinity of G_o alpha for GTP gamma S [20].

In addition to the ionic components, there are other molecules that are incorporated into the internal solution. Included are molecules such as ATP/GTP, intracellular protease inhibitors, and Ca^{2+} chelators. The function of each component is discussed below, but we refer the reader to the references for more in-depth reading.

5 ATP and GTP (Internal Containing ATP and GTP Must Be Kept on Ice throughout the Experiment to Prevent Degradation)

- ATP and GTP are often added to maintain basic activity of protein kinases as well as GTP-bound proteins to stay active minimizing channel dephosphorylation [21].
- Mitochondrial morphology can be affected by the perfusion of internal solution during whole-cell recordings. ATP minimizes the effect that this perfusion has on mitochondrial morphology [22].
- ATP preserves Ca^{2+} channels and $GABA_A$ channels.
- ATP augments NMDA receptor response.

6 Creatine/Phosphocreatine

- Creating/phosphocreatine contributes to the ATP regenerating system (creatine phosphate/creatine phosphokinase) preserving Ca^{2+} channels.

7 Intracellular Protease Inhibitors (i.e., Leupeptin)

Whole-cell recordings may induce intracellular protease release causing cell destabilization. As such, intracellular protease inhibitors are used for the following purposes:

- Leupeptin is a calcium dependent protease inhibitor and is used to prevent proteolysis-dependent rundown of calcium currents and other current [21].
- Conserves Ca^{2+} channels for intracellular proteolysis.
- Prevents cellular morphological changes in dissociated neurons (prevents spherical morphology).

8 Cations

- Cs^+ blocks K^+ channels. Substituting Cs^+ for K^+ is commonly used for better voltage control, particularly in whole-cell voltage-clamp recordings at positive potentials. This is because at positive holding potentials, K^+ channels are activated bringing the neuron back toward its resting membrane potential. Replacing K^+ with Cs^+ blocks K^+ channels, making it easier to hold the cell at positive holding potentials. However, in current clamp recordings substituting K^+ for Cs^+ alters the physiological properties of action potentials. Therefore, this substitution is not recommended for current clamping cells.

- Tetraethylammonium (TEA) blocks K^+ channels.
- QX314 blocks Na^+ channels preventing Na^+ current spikes, which can contaminate the desired signal.

9 Anions

- HEPES is used as a biological buffer [22].
- Methanesulfonate is inert and does not interact with divalent ions, making it a good substitute for chloride.

10 Ca^{2+} Chelators (BAPTA, EGTA, EDTA)

- Prevents Ca^{2+}-dependent plasticity [23].
- Prevents excess Ca^{2+} causing apoptosis.
- Prevents the internal membrane from reforming in whole-cell patch recordings [24].
- Prevents Ca^{2+} inactivation, thus augmenting Ca^{2+}-activated inward currents. Some Ca^{2+} channels are inactivated when internal Ca^{2+} rises above 10^{-7} M. Ca^{2+} chelators keep the Ca^{2+} concentration low.
- Suppresses activation of certain types of K^+ channels.
- Slows down calcium current rundown [25, 26].

11 Additional Information

- EDTA has a high affinity for Mg^{2+} so if Mg^{2+}-ATP is used, excess Mg should be used.
- Metal hypodermic syringes should be avoided when filling recording electrodes with EGTA containing internal solution because EGTA chelates metals resulting in acidification of the solutions.

12 Typical Intracellular Ionic Concentrations

13 Solutions Used for Extracellular Field Recordings-In Vitro

Micropipettes can be filled with 2–4 M NaCl.

Table 5
Intracellular ionic concentrations [22]

Ion	Concentration (mM)	Reference
Na^+	10–30	3
K^+	130–140	
Cl^-	5–15	1
HCO_3^-	10–16	1
A^-	109–155	[a]
Mg^{2+}	0.3–0.5	4
Ca^{2+}	50–100	3
ATP	2–3	3
Pi	1.7	3
Phosphocreatine	5	3
Creatine	6	3
pH	7.1–7.3	2

1 [16]
2 [17]
3 [18]
4 [19]
[a] A^- = membrane impermeant ion concentration. This value was calculated as the number of negative charges required to achieve electroneutrality

Table 6
Ionic concentrations used for EPSC or IPSC[a] recordings

Compound	Concentration (mM)
Cs-methanesulfonate	134.63
CsCl	5
TEA-Cl	5
EGTA (Cs)	0.4
HEPES	20
Mg-ATP	2.5
Na-GTP	0.25
QX-314 (Br)	1

pH 7.2–7.3 (CsOH)
Osmolality = 290 mOsm
Aliquot and freeze (−20 °C for 1 month or −80 °C for 1 year)
[a] For IPSC recordings the neuron would need to sit at positive holding potentials since the reversal potential for Cl– would be near −70 mV. Meaning no current would be detected if you were to hold the neuron at −70 mV

Table 7
Ionic concentrations used for IPSC recordings (depolarized Cl^- reversal potential ~0 mV)

Compound	Concentration (mM)
Cs-methanesulfonate	15
CsCl	120
NaCl	8
EGTA (Cs)	0.5
HEPES	10
Mg-ATP	2
Na-GTP	0.3
QX-314 (Br)	5

pH 7.2–7.3 (CsOH)
Osmolality = 290 mOsm
Aliquot and freeze (−20 °C for 1 month or −80 °C for 1 year)

Table 8
Ionic concentrations used for current clamp recordings

Compound	Concentration (mM)
K-Methanesulfonate	130
KCl	10
HEPES	10
EGTA (K)	0.4
$MgCl_2{\cdot}6H_2O$	2.0
Mg-ATP	3
Na-GTP	0.5

pH = 7.2–7.3 (KOH)
Osmolality = 290 mOsm
Aliquot and freeze (−20 °C for 1 month or −80 °C for 1 year)

14 Solutions for Patch Preparations

14.1 Cell-Attached Patch

Fill the electrode with solution compatible to the extracellular solution. Emit Ca^{2+} chelators since free Ca^{2+} is important for membrane integrity. Osmolality should be similar as seals become increasingly difficult when the osmolality deviates. Hypotonic pipette solutions are recommended.

14.2 Whole-Cell Path

See above tables for intracellular solutions.

14.3 Inside-Out Patch

Pipette solution should be similar to aCSF, while bath solution should be switched to solutions similar to intracellular conditions after formations of the inside-out patch.

14.4 Outside-Out Patch

Low levels of Ca^{2+} should be used so that the internal membrane does not reform after being ruptured. Substituting fluoride for chloride stabilizes outside out patches, but fluoride changes receptor properties and activates adenylate cyclase [22].

15 Pipette Offset

As discussed in Chap. 3, the pipette offset resets the pipette potential to zero with respect to the external solution or the bath (determined by the ground). In other words, an opposing voltage is applied to offset the potential difference between the internal and external solutions. A problem with either external or internal solutions can cause inconsistencies from day to day in the pipette offset value. For example, if using the external solution recipe in Table 2 with the internal recipe in Table 6, the pipette offset should be around +30 to +50 mV when the recording and ground electrodes have been properly coated with chloride. Pipette offset potentials far lower (−80 to 0 mV) can indicate that an important component has been left out of the solution. In this example, an anion would most likely have been left out due to the excess positive charge of the pipette potential causing the pipette offset to be in a subzero range. To avoid subzero potential ranges in your pipette offset, be sure to mix the internal solution well during the preparation process before aliquoting. In addition, vortexing each aliquot before use may also prevent inaccuracies in the pipette offset.

16 Troubleshooting

It is very easy to blame the solutions during a bad day of recording and very often poorly made solutions are the problem. Below we listed identifiable problems that are typically associated with poor external or internal solutions. Additional sources to blame are also included.

1. Unhealthy or apoptotic neurons noticeable even at 50 μm deep within a slice. This problem can be caused by inaccurate osmolality measurements, >1 h incubation time at 37 °C, unfiltered aCSF (bacterial buildup; filter with a 0.22 μm sized filter), or a dirty incubation chamber (incubation chambers can be cleaned with 70 % EtOH followed by a thorough rinse with distilled water). Unhealthy cells can also be caused by the

machinery used to cut the slices. Often the vibrating amplitude along the z-axis can be large or the clearance angle of the cutting blade can be impaired causing damage to the neurons (consult the cutting machine manual or technical service for support).

2. Difficulty obtaining a cell-attached patch or acceptable access to the cell membrane while attempting a whole-cell patch. This is very often due to unhealthy neurons. See (1) for troubleshooting strategies.
3. Cells consistently dying within 5–15 min after achieving a whole-cell patch. If the health condition of the cells is fine, a possible reason is the internal solution. Check the osmolality of the internal solution first. When stored at subzero temperatures, water evaporates from the solution causing an increase in the osmolality, which can affect recording quality.
4. The internal solution contains solid dark particles that buildup at the micropipette tip when positive pressure is applied. When this occurs, the internal solution needs to be filtered using a 0.22 μm filter. In addition, the apparatus used to fill the micropipette may also need to be cleaned with 70 % EtOH followed by a thorough rinse with double distilled water.

17 Summary

Successful practices in in vitro brain slice electrophysiology require careful selection and preparation of ionic-salt solutions. There are external solution recipes available for brain slice preparation, recovery, and recording; each can be used to enhance cellular viability. Importantly, these external solutions do not alter receptor expression or function, which have been proven empirically. Internal solutions are composed of anions and cations as well as other components slowing down cell death, current run-down, and preservation of Ca^{2+} concentrations. With properly made solutions, viable cells should be plentiful. However, problems with cellular viability are common in electrophysiology so we hope that the troubleshooting guide included in this chapter, will assist the reader.

References

1. Polderman KH, van de Kraats G, Dixon JM, Vandertop WP, Girbes AR (2003) Increases in spinal fluid osmolarity induced by mannitol. Crit Care Med 31(2):584–590
2. Sharp PE, Regina MCL (1998) The laboratory rat. Taylor & Francis, Boston, MA
3. Irani DN (2009) Cerebrospinal fluid in clinical practice. Saunders/Elsevier, Philadelphia, PA
4. Reed D, Withrow CD, Woodbury D (1967) Electrolyte and acid-base parameters of rat cerebrospinal fluid. Exp Brain Res 3(3):212–219
5. Jeong SM, Hahm KD, Shin JW, Leem JG, Lee C, Han SM (2006) Changes in magnesium concentration in the serum and cerebrospinal fluid of neuropathic rats. Acta Anaesthesiol Scand 50(2):211–216

6. LeVine SM, Wulser MJ, Lynch SG (1998) Iron quantification in cerebrospinal fluid. Anal Biochem 265(1):74–78
7. Espino A, Ambrosio S, Bartrons R, Bendahan G, Calopa M (1994) Cerebrospinal monoamine metabolites and amino acid content in patients with parkinsonian syndrome and rats lesioned with MPP+. J Neural Transm Park Dis Dement Sect 7(3):167–176
8. Ganrot K, Laurell C-B (1974) Measurement of IgG and albumin content of cerebrospinal fluid, and its interpretation. Clin Chem 20(5):571–573
9. Habgood MD, Sedgwick JE, Dziegielewska KM, Saunders NR (1992) A developmentally regulated blood-cerebrospinal fluid transfer mechanism for albumin in immature rats. J Physiol 456:181–192
10. Hutchesson A, Preece MA, Gray G, Green A (1997) Measurement of lactate in cerebrospinal fluid in investigation of inherited metabolic disease. Clin Chem 43(1):158–161
11. Swahn CG, Sedvall G (1988) CSF creatinine in schizophrenia. Biol Psychiatry 23(6):586–594
12. Martina M, Taverna S (2014) Patch-clamp methods and protocols. Springer, New York
13. Huang S, Uusisaari MY (2013) Physiological temperature during brain slicing enhances the quality of acute slice preparations. Front Cell Neurosci 7:48
14. Lipton P, Aitken PG, Dudek FE, Eskessen K, Espanol MT, Ferchmin PA, Kelly JB, Kreisman NR, Landfield PW, Larkman PM et al (1995) Making the best of brain slices: comparing preparative methods. J Neurosci Methods 59(1): 151–156
15. Lee BR, Ma YY, Huang YH, Wang X, Otaka M, Ishikawa M, Neumann PA, Graziane NM, Brown TE, Suska A, Guo C, Lobo MK, Sesack SR, Wolf ME, Nestler EJ, Shaham Y, Schluter OM, Dong Y (2013) Maturation of silent synapses in amygdala-accumbens projection contributes to incubation of cocaine craving. Nat Neurosci 16(11):1644–1651
16. Alvarez-Leefmans F (1990) Intracellular Cl– regulation and synaptic inhibition in vertebrate and invertebrate neurons. In: Alvarez-Leefmans F, Russell J (eds) Chloride channels and carriers in nerve, muscle, and glial cells. Springer, USA, pp 109–158
17. Chesler M (1990) The regulation and modulation of pH in the nervous system. Prog Neurobiol 34(5):401–427
18. Erecinska M, Silver IA (1989) ATP and brain function. J Cereb Blood Flow Metab 9(1): 2–19
19. Taylor JS, Vigneron DB, Murphy-Boesch J, Nelson SJ, Kessler HB, Coia L, Curran W, Brown TR (1991) Free magnesium levels in normal human brain and brain tumors: 31P chemical-shift imaging measurements at 1.5 T. Proc Natl Acad Sci U S A 88(15):6810–6814
20. Higashijima T, Ferguson KM, Sternweis PC (1987) Regulation of hormone-sensitive GTP-dependent regulatory proteins by chloride. J Biol Chem 262(8):3597–3602
21. Sarantopoulos C (2007) Perforated patch-clamp techniques. Neuromethods 38:253–293
22. Kay AR (1992) An intracellular medium formulary. J Neurosci Methods 44(2–3):91–100
23. Liao D, Hessler NA, Malinow R (1995) Activation of postsynaptically silent synapses during pairing-induced LTP in CA1 region of hippocampal slice. Nature 375 (6530):400–404
24. Kettenmann H, Grantyn R (1992) Practical electrophysiological methods: a guide for in vitro studies in vertebrate neurobiology. Wiley, Chichester
25. Belles B, Malécot CO, Hescheler J, Trautwein W (1988) "Run-down" of the Ca current during long whole-cell recordings in guinea pig heart cells: role of phosphorylation and intracellular calcium. Pflugers Arch 411(4):353–360
26. Horn R, Korn SJ (1992) Prevention of rundown in electrophysiological recording. Methods Enzymol 207:149–155

Chapter 5

Patch Pipettes (Micropipettes)

Nicholas Graziane and Yan Dong

Abstract

Joining an established electrophysiology laboratory typically means that the micropipette fabrication process has already been perfected allowing newcomers to be trained without a complete understanding of how the current method was developed. This chapter looks to fill that knowledge gap by discussing available micropipette options and fabrication methods currently used for in vitro electrophysiology (e.g., inside-out, outside-out, whole-cell, cell-attached, and sharp-electrode patches). We also discuss the importance of the micropipette's shape in relation to the in vitro electrophysiology model used (e.g., tissue slice, cell culture, dissociated neurons). For laboratories setting-up electrophysiology equipment, we have provided information regarding tools needed for pulling micropipettes as well as post-pulling fabrication processes (e.g., fire polishing). Finally, we discuss micropipette drift, which can occur during electrophysiological recordings accompanied by a micropipette-drift troubleshooting guide. Surprisingly, the process of micropipette fabrication can play a significant role in data collection. With this in mind, we anticipate by the end of this chapter that the reader will have the information necessary to fabricate the appropriate micropipette for their desired electrophysiology preparation allowing for successful data collection.

Key words Micropipette fabrication, Micropipette equipment, Micropipette resistance, Electrode holders, Pipette drift

1 Introduction

Joining an established electrophysiology laboratory typically means that the micropipette fabrication process has already been perfected allowing newcomers to be trained without a complete understanding of how the current method was developed. This chapter looks to fill that knowledge gap by discussing available micropipette options and fabrication methods currently used for in vitro electrophysiology (e.g., inside-out, outside-out, whole-cell, cell-attached, and sharp-electrode patches). We also discuss the importance of the micropipette's shape in relation to the in vitro electrophysiology model used (e.g., tissue slice, cell culture, dissociated neurons). For laboratories setting-up electrophysiology equipment, we have provided information regarding tools needed for pulling micropipettes as well as post-pulling fabrication

Nicholas Graziane and Yan Dong, *Electrophysiological Analysis of Synaptic Transmission*, Neuromethods, vol. 112,
DOI 10.1007/978-1-4939-3274-0_5,

processes (e.g., fire polishing). Finally, we discuss micropipette drift, which can occur during electrophysiological recordings accompanied by a micropipette-drift troubleshooting guide. Surprisingly, the process of micropipette fabrication can play a significant role in data collection. With this in mind, we anticipate by the end of this chapter that the reader will have the information necessary to fabricate the appropriate micropipette for their desired electrophysiology preparation allowing for successful data collection.

2 Micropipette Material

There are many glass options that can be used for micropipette fabrication. The most common material used by electrophysiologists is borosilicate. Borosilicate glass is an affordable option that produces low noise recordings due to its low dielectric constant* and low dissipation factor* (dielectric constant = 4.5–6, dissipation factor = 0.002–0.005). Other options are quartz and alumina silicate. Quartz glass is considered the gold standard in that it is superior to borosilicate in strength and stiffness. These properties make it easier to fabricate small tip micropipettes. More importantly, quartz has an even lower dielectric constant and dissipation factor in comparison to borosilicate (dielectric constant = 3.86, dissipation factor = 10^{-5} to 10^{-4}) [1] making it an ideal option for single channel recordings where low noise is a must. Unfortunately, the cost of quartz is a major drawback. In addition, quartz glass softens at ~1600 °C so platinum filaments (≤800 °C is acceptable for platinum filaments) or coils typically used for pulling pipettes are impractical for quartz glass fabrication. Instead, expensive laser-based pullers are needed. Another glass option, alumina silicate glass, can be useful for low noise recordings. However, alumina silicate glass softens at ~900 °C, limiting the platinum filament's lifetime [2].

When selecting the type of glass the dielectric constant (relative permittivity) and dissipation factor should be considered. The dielectric constant and dissipation factor are proportional to capacitance. In other words, a micropipette pulled from glass with a high dielectric constant has a high pipette capacitance (Chap. 3: pipette capacitance can increase root mean square noise and distort rapid-onset ionic currents). Electrophysiologists looking for cheap glass options should do so cautiously. Soda and potash glass are both cheap, soft glass options, but produce high noise and have also been shown to change channel properties [3, 4]. Both of these problems most likely stem from the impurities in the glass. The impurities lead to the high dielectric constant (causing increased capacitance) and could leach from the glass altering channel gating or conductance. The latter is avoidable when implementing the perforated patch technique (see Chap. 7).

Some manufacturers make an internal filament fused to the wall of the micropipette glass, which allows for easy filling of the micropipette and allows sealing to the cell with little difficulty. Despite these advantages the internal filament can create significant noise [2] and it is not advised to use if noise is a concern during the recording.

Micropipettes come in thick-or thin-walled glass. Thick-walled glass (outer diameter (OD) = 1.5 mm, inner diameter (ID) = 0.86 mm) is used for cultured or dissociated neuronal recordings and can be used to minimize pipette capacitance (capacitance is inversely proportional to distance between parallel plates). Thin-walled glass (OD = 1.5 mm, ID = 1.1 mm) is typically used for tissue slice preparations and for low resistance.

3 Shape of the Micropipette

Pulling the right type micropipette shape is an art form and if done properly, can reduce potential problems that may arise during recording. For example, if an electrophysiologist were to use a micropipette with a minimum taper to patch onto dissociated or cultured neurons, they would find little trouble. However, using that same pipette for a tissue slice preparation would be far more difficult. This is because a minimum tapered pipette would not be able to penetrate the tissue slice resulting in "cell chasing." "Cell chasing" means that no matter how deep in the slice you advance your pipette, the neuron seems to stay just out of reach of the pipette. Therefore, it is critical to match the micropipette's shape to the cellular preparation. As a general rule, pipettes with 3–4 mm taper (2 μm tip) can be used for cultured or dissociated cells while pipettes with 4–5 mm taper (1–3 μm tip) can be used for tissue slice recording [5]. A micropipette resistance between 3 and 5 MΩ is acceptable for most in vitro recording techniques. Pipette taper should also be paid close attention to when performing sharp electrode recordings with high resistance pipettes (30–100 MΩ). Using gradually tapered electrodes (electrodes lacking an inflection) can effectively minimize excessive cellular damage when impaling the cell.

The shape of the electrode is also important for single-channel and whole-cell current measurements. As described in Chap. 3, pipette capacitance can increase root mean square noise. In addition, capacitance is inversely proportional to the distance separating the capacitor's parallel plates. Therefore, selecting the shortest possible pipettes with the largest possible wall thickness reduces noise, which is beneficial for recordings that are sensitive to noise (e.g., single-channel recordings) [2].

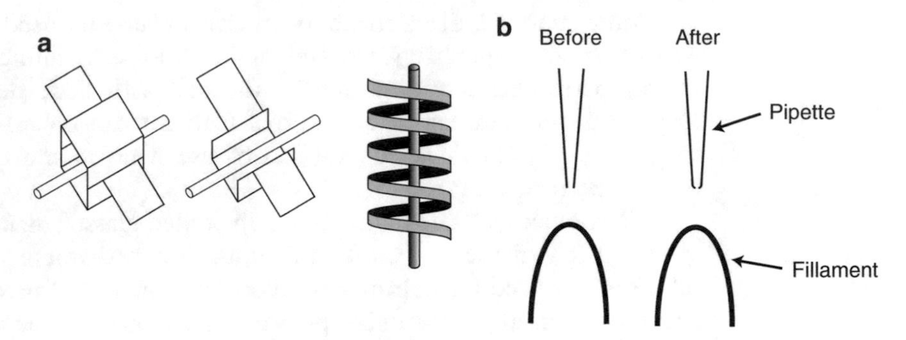

Fig. 1 Components used for fabricating micropipettes. (**a**) Diagram of a box, trough, and coil filaments with a glass capillary situated inside each. (**b**) A before and after image of a micropipette following fire polishing with a heated filament

4 Filaments and Coils

There are three options that can be used to heat the micropipette when pulling the electrode: (1) platinum'/iridium box filaments, (2) platinum'/iridium trough filaments, and (3) coils (Fig. 1a). Box filaments are recommended for slice preparations, which require long, parallel walls to enhance tissue slice penetration. When ordering a box filament, be sure that it is 1.0–1.5 mm larger that the OD of the glass. Trough filaments are recommended for micropipettes made from standard or thin-wall glass. A general rule of thumb for both box and trough filaments is that as filament width increases, the pipette taper becomes longer. The major advantage of using filaments over coils is that they heat the pipette evenly and are more efficient at heating and cooling. Coils produce uneven heat since one end of the coil has 180° more coil than the other end. On the other hand, the advantage of coils over platinum/iridium is that coils are far more durable (much less fragile) than the platinum/iridium.

5 Pulling Micropipettes

There are two types of pullers typically used for fabricating micropipettes; horizontal pullers and vertical pullers. Both pullers use the same basic principle. The glass is heated using orange heat from a platinum filament and pulled. Horizontal pullers pull the glass capillary using springs, pullers, and pull cables, while vertical pullers use gravity and weights. As far as comparisons, vertical pullers are easy to maintain, small so they occupy less lab bench space, and are far less expensive than horizontal pullers. However, horizontal pullers are known for their high reproducibility of micropipette shapes, which can save an electrophysiologist a lot of precious time during the fabrication process (for a video guide to pulling pipettes on a puller see [6].

In our experience, both types of pullers can produce consistent micropipettes in terms of resistance and taper. Inconsistencies arise when the filament becomes worn. This happens most often when the filament is hit while loading the glass capillary; an avoidable problem as long as careful attention is paid during glass loading. A second potential problem causing inconsistencies arises when the filament is not properly mounted. This is a common problem for horizontal pullers. The Pipette Cookbook [5] provides essential how-to instructions, including pictures, to assist any new electrophysiologist attempting to install a new filament. A third cause of inconsistently pulled micropipettes occurs when the pulling unit used has an air cooling system that is not working properly or has exhausted drierite (blue color = active, purple/pink color = exhausted (wet state). Regenerate drierite by placing it at 200 °C for 1 h). Other problems with reproducibility can occur if (1) dirt is on the glass capillary, (2) dirt is on the filament, (3) there are sudden changes in air currents, or (4) there are sudden changes in temperature. All of these problems can be avoided. Glass should be kept covered when not in use to prevent contaminants from sticking to the glass. Changes in air currents or temperature can be prevented since pullers come equipped with humidity control chambers (horizontal pullers) or cover plates (vertical pullers) that protect the filament.

6 Fire Polishing

To fire polish a pipette, a light microscope, a 10× or 15× (low magnification objective), a 40× or 100× objective (high magnification objective), and platinum wire can be used (Fig. 1b). The micropipette is moved toward the heated wire (heated by current) under low magnification until it is about 50 μm from the wire filament. Then switching to higher magnification the pipette can be moved into position and heated until the glass can flow and make a final tip [2]. The purpose of fire polishing is to make a final tip geometry that prevents the glass from penetrating the cell when pressed against it. Also, fire polishing the pipette cleans the pipette tip enhancing seal formation and creates a blunted edge leading to lower pipette resistance (Fig. 1b). For a detailed step-by-step instruction on how to fire polish a pipette as well as how to create your own fire-polishing apparatus see [2]. Otherwise, a microforge can be purchased and used to fire-polish the pipettes.

Pressure polishing a pipette is another option in which compressed air (40 PSI) is passed through the pipette while the pipette is fire polished [7, 8]. This technique is posited to be useful for very small cells (<5 μm) as well as for producing a blunted shape micropipette, which lowers pipette resistance.

7 Pipette Resistance

Often it is useful to visualize pulled pipettes using a microscope to determine acceptable tip sizes/tapers for the desired pipette resistance. An alternative to checking pipette resistance is to attach a 10 mL syringe to a piece of tubing, which can then attach to the pulled pipette. Dip the pipette tip into clean methanol. Starting at 10 mL apply pressure until bubbles are expelled from the pipette. A bubble number of 5 or less indicates tip with >10 MΩ resistance, while 6 or more indicates a low resistance pipette suitable for whole-cell recordings [9].

8 Electrode Holders

Electrode holders house the electrode (Ag/AgCl) and the micropipette. The holders are made of acrylic or polycarbonate and can be purchased directly from the manufacturer selling the headstage or from manufacturers selling electrophysiology equipment (e.g., Warner Instruments). The holders come in various styles from straight or curved (45° or 90°) (Fig. 2a–c) with the curved holders being useful for rigs with space limiting factors. Straight holders come in different lengths. The suitable length can be determined based on where the slice is mounted and where the headstage is positioned. Holders also come in different bore sizes depending on the OD of the micropipette being used. Holders also contain ports or vents. Tubing can be attached to the port or vent in order to apply positive/negative pressure to the micropipette (Fig. 2a). Once the micropipette is filled with solution, it can be inserted into the holder and the end cap of the holder can be tightened compressing a silicone rubber gasket. The purpose of the rubber gasket

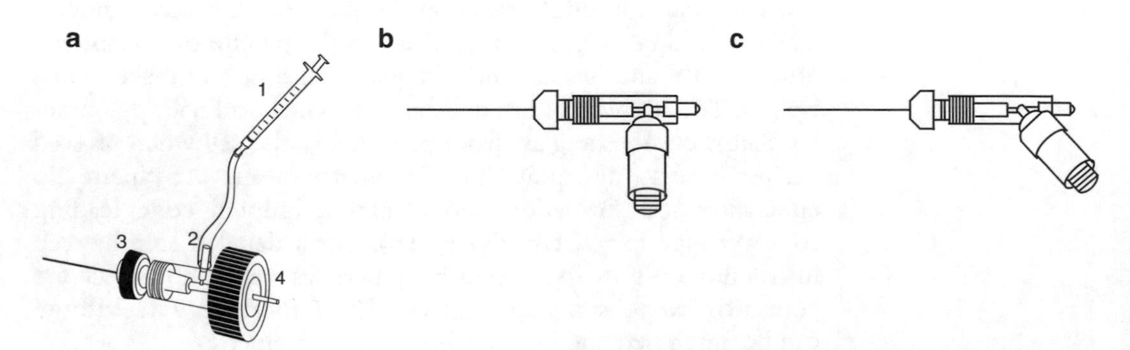

Fig. 2 Micropipette holders containing an electrode (typically AgCl), which changes in ionic composition. (**a**) A micropipette holder consisting of an electrode (*3*), a port for tubing (*2*), which is connected to a syringe (*1*), and a gold pin (*4*), which connects to a headstage. (**b**) A micropipette holder with a 90° curve. (**c**) A micropipette with a 45° curve

is to form a tight seal around the micropipette. This rubber gasket should be monitored for wear as worn down gaskets can cause pipette drift or loss of pressure control. Loss of positive pressure (easily noticeable when lowering the micropipette through the tissue slice) can also be due to the rubber gasket on the opposite end of the holder (the side near the gold pin, which attaches to the headstage). If there is a loss of pressure control and you need to identify the leak, remove the holder from the headstage, insert a micropipette, and then submerge the holder + micropipette into water. To expose any leaks, blow positive pressure through the tube attached to the port and locate where the bubbles are formed.

A lot of troubleshooting in electrophysiology deals with noise control. If there are noticeable increases in noise, the electrode holder should be checked for fluid or checked to be sure that it is not dirty. Noise from a dirty holder can contribute to noise across a broad range of frequencies [1]. If fluid or dirt in the holder is seen, it can be cleaned by disassembly and then washed with ethanol. Do a thorough, final wash with distilled water and then to dry, blow clean air through (alternatively dry for several hours in an oven at 60–70 °C) [1].

9 Pipette Drift (Mechanical)

A common issue to deal with for an electrophysiologist is pipette drift. A drifting pipette prevents long recordings as the pipette moves off of the recorded neuron causing intolerable "current leak". Pipette drift should be checked if recordings are regularly being lost within 30 min of patching. With clean internal and external solution at the proper pH and osmolality, healthy neurons, and no pipette drift, whole-cell recordings should easily last 1–2 h. Most micromanipulators are manufactured to have no more than 1 μm drift/h so any pipette movement >1 μm in 30 min should be corrected immediately. If pipette drift is detected, the following list may help correct the problem:

1. Replace the rubber gasket in the electrode holder if it looks worn. The rubber gasket forms a tight seal on the glass pipette. If worn, this tight seal is lost, which can cause the pipette to move.
2. Do not over-tighten the pipette end caps on either end of the electrode holder (i.e., the rubber-gasket side or the headstage-attached side) as it can cause relaxation in the Teflon threads exacerbating drift. Finger-tight is sufficient.
3. Check cables throughout the rig especially on the headstage and micromanipulator. Make sure that they are not under tension potentially pulling what they are connected to. It is recommended to use large service loops (Fig. 3a). Also, make

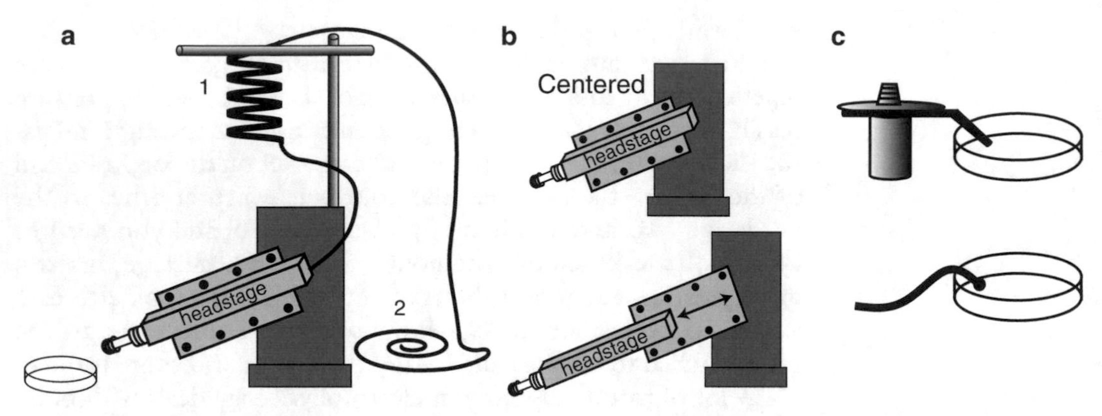

Fig. 3 Troubleshooting strategies for pipette drift. (**a**) Configuration strategies for cable wires connecting components nearby the recording chamber with the amplifier. Large service loops (*1*) and/or spirally the cables on the air table can help prevent manual pipette drift. (**b**) Keeping the headstage centered on the mounting plate can help prevent pipette drift. Often, a drifting pipette can be caused by the headstage being too far from the center of the mounting plate, thus creating a lack of headstage support (*bottom*). (**c**) Components placed inside the recording chamber (e.g., suction, temperature probe, grounding wires) are placed so that they do not come into contact with the recording chamber, thus preventing the recording chamber from moving (*top*). Movement of the chamber or vibrations can be introduced if the suction rests on the chamber (*bottom*)

sure that the cables are secured to the air table so that little vibration is introduced through the cables to vulnerable parts near the stage. The air table can dampen potential vibrations that may come from outside sources.

4. Secure the tubing (attached to the electrode holder port) to the headstage.
5. If possible, try to keep the headstage centered with the mounting plate (Fig. 3b). This keeps the headstage balanced preventing unnecessary shifts during recording.
6. Make sure that all bolts, nuts, screws are secure on the stage that mounts the headstage and micromanipulator.
7. Check for movement in the dish where cells/tissue slices are mounted. This can be caused by ground pellets, perfusion tubing, suction needles, or temperature probes, which can introduce movement through tensed wires or vibrations. The best solution is to make sure that everything placed in the dish does not make physical contact with the dish (Fig. 3c).
8. Check the micromanipulator for drift by clamping a pipette directly to the mounting plate. Unfortunately, if the micromanipulator is the problem, contact the manufacturer for a trained service engineer.
9. Temperature changes can be another potential problem. It is recommended to place rigs away from air conditioning vents or any heating/cooling system. For temperature-sensitive manipulators, it is helpful if the pipette is placed in warm

recording solution for a few minutes before patching (conclusion based on a temperature analysis performed by a company using borosilicate glass).

10. Check to be sure the air table is working and balanced properly.

If the problem cannot be corrected after exhausting this list, a low-drift pipette holder can be purchased from G23 Instruments called the ISO range. Additional readings can be found in "A low drift micropipette holder" written by Frederick Sachs in 1995 [10].

10 Summary

Micropipette fabrication can be quite complex when considering the various components that are involved in the process. This includes the different glass options available and the different micropipette shapes that can be formed from pullers or from fire-polishing techniques. However, once the appropriate options have been identified and proper care and handling of instruments has been taken into consideration, micropipette fabrication takes <30 min of daily preparation time, which is achievable at any experience level. With the knowledge gained from this chapter, we anticipate that beginning electrophysiologists, who run into the inevitable problems during the fabrication process, can troubleshoot the problems quickly preventing mounting frustrations.

*Dielectrics are insulators that fill the space between capacitors. These insulators can contain polar molecules that are in random orientations when no electric field is present, but polarizes when exposed to an electric field. What results is a decrease in the electric field between the parallel plates of a capacitor. As a result the capacitance is increased (remember distance between capacitor plates is inversely proportional to its capacitance) (Fig. 4. For more information on dielectrics and dissipation factors see [2]).

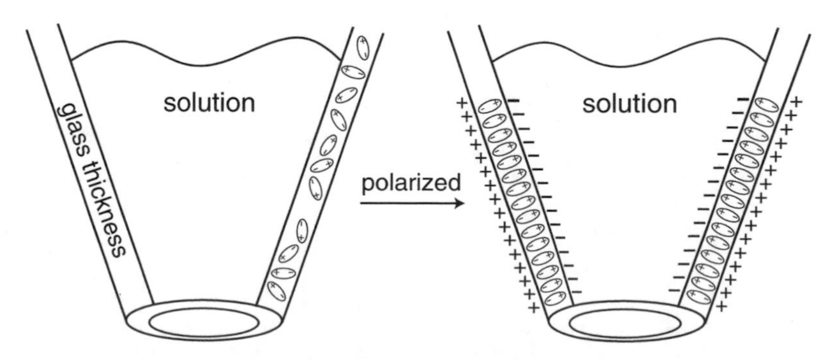

Fig. 4 Glass micropipettes are capacitors with a capacitance dependent upon the wall thickness and the glass components. The glass contains polar molecules that become polarized when exposed to an electric field

*Dissipation factor defines how much electrical potential energy is dissipated by an insulator. The more efficient an insulator is, the lower the dissipation factor.

References

1. Levis R, Rae J (2002) Technology of patch-clamp electrodes. In: Walz W, Boulton A, Baker G (eds) Patch-clamp analysis, 35. Humana, Totowa, pp 1–34
2. Rae JL, Levis RA (2004) Fabrication of patch pipets. Curr Protoc Neurosci. Chapter 6:Unit 6.3
3. Copello J, Simon B, Segal Y, Wehner F, Ramanujam VM, Alcock N, Reuss L (1991) Ba2+ release from soda glass modifies single maxi K+ channel activity in patch clamp experiments. Biophys J 60(4):931–941
4. Furman RE, Tanaka JC (1988) Patch electrode glass composition affects ion channel currents. Biophys J 53(2):287–292
5. Oesterle A (2008) P-97 pipette cookbook. Sutter Instruments, Novato, CA
6. Brown AL, Johnson BE, Goodman MB (2008) Making patch-pipettes and sharp electrodes with a programmable puller. J Vis Exp(20)
7. Goodman MB, Lockery SR (2000) Pressure polishing: a method for re-shaping patch pipettes during fire polishing. J Neurosci Methods 100(1–2):13–15
8. Johnson BE, Brown AL, Goodman MB (2008) Pressure-polishing pipettes for improved patch-clamp recording. J Vis Exp(20)
9. Ogden D, Stanfield P (1994) Patch clamp techniques for single channel and whole-cell recording Microelectrode techniques: the Plymouth workshop handbook. Company of Biologists, Cambridge, UK
10. Sachs F (1995) A low drift micropipette holder. Pflugers Arch 429(3):434–435

Chapter 6

Spatiotemporal Effects of Synaptic Current

Nicholas Graziane and Yan Dong

Abstract

A neuron receives excitatory and inhibitory electrical signals from other neurons via synaptic connections on dendritic spines, branches, or somatic membranes. These signals travel from different locations to the soma depolarizing or hyperpolarizing the membrane, thus regulating cell excitability. Although this is a straightforward concept, there are a multitude of intricate details that occur at the neuronal level, which regulate this process. This chapter looks to describe these spatiotemporal details of synaptic input so that the beginning electrophysiologist can become aware of the full potential of neuronal communication. To begin, we discuss Wilfrid Rall's cable theory followed by empirical evidence supporting or refuting Rall's predictions. This is followed by an introduction into Hebbian plasticity or spike-timing dependent plasticity (STDP), which is a process critically dependent upon the temporal firing activity of both presynaptic and postsynaptic neurons.

Key words Cable filtering theory, Membrane responses to synaptic input, Hebbian plasticity, Space clamp artifacts

1 Introduction

A neuron receives excitatory and inhibitory electrical signals from other neurons via synaptic connections on dendritic spines, branches, or somatic membranes. These signals travel from different locations to the soma depolarizing or hyperpolarizing the membrane, thus regulating cell excitability. Although this is a straightforward concept, there are a multitude of intricate details that occur at the neuronal level, which regulate this process. This chapter looks to describe these spatiotemporal details of synaptic input so that the beginning electrophysiologist can become aware of the full potential of neuronal communication. To begin, we discuss Wilfrid Rall's cable theory followed by empirical evidence supporting or refuting Rall's predictions. This is followed by an introduction into Hebbian plasticity or spike-timing dependent plasticity (STDP), which is a process critically dependent upon the temporal firing activity of both presynaptic and postsynaptic neurons.

Nicholas Graziane and Yan Dong, *Electrophysiological Analysis of Synaptic Transmission*, Neuromethods, vol. 112,
DOI 10.1007/978-1-4939-3274-0_6,

2 Theoretical Passive Membrane Responses to Localized Synaptic Input (Cable Filtering)

In order to model the passive electrical properties of dendrites, Wilfrid Rall used three cable parameters: axial resistance (r_i), membrane capacitance (C_m), and membrane conductance (G_m) [1, 2]. These three parameters along with the neuron's morphology control the propagation of membrane potentials generated from synaptic currents (Fig. 1a). The axial resistance causes a voltage drop between the synaptic input site and the soma causing the current driving force to decrease as the distance from the cell body increases (Fig. 1a). The capacitance stores synaptic charge, thus slowing kinetic properties of membrane potentials. Therefore, potential's measured from the soma have varying amplitude and kinetics depending upon where along the dendrite the current is generated. Since r_i increases depending on length and C_m increases, which is determined by membrane area, somatic potentials generated at distal dendrites have smaller

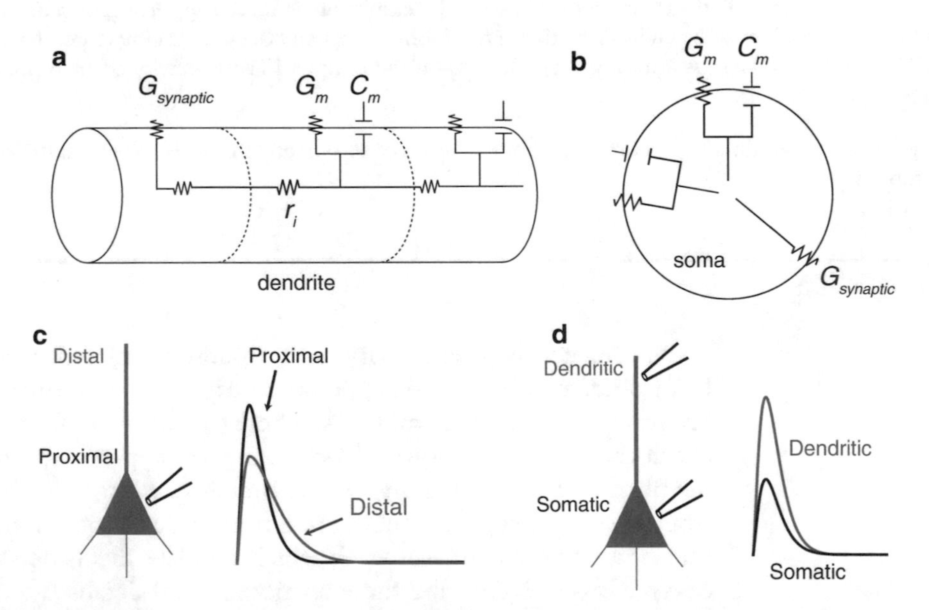

Fig. 1 Dendritic membrane properties and cable filtering. (**a**) Diagram of the dendritic electrical circuit made up of multiple compartments separated by the axial resistance (r_i). The membrane capacitance (C_m) is determined by the membrane area and the membrane conductance (G_m) is controlled by the transmembrane channels or receptors along the dendrite or at the synapse ($G_{synaptic}$). (**b**) Diagram of the somatic electrical circuit. Axial resistance is not present in the soma. Only somatic C_m and G_m along with the conductance from synaptic inputs ($G_{synaptic}$). (**c**) Somatic recordings from a neuron (*left*) produce different waveforms (*right*) depending on the site of synaptic input. When channels or receptors at distal sites are activated, the potentials recorded from the soma have smaller amplitudes and slower kinetics than those activated at proximal locations. (**d**) Dendritic recordings of channel/receptor activation at the dendrite produce larger amplitudes than somatic recordings of channel/receptor activation at the soma. This is due to the small diameter, high resistance properties of dendrites

amplitudes and slower kinetics than those generated at proximal sites [3] (Fig. 1c).

The differences in morphology between the soma and the dendrite can alter the current flow at synaptic input sites. The spherical soma has only one compartment so current flows into the somatic G_m and C_m. However, the dendrite has multiple compartments separated by r_i (Fig. 1a). Therefore, similar to the soma, synaptic current flows through the membrane into G_m and C_m. However, unlike the soma, the current must then travel through the axial resistance to the next compartment.

With these theoretical principles in mind, Rall predicted the functional impact that distal vs. proximal synaptic inputs had on somatic potentials. The distal synaptic inputs with slower current kinetics and smaller current amplitudes would only cause subthreshold changes in somatic membrane potentials (temporal integration-action potentials generated from summed temporally dispersed synaptic activity), while proximal synapses would produce action-potential-triggering depolarizations (coincidence detection-action potentials generated from short duration inputs arriving simultaneously and producing summed depolarizations) [4, 5].

3 Empirically Derived Membrane Responses to Synaptic Inputs (In Vitro)

Rall's theoretical points on passive cable filtering can be summarized as follows: (1) local potentials derived at dendritic synaptic inputs have larger amplitudes than somatic inputs due to the small diameter, high resistance dendritic properties (Fig. 1d), (2) somatic potentials generated at proximal synapses have larger amplitudes than those generated at distal synapses causing coincidence detection at the soma (Fig. 1c), and (3) somatic potentials generated at distal synapses have slower kinetics (Fig. 1c) than those generated at proximal synapses causing temporal summation at the soma. These points have been empirically tested and are summarized below:

1. Local potentials generated at dendritic synaptic inputs have larger amplitudes than potentials at somatic inputs due to the small diameter, high resistance dendritic properties.

Empirical evidence in CA1 hippocampal pyramidal neurons and in layer 5 neocortical pyramidal neurons supports Rall's prediction [6, 7]. In fact, empirical evidence suggests that as the site of generation increases away from the soma, the excitatory postsynaptic potential (EPSP) amplitude increases (Fig. 2a).

2. Somatic potentials generated at proximal dendritic synapses have larger amplitudes than those generated at distal dendritic synapses causing coincidence detection at the soma.

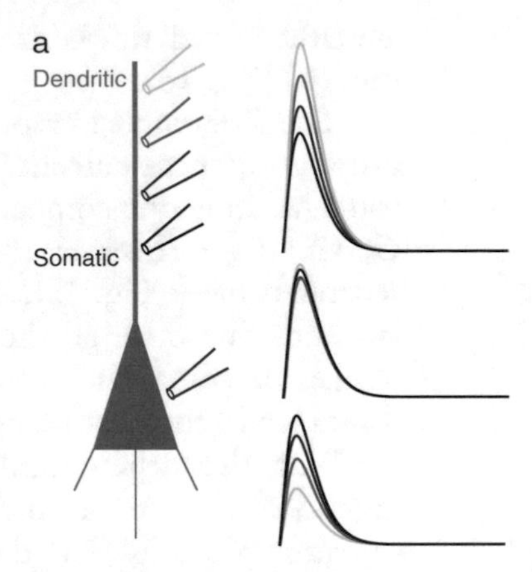

Fig. 2 (**a**) As the excitatory input increases away from the soma, the excitatory potentials become larger (measured at the input site). At the soma, in some cell types, despite the change in distance of activated inputs away from the soma, there is no significant difference in the potential amplitude (*middle* traces). However, in other cell types, the measured potential amplitude at the soma is inversely related to the activated input's distance from the soma

The difference between point number 1 and 2 is that point number 1 discusses local potentials recorded either at the dendrite or at the soma, while point number 2 focuses on potentials recorded only from the soma.

Empirical evidence in CA1 pyramidal neurons suggests that Rall's theoretical principle is not accurate (but read on). Magee and Cook [6] show that EPSPs generated at proximal synapses have similar somatic amplitudes when compared EPSPs generated at distal synapses. However, CA1 pyramidal neurons possess increased numbers of AMPA-type glutamate receptors at distal dendritic synapses [8–10]. This enables greater potential amplitudes to be generated distally, thus counterbalancing current loss via axial resistance or capacitance as the depolarizing potential travels to the soma.

In support of Rall's second principle, empirical evidence from layer 5 pyramidal neurons in the neocortex indicate that EPSPs generated at more distal sites show decreases in somatic amplitude. This is most likely due to layer 5 pyramidal neurons lacking the increased AMPA-type glutamate receptor expression at distal synapses [7, 11].

3. Somatic potentials generated at distal synapses have slower kinetics than those generated at proximal synapses causing temporal summation at the soma.

Experimental evidence indicates that proximal and distal synaptic inputs have similar EPSP kinetics when measured at the soma [12–14]. This is due to dendritic expression of voltage-activated channels, such as I_H channels (Na^+ and K^+ conducting channels), which are expressed at higher densities as the distance increases

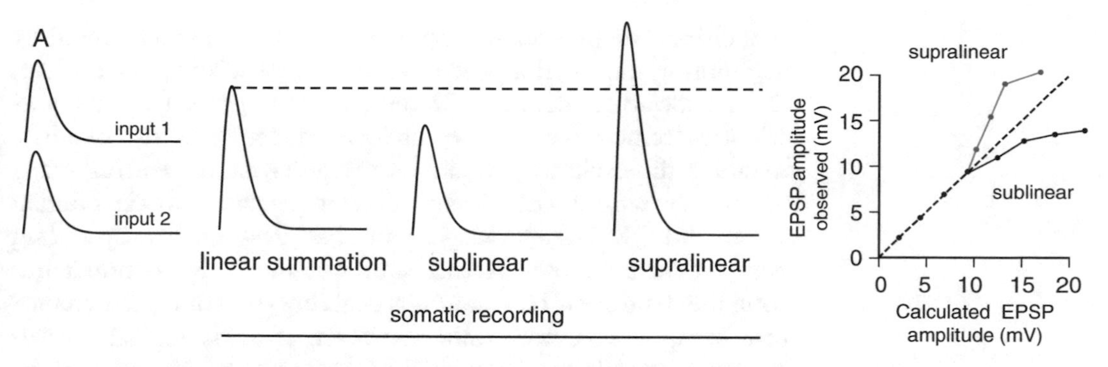

Fig. 3 (**a**) Two excitatory synaptic inputs undergo linear, sublinear, or supralinear summation patterns at the soma depending upon the active dendritic properties of the neuron. (*Right*) Graph showing the potential summation of synaptic inputs at the soma as the stimulus intensity increases

from the soma. I_H channels are active at rest and have a deactivation rate that is slow, thus affecting EPSP decay and not amplitude [13–15]. When I_H channels deactivate they produce a hyperpolarizing current that shortens the local EPSP duration, thus blocking temporal summation. However, the above example is just one possible scenario. Subthreshold EPSPs are likely to affect several voltage-gated currents such as Na^+ channels, low-voltage Ca^{2+} channels, and A-type K^+ channels (see Chap. 16). The balance between the inward and outward currents produced by these channels dictates the EPSP shape and temporal summation at the soma [16–18].

For example, experiments show that EPSPs elicited on different dendritic branches sum linearly at the soma [19, 20], while other experiments demonstrate sublinear summation of EPSPs at the soma when the synaptic inputs converge on the same dendritic branch (Fig. 3a) [21]. Other experiments show supralinear summation of EPSPs when calcium currents and sodium currents are activated [22–24, 53], while other experiments show EPSP supralinear summation in hyperpolarized neurons (attributed to the hyperpolarization-induced activation of the T-type Ca^{2+} channel [25]. To sum up, Rall's theory of EPSP summation is not entirely false despite only taking into consideration passive dendritic properties. After factoring in active dendritic properties, EPSP summation can occur depending upon the voltage-gated inward and outward currents produced.

There are a couple of more points to make concerning this topic that may benefit the beginning electrophysiologist. First, concerning anatomical organization of synaptic inputs, most neurons receive multiple afferents from different brain regions and evidence suggests that these afferents are specifically organized along the dendritic tree [26]. Therefore, it is plausible that EPSP summation may depend not only on active conductance along the dendrite, but also upon the location of the presynaptic inputs.

In addition, the presynaptic inputs may then be further altered by neighboring inputs in a process known as associativity (see Chap. 12 for associativity defined). Second, concerning electrophysiological measurements, since the active conductance at the dendrite can affect the excitatory inputs on the postsynaptic neuron, many synaptic current measurements in voltage-clamp mode contain cesium and QX-314 blocking K^+ and Na^+, respectively. By adding these blockers to the internal solution, the active conductance along the dendrite is blocked, thus enabling current measurements from synaptic receptors only. However, if physiological measurements of membrane potentials are desired, these blockers can be left out of the internal solution enabling EPSP measurements generated from synaptically activated receptors and voltage-gated channels.

4 Hebbian Plasticity (Spike-Timing-Dependent Plasticity (STDP))

Hebbian plasticity [27], developed by Donald Hebb, refers to synaptic strengthening, which is facilitated by the precise temporal activity of both the presynaptic and postsynaptic neuron [28, 29]. This temporal specificity is in the order of milliseconds whereby the presynaptic neuron precisely fires before the postsynaptic neuron causing synaptic depression. However, if postsynaptic excitation precedes presynaptic action potentials, synaptic strengthening occurs (Fig. 4a) [30–37]. In contrast to Hebbian plasticity, there is anti-Hebbian plasticity, which refers to synaptic weakening facilitated by coordinated presynaptic and postsynaptic firing (Fig. 4b). When the presynaptic action potential precedes the postsynaptic excitatory potential, long-term depression is observed. Anti-Hebbian plasticity can also refer to synaptic strengthening that is facilitated by presynaptic action potentials preceding postsynaptic firing.

The research on STDP is vast including mechanisms mediating specific responses in neuronal subtypes throughout the brain. A literature review is beyond the scope of this chapter, but for those interested, there are great reviews cited here [38, 39]. For the purpose of this chapter we want to touch on a few interesting ideas behind temporal presynaptic and postsynaptic activity to provide the reader with an appreciation for the many electrophysiological properties that mediate neuronal activity:

1. *Back propagating action potentials (BAPs) and EPSPs.* For those unfamiliar, BAP refers to the action potentials that travel from the soma to the dendrites (Fig. 4c). This is opposite to the conventional action potential propagation, which is initiated in the axon hillock and travels down the axon. BAPs are critically involved in synaptic modifications because they facilitate Ca^{2+} influx [40] through NMDA receptors and voltage-dependent

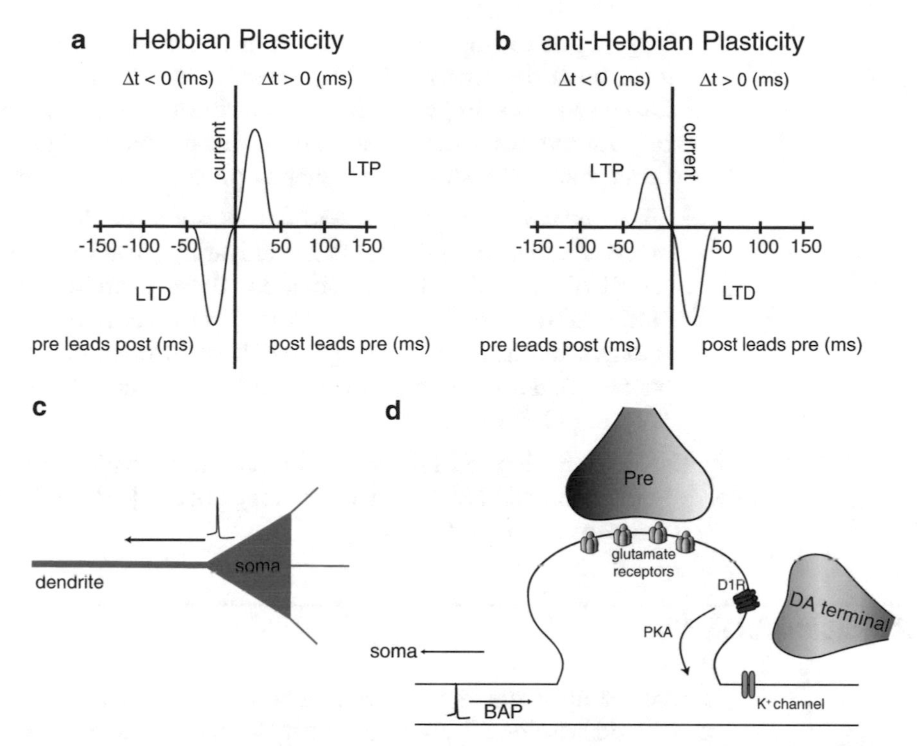

Fig. 4 (**a**) An illustration of Hebbian plasticity. When the presynaptic neuron fires an action potential ≤50 ms before the postsynaptic neuron depolarizes, long-term depression (LTD) occurs at the presynaptic to postsynaptic connection. However, if the postsynaptic neuron is depolarized ≤50 ms prior to a presynaptic action potential, long-term potentiation (LTP) occurs. (**b**) An illustration of anti-Hebbian plasticity showing LTP or LTD at synaptic connections caused by pre-before postsynaptic excitation or post-before presynaptic excitation, respectively. (**c**) Illustration of a backpropagating action potential generated at the soma and traveling retrogradely toward the dendrites. (**d**) A tripartite excitatory synapse containing a presynaptic terminal which releases glutamate, thus activating glutamate receptors postsynaptically. Dopamine terminals synapse onto the postsynaptic neuron activating dopamine receptors, which can alter K+ channels through G protein signaling pathways, thus affecting back propagating action potentials

Ca^{2+} channels. The influx of Ca^{2+} initiates signaling mechanisms necessary for synaptic strengthening.
Timing an EPSP with a BAP in a dendrite can inactivate hyperpolarizing channels (e.g., A-type K^+ channels) [35, 41] or activate Na^+ channels in distal dendrites [42]. This increases Ca^{2+} influx and boosts the BAP speed, which enhances BAP arrival times at synapses.

2. *Dendritic location and spike-timing-dependent plasticity (spatial effects)*. Spike-timing-dependent synaptic modifications are dependent upon the synaptic location along the dendrite. This is most likely regulated by dendritic resistance and capacitance, which attenuate the BAP as it travels distally. Empirically tested, the magnitude of synaptic strengthening (commonly referred to as long-term potentiation (LTP)) is smaller at the distal vs. proximal apical dendrite [43].

Synaptic location along the dendrite can also dictate LTP or long-term depression (LTD) (synaptic weakening) induction. Distal synapses on pyramidal neurons in the somatosensory cortex potentiates when postsynaptic firing precedes presynaptic firing, but depress when presynaptic precedes postsynaptic [44].

3. *Neuromodulators.* Neuromodulators such as dopamine and acetylcholine regulate STDP by activating protein kinases such as PKA and PKC. Once activated, these protein kinases can activate Na^+ and K^+ channels, thus affecting conductance states along a dendrite (Fig. 4d) [45–48]. By affecting conductance states, BAPs traveling through dendrites vary in speed, thus altering Ca^{2+} influx.
4. *STDP* in vivo. STDP occurs in vivo with evidence in retinal ganglion cells [37], somatosensory cortex [49, 50], and the human motor system [51].

5 Space-Clamp Artefacts

Space clamp artifacts can occur when voltage clamping a cell in a somatic whole-cell configuration. We have previously discussed that dendrites contain compartments that are separated by axial resistances. This axial resistance runs in series along the dendrite (Fig. 1a) causing a significant voltage drop as the distance from the soma increases. In addition to axial resistance, the cell capacitance can add to space clamp artifacts by delaying changes in the clamped voltage. The thin-branched dendritic compartments generate high capacitance. Therefore, dendritic compartments extending far away from the soma will be poorly voltage clamped leading to sluggish, incomplete current responses [52] (Fig. 5a).

To limit space-clamp artifacts voltage-clamp recordings from highly branched neurons in slices or highly developed cultured neurons should be avoided. However, this mostly pertains to

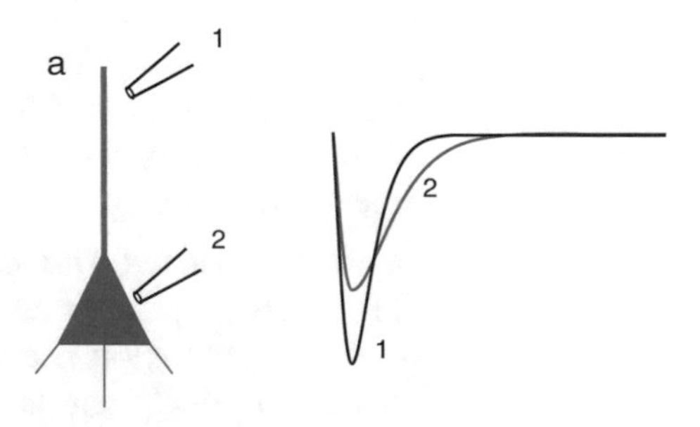

Fig. 5 (**a**) An example of how space clamping can alter the currents recorded in voltage-clamp mode. A whole-cell recording at site 1 near the synaptic input generates a large amplitude, fast deactivating inward current relative to an inward current generated by the same synaptic input, but recorded at site 2

receptors under study that extend far away from the soma. If the receptor is expressed near or on the soma, space-clamp artifacts should be relatively small. Finally, we should mention that space clamp is restricted to voltage-clamp recordings.

6 Summary

The complex nature of neuronal communications relies in part on synaptic location, cellular morphology, receptor expression, dendritic active properties (e.g., voltage-gated channels), and temporal-dependent firing of both presynaptic and postsynaptic neurons. Each component can affect postsynaptic membrane potentials both in the short-term and over long-term periods. Electrophysiologists interested in synaptic plasticity should take each of these components into consideration as it demonstrates how intracellular signaling mechanisms can be regulated by neuronal morphology, receptor expression, and firing properties.

The neuronal properties of dendritic branches may be invaluable in regulating membrane excitability. However, these same properties have the disadvantage of producing errors during electrophysiology measurements. Care should be taken when recording from highly branched neurons or from over-developed cultured neurons in order to reduce space-clamp artifacts.

References

1. Rall W (1967) Distinguishing theoretical synaptic potentials computed for different soma-dendritic distributions of synaptic input. J Neurophysiol 30(5):1138–1168
2. Rall W, Burke RE, Smith TG, Nelson PG, Frank K (1967) Dendritic location of synapses and possible mechanisms for the monosynaptic EPSP in motoneurons. J Neurophysiol 30(5): 1169–1193
3. Magee JC (2000) Dendritic integration of excitatory synaptic input. Nat Rev Neurosci 1(3):181–190
4. Rall W, Segev I, Rinzel J, Shepherd GM (1995) The theoretical foundation of dendritic function: selected papers of Wilfrid Rall with commentaries. Bradford Books, London
5. Reiss RF, U S A F O o S Research and I. General Precision (1964). Neural theory and modeling: proceedings of the 1962 Ojai symposium, Stanford University Press
6. Magee JC, Cook EP (2000) Somatic EPSP amplitude is independent of synapse location in hippocampal pyramidal neurons. Nat Neurosci 3(9):895–903
7. Williams SR, Stuart GJ (2002) Dependence of EPSP efficacy on synapse location in neocortical pyramidal neurons. Science 295(5561): 1907–1910
8. Andrasfalvy BK, Magee JC (2001) Distance-dependent increase in AMPA receptor number in the dendrites of adult hippocampal CA1 pyramidal neurons. J Neurosci 21(23): 9151–9159
9. Matsuzaki M, Ellis-Davies GC, Nemoto T, Miyashita Y, Iino M, Kasai H (2001) Dendritic spine geometry is critical for AMPA receptor expression in hippocampal CA1 pyramidal neurons. Nat Neurosci 4(11):1086–1092
10. Spruston N, Jonas P, Sakmann B (1995) Dendritic glutamate receptor channels in rat hippocampal CA3 and CA1 pyramidal neurons. J Physiol 482(Pt 2):325–352
11. Williams SR, Stuart GJ (2003) Role of dendritic synapse location in the control of action potential output. Trends Neurosci 26(3): 147–154
12. Berger T, Larkum ME, Luscher HR (2001) High I(h) channel density in the distal apical

dendrite of layer V pyramidal cells increases bidirectional attenuation of EPSPs. J Neurophysiol 85(2):855–868
13. Magee JC (1999) Dendritic lh normalizes temporal summation in hippocampal CA1 neurons. Nat Neurosci 2(6):508–514
14. Williams SR, Stuart GJ (2000) Site independence of EPSP time course is mediated by dendritic I(h) in neocortical pyramidal neurons. J Neurophysiol 83(5):3177–3182
15. Magee JC (1998) Dendritic hyperpolarization-activated currents modify the integrative properties of hippocampal CA1 pyramidal neurons. J Neurosci 18(19):7613–7624
16. Hoffman DA, Magee JC, Colbert CM, Johnston D (1997) K+ channel regulation of signal propagation in dendrites of hippocampal pyramidal neurons. Nature 387(6636): 869–875
17. Schwindt PC, Crill WE (1995) Amplification of synaptic current by persistent sodium conductance in apical dendrite of neocortical neurons. J Neurophysiol 74(5):2220–2224
18. Stuart G, Sakmann B (1995) Amplification of EPSPs by axosomatic sodium channels in neocortical pyramidal neurons. Neuron 15(5): 1065–1076
19. Cash S, Yuste R (1999) Linear summation of excitatory inputs by CA1 pyramidal neurons. Neuron 22(2):383–394
20. Skydsgaard M, Hounsgaard J (1994) Spatial integration of local transmitter responses in motoneurones of the turtle spinal cord in vitro. J Physiol 479(Pt 2):233–246
21. Reyes A (2001) Influence of dendritic conductances on the input-output properties of neurons. Annu Rev Neurosci 24:653–675
22. Golding NL, Jung HY, Mickus T, Spruston N (1999) Dendritic calcium spike initiation and repolarization are controlled by distinct potassium channel subtypes in CA1 pyramidal neurons. J Neurosci 19(20):8789–8798
23. Schiller J, Schiller Y, Stuart G, Sakmann B (1997) Calcium action potentials restricted to distal apical dendrites of rat neocortical pyramidal neurons. J Physiol 505(Pt 3):605–616
24. Schwindt P, Crill W (1999) Mechanisms underlying burst and regular spiking evoked by dendritic depolarization in layer 5 cortical pyramidal neurons. J Neurophysiol 81(3): 1341–1354
25. Nettleton JS, Spain WJ (2000) Linear to supralinear summation of AMPA-mediated EPSPs in neocortical pyramidal neurons. J Neurophysiol 83(6):3310–3322
26. Britt JP, Benaliouad F, McDevitt RA, Stuber GD, Wise RA, Bonci A (2012) Synaptic and behavioral profile of multiple glutamatergic inputs to the nucleus accumbens. Neuron 76(4):790–803
27. Hebb DO (1949) The organization of behavior: a neuropsychological theory. Wiley, New York
28. Levy WB, Steward O (1983) Temporal contiguity requirements for long-term associative potentiation/depression in the hippocampus. Neuroscience 8(4):791–797
29. Walters ET, Byrne JH (1983) Associative conditioning of single sensory neurons suggests a cellular mechanism for learning. Science 219(4583):405–408
30. Bell CC, Han VZ, Sugawara Y, Grant K (1997) Synaptic plasticity in a cerebellum-like structure depends on temporal order. Nature 387(6630):278–281
31. Bi GQ, Poo MM (1998) Synaptic modifications in cultured hippocampal neurons: dependence on spike timing, synaptic strength, and postsynaptic cell type. J Neurosci 18(24): 10464–10472
32. Debanne D, Gahwiler BH, Thompson SM (1998) Long-term synaptic plasticity between pairs of individual CA3 pyramidal cells in rat hippocampal slice cultures. J Physiol 507 (Pt 1):237–247
33. Egger V, Feldmeyer D, Sakmann B (1999) Coincidence detection and changes of synaptic efficacy in spiny stellate neurons in rat barrel cortex. Nat Neurosci 2(12):1098–1105
34. Fino E, Deniau JM, Venance L (2008) Cell-specific spike-timing-dependent plasticity in GABAergic and cholinergic interneurons in corticostriatal rat brain slices. J Physiol 586(1): 265–282
35. Magee JC, Johnston D (1997) A synaptically controlled, associative signal for Hebbian plasticity in hippocampal neurons. Science 275(5297):209–213
36. Markram H, Lubke J, Frotscher M, Sakmann B (1997) Regulation of synaptic efficacy by coincidence of postsynaptic APs and EPSPs. Science 275(5297):213–215
37. Zhang LI, Tao HW, Holt CE, Harris WA, Poo M (1998) A critical window for cooperation and competition among developing retinotectal synapses. Nature 395(6697):37–44
38. Caporale N, Dan Y (2008) Spike timing-dependent plasticity: a Hebbian learning rule. Annu Rev Neurosci 31:25–46
39. Feldman DE (2012) The spike-timing dependence of plasticity. Neuron 75(4):556–571
40. Tsubokawa H, Offermanns S, Simon M, Kano M (2000) Calcium-dependent persistent facilitation of spike backpropagation in the

CA1 pyramidal neurons. J Neurosci 20(13): 4878–4884

41. Watanabe S, Hoffman DA, Migliore M, Johnston D (2002) Dendritic K+ channels contribute to spike-timing dependent long-term potentiation in hippocampal pyramidal neurons. Proc Natl Acad Sci U S A 99(12): 8366–8371
42. Stuart GJ, Hausser M (2001) Dendritic coincidence detection of EPSPs and action potentials. Nat Neurosci 4(1):63–71
43. Froemke RC, Poo MM, Dan Y (2005) Spike-timing-dependent synaptic plasticity depends on dendritic location. Nature 434(7030): 221–225
44. Letzkus JJ, Kampa BM, Stuart GJ (2006) Learning rules for spike timing-dependent plasticity depend on dendritic synapse location. J Neurosci 26(41):10420–10429
45. Hoffman DA, Johnston D (1998) Downregulation of transient K+ channels in dendrites of hippocampal CA1 pyramidal neurons by activation of PKA and PKC. J Neurosci 18(10):3521–3528
46. Hoffman DA, Johnston D (1999) Neuromodulation of dendritic action potentials. J Neurophysiol 81(1):408–411
47. Ruan H, Saur T, Yao WD (2014) Dopamine-enabled anti-Hebbian timing-dependent plasticity in prefrontal circuitry. Front Neural Circuits 8:38
48. Tsubokawa H, Ross WN (1997) Muscarinic modulation of spike backpropagation in the apical dendrites of hippocampal CA1 pyramidal neurons. J Neurosci 17(15):5782–5791
49. Cassenaer S, Laurent G (2007) Hebbian STDP in mushroom bodies facilitates the synchronous flow of olfactory information in locusts. Nature 448(7154):709–713
50. Jacob V, Brasier DJ, Erchova I, Feldman D, Shulz DE (2007) Spike timing-dependent synaptic depression in the in vivo barrel cortex of the rat. J Neurosci 27(6):1271–1284
51. Wolters A, Sandbrink F, Schlottmann A, Kunesch E, Stefan K, Cohen LG, Benecke R, Classen J (2003) A temporally asymmetric Hebbian rule governing plasticity in the human motor cortex. J Neurophysiol 89(5):2339–2345
52. Molleman A (2003) Basic theoretical principles. Patch clamping. Wiley, New York, pp 5–42
53. Stuart G, Schiller J, Sakmann B (1997) Action potential initiation and propagation in rat neocortical pyramidal neurons. J Physiol 505 (Pt 3):617–632

Chapter 7

Perforated Patch

Nicholas Graziane and Yan Dong

Abstract

In whole-cell patch clamp mode the internal solution of the micropipette perfuses the cell replacing the much smaller cytosolic solution. Because of this, some soluble factors that modulate cellular excitability and influence signaling pathways are washed out via the micropipette causing altered intracellular signaling, cellular function, or the active state of ion channels. One of the commonly observed consequences is current run-down, which refers to the gradual loss of current over time. Key molecules have been added to the micropipette's intracellular solution in order to impede current run-down. ATP and/or creatine/phosphocreatine are added to prevent channel dephosphorylation and protease inhibitors are added to prevent proteolytic degradation of channel proteins [1]. However, these components are not always successful in preventing current run-down as other factors can elicit the slow demise of current recordings in whole-cell patch through the disruption of the actin cytoskeleton [2].

Key words Polyene antibiotics, Gramicidin D, β-Escin

1 Introduction

In whole-cell patch clamp mode the internal solution of the micropipette perfuses the cell replacing the much smaller cytosolic solution. Because of this, some soluble factors that modulate cellular excitability and influence signaling pathways are washed out via the micropipette causing altered intracellular signaling, cellular function, or the active state of ion channels. One of the commonly observed consequences is current run-down, which refers to the gradual loss of current over time. Key molecules have been added to the micropipette's intracellular solution in order to impede current run-down. ATP and/or creatine/phosphocreatine are added to prevent channel dephosphorylation and protease inhibitors are added to prevent proteolytic degradation of channel proteins [1]. However, these components are not always successful in preventing current run-down as other factors can elicit the slow demise of current recordings in whole-cell patch through the disruption of the actin cytoskeleton [2].

Nicholas Graziane and Yan Dong, *Electrophysiological Analysis of Synaptic Transmission*, Neuromethods, vol. 112,
DOI 10.1007/978-1-4939-3274-0_7, © Springer Science+Business Media New York 2016

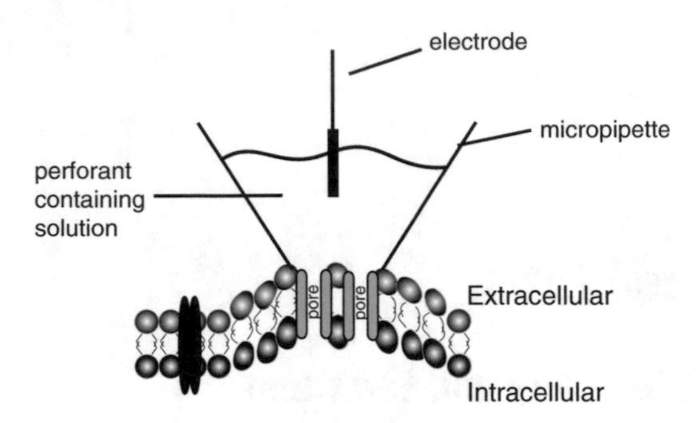

Fig. 1 An illustration of a perforated patch. The micropipette is filled with solution containing a perforant, which induces pore formation in the cellular membrane

The disadvantages of whole-cell recordings discussed above led to the invention of the perforated-patch clamp technique (Fig. 1) by Lindau and Fernandez in 1986. Unable to study cellular responses to antigens due to washout of bioactive molecules, Lindau and Fernandez used ATP (400 μM), which was previously known to permeabilize membranes, in the intracellular solution [3]. By using this technique, Lindau and Fernandez were able to minimize proteolysis and prevent the loss of bioactive molecules. Today, ATP is rarely used as a membrane permeabling substrate for forming perforation because of the high access resistance (0.1–5.0 GΩ) and complication of ATP receptors on the cell surface. Instead, nystatin, amphotericin B, gramicidin D, and β-escin are commonly used as alternatives to ATP.

2 Polyene Antibiotics

Polyene antibiotics are derived from the bacteria genus *Streptomyces*. These antibiotics include nystatin and amphotericin B, which are commonly used medicinally as antifungal agents. Both nystatin and amphotericin B bind to lipid bilayers containing sterols, in particular ergosterol [4]. However, it has been shown that polyene antibiotics can bind and form pores in sterol-free membranes [5–7]. The pores are formed from the large lactone ring which contains hydrophilic groups on one side and hydrophobic groups on the other (Fig. 2a). These barrel-shaped pores (roughly 2.8 nm in length) are stabilized by sterols (Fig. 2b) and are permeable to water, small hydrophilic nonelectrolytic molecules, and univalent cations (i.e., Li^+, Na^+, K^+, Cs^+). Polyene pores are 10× less permeable to univalent anions like Cl^- and are impermeable to large molecules and divalent ions (e.g., Ca^{2+}, Mg^{2+}) [1]. Because of this, intracellular Ca^{2+} stores remain intact [8, 9]. Since nystatin and amphotericin B are added to the micropipette, they remain localized to the membrane in the patch and are unable to traverse

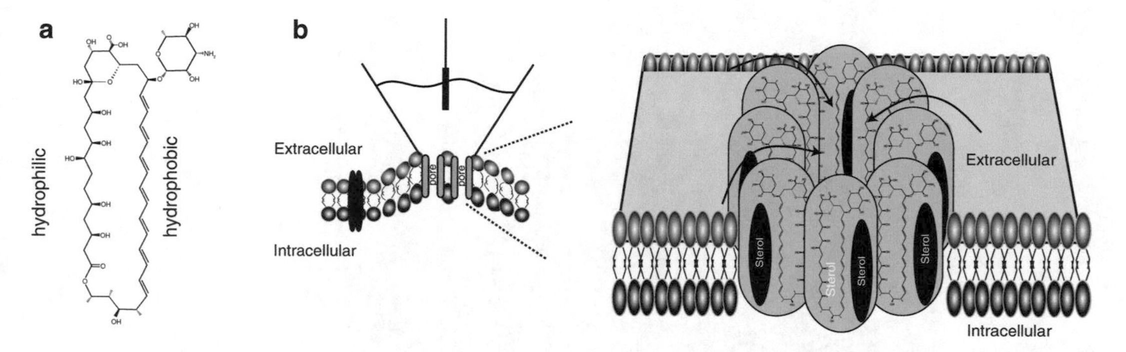

Fig. 2 An illustration showing how amphotericin B forms channel pores in a cellular membrane. (**a**) The structure of amphotericin B contains a hydrophilic side, which faces the pore opening and a hydrophobic side, which binds to sterols in the cellular membrane. (**b**) Amphotericin B-containing pores are formed from 8 to 10 molecules arranged as shown. The pore depicted by the *arrows* shows the location of ion permeability

the membrane unless the membrane ruptures causing pores to switch from predominantly cation selective to anion selective [10]. Table 1 summarizes the characteristics of both nystatin and amphotericin B. For solution preparation and protocols useful for polyene antibiotic methods for perforated-patch clamp see Sarantopoulos [1] and Ishibashi et al. [21].

3 Problems with Polyene Antibiotics

1. Solubility is difficult due to the hydrophobic characteristics of polyenes.
2. They are sensitive to light and heat and they lose potency over time. Therefore, fresh solutions need to be made daily and used within a couple of hours.
3. Inadequate gigaseal formation when polyene antibiotics are at the pipette tip. This can be partially circumvented by prefilling the tip of the pipettes with polyene free solution (e.g., a second immersion regular internal solution is sufficient), then backfilling with polyene containing solution.
4. Donnan potentials are formed leading to voltage differences between the cell's interior and interior of the pipette (potentials as high as 10 mV can be detected). Donnan potentials are caused by Cl^- fluxes generated by the permeability of Cl^- while the simultaneous impermeability to larger anions. Therefore, Cl^- redistributes between intracellular compartment and the micropipette.
5. Cell shrinking and swelling can occur due to Cl^- redistribution. This is preventable using appropriate filling solutions [23] or replacing some Cl^- with large non-permeable anions such as SO_4^{2-}.

Table 1 The properties of different perforants used for the perforated patch technique

Perforant	Class	Pore structure	Pore radius (nm)	Selectivity	Conductance (pS)	Access resistance (MΩ)	Pore forming duration (min)
Nystatin	Polyene antibiotic, *Streptomyces nousei*	Lactone ring forming barrel like pore	0.4[a]	Monovalent cations, Cl^-, urea (0.18 nm), glycerol (0.31 nm)[a]	2[b]	7–10	30
Amphotericin B	Polyene antibiotic, *Streptomyces nodosus*	Lactone ring forming barrel like pore	0.8[c]	Monovalent cations, Cl^-, urea (0.18 nm), glycerol (0.31 nm), glucose (0.42 nm), sucrose (0.52 nm)	6[b]	3–10	30
Gramicidin	Linear polypeptide antibiotic, *Bacillus brevis*	β-Sheet connected as end-to-end dimers	≤3.5	Monovalent cations, P_{Cl} ~0 $Rb^+ > K^+ > Cs^+ > Na^+ > Li$[d]	50[e]	20–40	40
β-Escin	Saponin derivative	Unknown	8[d]	Nonselective, Ca^{2+} permeable	Unknown	20–30[f]	<10

[a][4, 9, 11, 12]
[b][4, 13, 14]
[c][15–17]
[d][18–20]
[e][21]
[f][22]

6. Perforated-patch clamping is a slow process taking ≥30 min to gain sufficient electrophysiological access to the intracellular compartment. Since polyenes are backfilled in the micropipette, it takes time for these molecules to diffuse to the pipette tip and form pores in the cellular membrane. In addition, similar to whole-cell patch clamp, failure to achieve an acceptable access resistance can occur.
7. Inability to apply positive pressure to the micropipette increases the risk of blockage at the tip from debris in the bath or from debris in the tissue slice. Positive pressure cannot be applied as polyene antibiotics will be pushed to the micropipette tip impeding gigaseal formation.

4 Gramicidin D

One of the disadvantages of using polyenes as a perforant is that they are permeable to Cl^- causing Donnan potentials and introducing some bias in investigations of Cl^- permeable channels (e.g., GABAR, glycine). With this disadvantage in mind, researchers have used gramicidin D as a performant, which has negligible Cl^- permeability. Gramicidin D is a mixture of gramicidin A, B, and C. It forms voltage-insensitive membrane pores from molecules tied up as head-to-head dimers and folding as right hand helices [1]. The lipid bilayer interacts with the hydrophobic outer surface, while the hydrophilic inner surface of the dimer forms the channel. Properties of gramicidin D are listed in Table 1.

In addition to lacking Cl^- permeability, gramicidin D in low concentrations (<5 μg/mL) does not interfere with gigaseal formation, while achieving an access resistance of 25–50 MΩ [1]. This negates the need to prefill and backfill the micropipette and allows for the application of positive pressure in the pipette. However, gramicidin D does have the disadvantage of lacking solubility, which is why another perforant alternative was developed (i.e., β-escin).

5 β-Escin

To overcome the disadvantages of polyenes and gramicidin D discussed above, β-escin was developed providing many advantages over the former mentioned perforants. β-Escin has a higher success rate (65 %) in achieving low access resistance (20–30 MΩ) compared with nystatin (50 %), gramicidin D (33 %), and amphotericin B (22 %) as demonstrated work done in cardiac myocytes [1]. In fact 94 % success rate has been reported in ventricular myocytes [22].

β-Escin's are soluble in water and can be aliquoted and stored at −20 °C for up to 1 week. β-Escin achieves maximum permeability quickly, previously reported as short as <10 min [1]. Finally, β-Escin does not affect gigaseal formation at concentrations of 25 μmol/L [22].

β-Escins (saponins) are glycosides containing a steroid, a triterpenoid nucleus, and one or more carbohydrate branches [1]. They form complexes with cell membranes linking to sterols, consequently causing pore formation and cell permeabilization, while causing additional alterations in negatively charged carbohydrate portions on the cell surface [24–27].

6 Summary

The whole-cell patch clamp technique has a major disadvantage in that certain ionic channels, such as Ca^{2+} channels, show decreased conductance with increased recording time. The reason is because the pipette solution perfuses the cell decreasing intracellular bioactive molecular concentrations needed for channel function. To overcome this problem, the perforated patch clamp technique was developed. In this configuration, pores are formed in the cellular membrane providing electrical access to the recorded cell, while preventing "wash-out" of soluble intracellular components. There are three common pore forming agents used today. Polyene antibiotics are widely used, but possess limitations such as poor solubility, short half-life, Cl^- permeability, and relatively long period of time (≥30 min) to achieve sufficient electrical access to the cell. Gramicidin D generated pores lack Cl^- permeability reducing the potential risk of Donnan potentials as well as cell swelling and shrinking. Gramicidin D also does not interfere with gigaseal formation. However, solubility of Gramicidin D is poor. The most recent development in pore forming agents has been the β-escins. β-Escins take <10 min to achieve cellular access and have a high success rate. They are soluble in water and have a longer duration of potency compared to polyene antibiotics and gramicidin D.

Through perforated patch clamp, channel function can be tested empirically without altering intracellular concentrations of bioactive molecules. Electrophysiologists eager to study channels susceptible to run-down or dependent upon intracellular calcium stores should consider using perforated patches as an alternative to whole-cell configurations. If interested, the reader can use Sarantopolous [1] and Ishibashi et al. [21] for protocols used for perforated patches.

References

1. Sarantopoulos C (2007) Perforated patch-clamp techniques. Neuromethods 38:253–293
2. Furukawa T, Yamane T-i, Terai T, Katayama Y, Hiraoka M (1996) Functional linkage of the cardiac ATP-sensitive K+ channel to the actin cytoskeleton. Pflugers Arch 431(4):504–512
3. Arav R, Friedberg I (1985) ATP analogues induce membrane permeabilization in transformed mouse fibroblasts. Biochim Biophys Acta (BBA) Biomembr 820(2):183–188
4. Akaike N, Harata N (1994) Nystatin perforated patch recording and its applications to analyses of intracellular mechanisms. Jpn J Physiol 44(5):433–473
5. Cotero BV, Rebolledo-Antunez S, Ortega-Blake I (1998) On the role of sterol in the formation of the amphotericin B channel. Biochim Biophys Acta 1375(1 2):43–51
6. Gruszecki WI, Gagos M, Herec M, Kernen P (2003) Organization of antibiotic amphotericin B in model lipid membranes. A mini review. Cell Mol Biol Lett 8(1):161–170
7. Venegas B, Gonzalez-Damian J, Celis H, Ortega-Blake I (2003) Amphotericin B channels in the bacterial membrane: role of sterol and temperature. Biophys J 85(4):2323–2332
8. Horn R, Korn SJ (1992) Prevention of rundown in electrophysiological recording. Methods Enzymol 207:149–155
9. Horn R, Marty A (1988) Muscarinic activation of ionic currents measured by a new whole-cell recording method. J Gen Physiol 92(2): 145–159
10. Marty A, Finkelstein A (1975) Pores formed in lipid bilayer membranes by nystatin, Differences in its one-sided and two-sided action. J Gen Physiol 65(4):515–526
11. Hladky SB, Haydon DA (1970) Discreteness of conductance change in bimolecular lipid membranes in the presence of certain antibiotics. Nature 225(5231):451–453
12. Holz R, Finkelstein A (1970) The water and nonelectrolyte permeability induced in thin lipid membranes by the polyene antibiotics nystatin and amphotericin B. J Gen Physiol 56(1):125–145
13. Ermishkin LN, Kasumov KM, Potzeluyev VM (1976) Single ionic channels induced in lipid bilayers by polyene antibiotics amphotericin B and nystatine. Nature 262(5570):698–699
14. Kasumov KM, Borisova MP, Ermishkin LN, Potseluyev VM, Silberstein AY, Vainshtein VA (1979) How do ionic channel properties depend on the structure of polyene antibiotic molecules? Biochim Biophys Acta 551(2): 229–237
15. de Kruijff B, Demel RA (1974) Polyene antibiotic-sterol interactions in membranes of Acholeplasma laidlawii cells and lecithin liposomes. III. Molecular structure of the polyene antibiotic-cholesterol complexes. Biochim Biophys Acta (BBA) Biomembr 339(1): 57–70
16. de Kruijff B, Gerritsen WJ, Oerlemans A, Demel RA, van Deenen LL (1974) Polyene antibiotic-sterol interactions in membranes of Acholeplasma laidlawii cells and lecithin liposomes. I. Specificity of the membrane permeability changes induced by the polyene antibiotics. Biochim Biophys Acta 339(1):30–43
17. de Kruijff B, Gerritsen WJ, Oerlemans A, van Dijck PW, Demel RA, van Deenen LL (1974) Polyene antibiotic-sterol interactions in membranes of Acholesplasma laidlawii cells and lecithin liposomes. II. Temperature dependence of the polyene antibiotic-sterol complex formation. Biochim Biophys Acta 339(1): 44–56
18. Andreoli TE, Bangham JA, Tosteson DC (1967) The formation and properties of thin lipid membranes from HK and LK sheep red cell lipids. J Gen Physiol 50(6):1729–1749
19. Mueller P, Rudin DO (1967) Development of K+-Na+ discrimination in experimental bimolecular lipid membranes by macrocyclic antibiotics. Biochem Biophys Res Commun 26(4): 398–404
20. Tosteson DC, Andreoli TE, Tieffenberg M, Cook P (1968) The effects of macrocyclic compounds on cation transport in sheep red cells and thin and thick lipid membranes. J Gen Physiol 51(5):373–384
21. Ishibashi H, Moorhouse A, Nabekura J (2012) Perforated whole-cell patch-clamp technique: a user's guide. In: Okada Y (ed) Patch clamp techniques. Springer, Japan, pp 71–83
22. Fu LY, Wang F, Chen XS, Zhou HY, Yao WX, Xia GJ, Jiang MX (2003) Perforated patch recording of L-type calcium current with beta-escin in guinea pig ventricular myocytes. Acta Pharmacol Sin 24(11):1094–1098

23. Rae J, Cooper K, Gates P, Watsky M (1991) Low access resistance perforated patch recordings using amphotericin B. J Neurosci Methods 37(1):15–26
24. Abe H, Konishi H, Komiya H, Arichi S (1981) Effects of saikosaponins on biological membranes. 3. Ultrastructural studies on effects of saikosaponins on the cell surface. Planta Med 42(4):356–363
25. Gauthier C, Legault J, Girard-Lalancette K, Mshvildadze V, Pichette A (2009) Haemolytic activity, cytotoxicity and membrane cell permeabilization of semi-synthetic and natural lupane- and oleanane-type saponins. Bioorg Med Chem 17(5):2002–2008
26. Melzig MF, Bader G, Loose R (2001) Investigations of the mechanism of membrane activity of selected triterpenoid saponins. Planta Med 67(1):43–48
27. Podolak I, Galanty A, Sobolewska D (2010) Saponins as cytotoxic agents: a review. Phytochem Rev 9(3):425–474

Part II

Recording of Synaptic Current

Chapter 8

Isolation of Synaptic Current

Nicholas Graziane and Yan Dong

Abstract

Synaptic currents can be examined by a variety of different approaches such as isolating quantal currents or evoking currents using electrical or optical stimulation. It is important for a beginning electrophysiologist to fully comprehend each approach so that the right method is used for the intended scientific question. Below we provide conceptual and technical information for miniature, spontaneous, quantal, and evoked postsynaptic currents with useful references that provide additional and in some cases more detailed information.

Key words Miniature postsynaptic currents, Spontaneous postsynaptic currents, Quantal postsynaptic currents, Evoked currents, Optical stimulation, Electrical stimulation, DREADDs, Nanoparticles

1 Introduction

Synaptic currents can be examined by a variety of different approaches such as isolating quantal currents or evoking currents using electrical or optical stimulation. It is important for a beginning electrophysiologist to fully comprehend each approach so that the right method is used for the intended scientific question. Below we provide conceptual and technical information for miniature, spontaneous, quantal, and evoked postsynaptic currents with useful references that provide additional and in some cases more detailed information.

2 Miniature Postsynaptic Currents (mPSCs)

Miniature PSC refers to currents that are generated by neurotransmitters released independently of action potentials (Fig. 1). Constitutively, in the absence of an action potential, a single presynaptic vesicle can fuse to a presynaptic terminal. As a result, neurotransmitters are released (from this single vesicle) into the presynaptic cleft and subsequently bind to postsynaptic

Nicholas Graziane and Yan Dong, *Electrophysiological Analysis of Synaptic Transmission*, Neuromethods, vol. 112, DOI 10.1007/978-1-4939-3274-0_8, © Springer Science+Business Media New York 2016

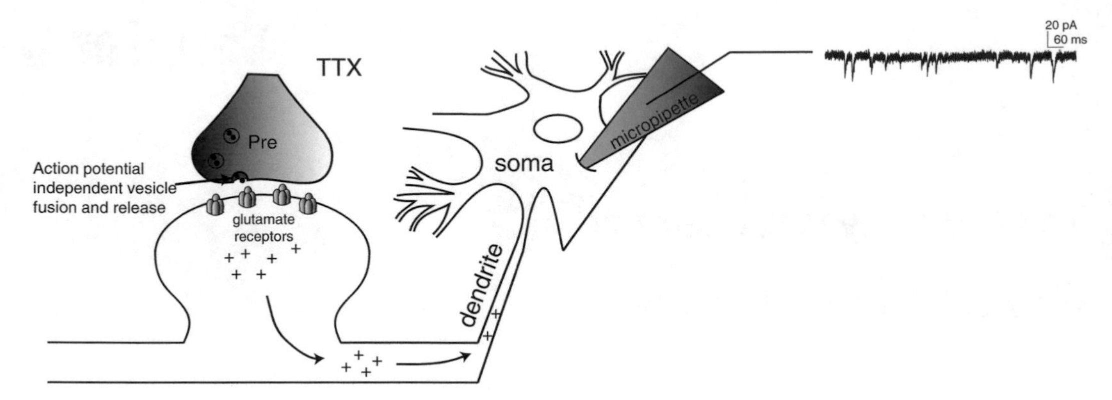

Fig. 1 A miniature postsynaptic current recording in the presence of TTX. Presynaptic vesicles spontaneously fuse to the presynaptic membrane causing the release of neurotransmitter via action potential independent means. Somatic whole cell recordings measure the EPSCs (traces illustrated)

receptors. The amount of neurotransmitters within one vesicle is called a quantum. The current produced in a quantum is referred to as quantal currents since it embodies the minimum current that can be generated from the contents of a single released vesicle.

Miniature PSCs can be recorded in the whole-cell configuration in the presence of tetrodotoxin (TTX) (a Na^+ channel blocker). Once the data is obtained there are many useful quantal analyses that can be measured such as mPSC amplitude and frequency. The current amplitude is associated with postsynaptic neurotransmitter sensitivity, which is regulated by postsynaptic receptor number and conductance. Changes in frequency are associated with presynaptic changes such as the probability of release and/or the number of release sites. Of course, these rules are not always true as changes in frequency can also be associated with postsynaptic modifications such as the unsilencing of synapses [1]. In addition to amplitude and frequency measurements, current kinetics can also be analyzed from mPSCs data such as rise times, decay times, and mean channel open times [2].

An interesting complication of mPSCs is that they occur in the absence of action potentials. Therefore, the data collected from mPSCs may not correlate with other current measurements that rely on action potentials. For example, mPSCS are small currents. Therefore, to be detected in a somatic whole-cell patch configuration, the site of neurotransmitter release must be in close proximity to the soma. Evoked currents, on the other hand, may be generated at distal sites causing changes in the size and time course of the recorded currents. Another complication is that spontaneous leak of synaptic vesicles may occur at distinct classes of synapses that are separate from action potential dependent release sites [3, 4]. Therefore, making correlations between action potential dependent currents and mPSCs should be done so with caution.

2.1 Technical Considerations

To record mPSCs in a whole-cell patch configuration, it is always good practice to wait 5 to 10 min before collecting the data of interest. During the 5–10 min, the internal solution has time to perfuse the cell allowing for equilibrium to be reached between intracellular and extracellular compartments. Consequently, analyzed currents will not be contaminated by potential effects generated by non-equilibrated solutions. Since mPSCs require bath-perfused TTX, it may be beneficial to determine the time it takes for TTX to reach and block Na^{2+} channels. This time can be determined by evoking currents while perfusing the cells with TTX. The time depends on the flow rate as well as the distance between the chamber and the reservoir, but typically it takes about 10 min. While collecting mPSC data, it is important to watch the access resistance as changes in the access resistance can lead to changes in mPSC current characteristics (e.g., amplitude, frequency, kinetics). In addition, the baseline noise should be no greater than 5–10 pA. Recording stable baselines with low noise enhances the accuracy of the data analysis as well as makes the overall analysis process much easier. Data analysis can be accomplished with computer software (e.g., Clampfit or Minianalysis), which can identify mPSCs and calculate important measurements such as amplitude, frequency, and kinetics. However, it is often necessary to double check the computer-identified mPSCs since it is possible that baseline noise can be inaccurately identified as mPSCs. The gain on the amplifier can be set to 5 and the filtering can be done using a bandpass filter set to 1 and 5 kHz with a Bessel filtered of 3 kHz to prevent aliasing.

3 Spontaneous Postsynaptic Currents (sPSCs)

Spontaneous PSCs refer to currents that are elicited either by spontaneous action potential dependent release of neurotransmitters or, like with mPSCs, by the spontaneous release of quanta. Since action potentials are not blocked during these recordings, differentiating action potential dependent and independent PSCs is not possible. The cause for the spontaneous activity of neurons in vitro is still not well understood, but potential causes could be based on the intrinsic properties of specific neurons or a subset of neurons.

Isolating these currents is very straightforward. Using the whole-cell patch clamp configuration the spontaneous activity can be recorded and analyzed using the same procedures for mPSC recordings (see Sect. 2.1). Since sPSCs are action potential dependent, often the current amplitude is far larger than that for mPSCs. Therefore, sPSC analysis and mPSC analysis should never be pooled.

4 Quantal Postsynaptic Currents (qPSCs)

Quantal postsynaptic current refers to the quantal release of neurotransmitter (see Sect. 2), but they differ from mPSCs in that they are evoked by either electrical or optical stimulation (Fig. 2). This technique is performed by replacing calcium in the bath solution with strontium. In the presence of strontium, there are increases in the asynchronous release of vesicles. This asynchronous release leads to the quantal release of neurotransmitters. There are two reasons why strontium generates asynchronous release. One, data suggests that strontium has poorer binding affinity for calcium-binding molecules, which regulate vesicular release in the presynaptic active zone [5] (see Chap. 11 for an explanation on the presynaptic active zone). This leads to inefficient vesicle docking, fusion, and release. Two, strontium vs. calcium has greater permeability through calcium channels causing higher intracellular strontium concentrations after an evoked stimulus. Coupled to this, strontium is less efficiently cleared from the presynaptic terminal leaving residual strontium to stimulate vesicle release long after the action potential has occurred [5].

The clear advantage to qPSCs over mPSCs is that the quantal events at electrically or optically evoked inputs can be measured. For example, the experimenter can measure quantal events at synapses that undergo plasticity [6]. Measuring quantal amplitude and/or frequency at these synapses can provide information about presynaptic and/or postsynaptic modifications that has occurred.

4.1 Technical Approach

In order to record qPSCs using evoked currents, it is important to elicit asynchronous release. This is done in the presence of extracellular strontium (2.5–4.0 mM) (absence of calcium). In addition,

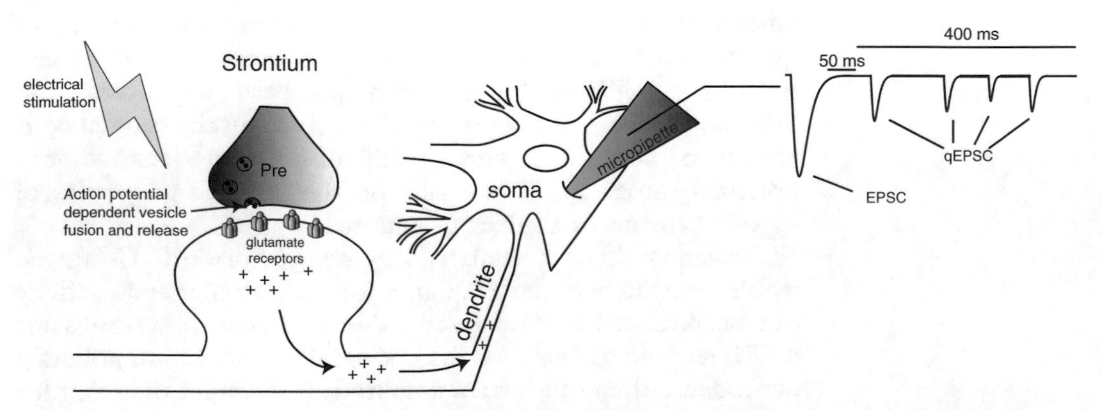

Fig. 2 Quantal postsynaptic currents are generated by electrical stimulation in the presence of strontium. The resulting current traces illustrate asynchronous release of neurotransmitter depicted by EPSCs occurring after the initial EPSC detected from somatic whole cell recordings. Measurements are taken 50 ms following the evoked EPSC for 400 ms

asynchronous release is elicited following a stimulus such as a 2-Hz, 10-pulse train delivered every 30 s [6]. The current analyzed is typically measured during a 400 ms period starting 50 ms after the evoked PSC (Fig. 2). This time window is chosen to minimize contamination by residual evoked PSCs and by spontaneous PSCs that predominate at long intervals after simulation [7]. Once the time window has been isolated, qPSCs can be detected using similar software that is used for mPSCs analysis.

5 Evoked Postsynaptic Currents (ePSCs)

5.1 Electrical

Evoked postsynaptic currents can be generated using electrical stimulation (Fig. 3). By passing a current, voltage-gated channels become activated via voltage sensing subunits [8]. This generates action potentials in the presynaptic neuron leading to neurotransmitter release and subsequent postsynaptic currents.

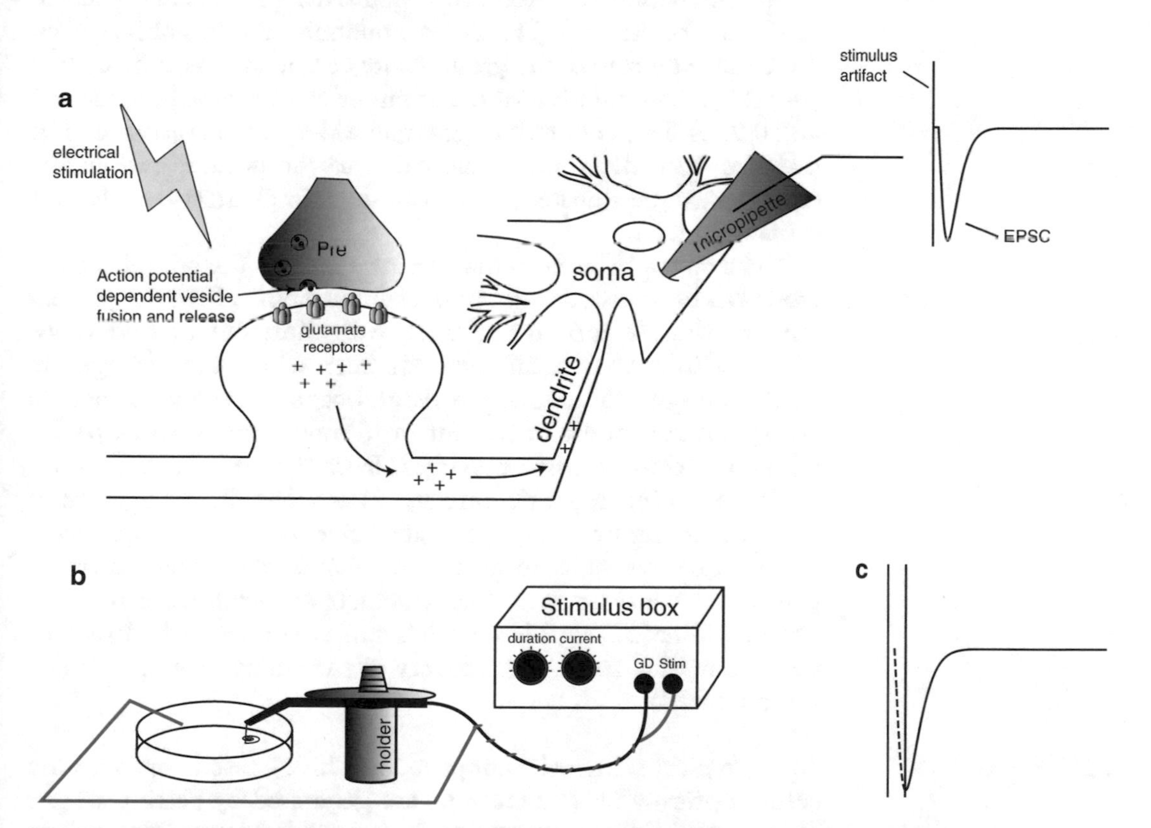

Fig. 3 (**a**) Electrically evoked currents measured from a somatic whole-cell recording showing an EPSC that follows a stimulus artifact produced by the electrical stimulus. (**b**) A ground wire is connected to a stimulus box. The part of the ground wire that is submerged in the external solution is coiled to increase the surface density and minimize the stimulus artifact. An active stimulus wire is wrapped around the ground wire and placed at the furthest point away from the ground wire. (**c**) A large stimulus artifact cuts off the activation phase of the EPSC causing difficulty identifying the true peak amplitude

5.2 Technical Considerations

Electrically evoked currents are generated using a stimulating electrode, which is connected to a stimulation box (e.g., Digitimer Research Instruments). The stimulating electrode can be made or purchased (e.g., FHC, Inc) in a variety of different ways. We prefer using an AgCl wire connected to copper wire, which is then connected to the stimulus box. A low resistance glass capillary is fabricated (see Chap. 5 for glass capillary fabrication procedures) and filled with the external solution. The AgCl wire is placed inside the glass capillary, which is then lowered down to the surface of the slice or submerged into the slice roughly by 10 μm. A ground wire, connected to the stimulus box, is submerged in the external solution, which bathes the brain slice (Fig. 3b). To minimize noise, the ground wire can be wrapped around the stimulus wire until right before they are positioned in the chamber (Fig. 3b). In performing these experiments, a stimulus artifact is visible a millisecond before the PSC. The artifact has large upward and downward currents. Care should be taken to minimize the time duration of the stimulus artifact because an excessively wide artifact could cut off activation phase of the PSC (Fig. 3c). To minimize the stimulus artifact, the surface density of the ground wire can be increased by coiling (Fig. 3b). The duration of the stimulus can be kept between 0.1 and 0.9 ms. Larger durations generate wider stimulus artifacts. The distance from the stimulus electrode to the ground wire in the chamber can be adjusted, which can sometimes affect the stimulus artifact (Fig. 3).

The strength of currents passing through the stimulus electrode can be adjusted through the stimulus box. However, passing current that is too strong can be detrimental in two ways. First, passing current can generate heat, which can damage the surrounding tissue. Damaged tissue becomes evident if there is consistent run-down of current or if brain tissue is stuck to the stimulus electrode when moving it from the brain slice. Second, when recording synaptic currents in whole-cell, voltage-clamp mode, it is important to generate currents that are not excessively large (typically no greater than 200 pA). Larger currents can lead to space-clamp effects, which can produce erroneous channel kinetics (e.g., large PSCs can cause overestimations of decay time due to slower recovery of the voltage-clamped holding potential).

5.3 Optical

In a typical channelrhodopsin-2 (ChR2)-based optogenetic setup, optically evoked currents are generated by passing a light at 470 nm (blue light) in order to activate the depolarizing channel ChR2 [9–11] (Fig. 4). ChR2 is isolated from the algae, *Chlamydomonas reinhardtii* and is a seven-transmembrane protein with a molecule of all-*trans* retinal bound to the core, which acts as a photosensor. The 470 nm light isomerizes all-*trans* retinal triggering a conformational change, thus creating a channel

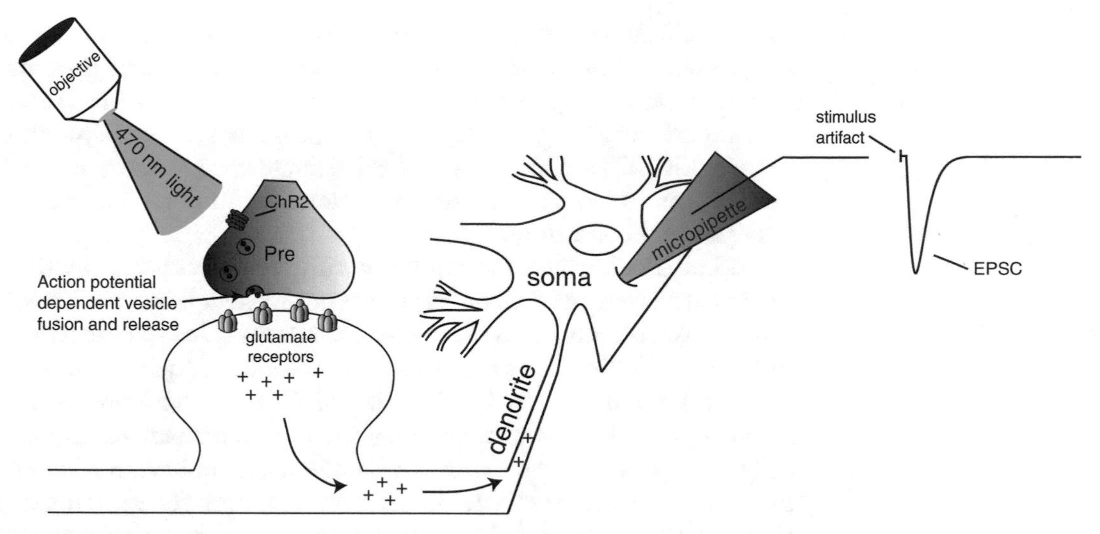

Fig. 4 Optically evoked currents measured from a somatic whole-cell recording showing an EPSC that follows an optically generated stimulus artifact (note that the stimulus artifact is much smaller relative to the electrical stimulation)

pore. This pore is a nonselective cation current passing H^+, Na^+, K^+, Ca^{2+} ions with a reversal potential near 0 mV and a conductance between 50 and 250 fS [12–15]. The photosensitivity is fast and can happen within 50 μs of illumination [14]. Since ChR2 development, multiple variants have been generated with varying opening and closing rates, light sensitivity, desensitization rates, and responses to spectral peaks (for a review see [16]). Briefly, these variants include ChR2 mutated H134R, ChETA, VChR1, ChD, ChEF, and ChIEF.

Since ChR2 needs to be introduced into the neurons, it is delivered, most commonly, using viral vectors and stereotaxic procedures in vivo (for a detailed protocol for delivery see Zhang et al. [17]). Lentivirus and adeno-associated virus (AAV) are widely used viral vectors with different expression time requirements. Lentivirus is incorporated directly into the DNA, leading to expression in 3–5 days, while AAV takes 3–4 weeks before expression is achieved. The advantage of using AAV is that there is greater area of infected neurons, but the selection of the proper serotype is required (see Zincarelli et al. [18] for empirical test of different AAV serotypes [18]). In order to visualize the level of expression, the viral vector is fused with a fluorescent reporter protein such as mCherry or tdTomato [19–21].

To activate ChR2 with 470 nm light in vitro, halogen/xenon lamp, a light emitting diode (LED), or two-photon excitation laser scanning methods can be used. Optically evoked currents have a major advantage in that specific neurocircuits can be isolated. For example, ChR2 can be injected into neuronal cell bodies and light can be delivered to the targeted region where ChR2 processes have innervated. PSCs recorded from the innervated site can then be said to have been generated from the ChR2 injected site (for a review on targeting strategies for optogenetics see Tye and Deisseroth [22]). This allows an experimenter to determine how connections from specific afferents function under control and experimental conditions.

Some disadvantages to using optically evoked currents are that expression levels need to be high in order to depolarize the ChR2 infected membrane. This is because the relatively low conductance of ChR2 (50–250 fS). In addition, the stimulus frequency desired may be limited due to the kinetics of ChR2. Therefore, when selecting a ChR2 for experimentation, it is important to empirically check whether the ChR2 can handle the stimulus train (for an example see Suska et al. [23]). Furthermore, optogenetic stimulation can potentially depolarize the presynaptic neuron to threshold inducing back propagating action potential generation and activation of other downstream targets, thus reducing the specificity of direct presynaptic to postsynaptic stimulation. Also, the presynaptic release is induced by experimenter controlled light stimulation. These patterns of light stimulation may not be comparable to endogenous firing patterns of the presynaptic neuron, which can potentially provide inaccurate interpretations of the postsynaptic responses.

5.4 Designer Receptor Exclusively Activated by a Designer Drug (DREADDs)

DREADDs are engineered G-protein coupled receptors that are activated by specific drug molecules [24, 25]. They are introduced into the biological system using viral vectors or expressed in transgenic mouse models. Once introduced biologically, DREADDs are activated by clozapine-N-oxide (CNO), which can be bath applied in vitro or locally perfused in vivo leading to changes in cellular activity [26]. Onset of activation is slow beginning at about 10 min post-injection and peaks within 45 min [24]. DREADDs enable the investigation of many different G-protein coupled receptors (GPCRs). These include GPCR subtypes such as G_q, G_i, G_s, Golf, and β-arrestin [27].

The limitations of using DREADDs include temporal control. As already stated, DREADD-induced modulation of neurons takes minutes versus optical methods, which takes only milliseconds. In addition, the clearance of CNO from plasma can take up to 2 h [28]. Additionally, when using DREADDs, it is difficult to deliver concentrations of CNO capable of precisely modulating a given circuit, which can cause difficulty in controlling behaviors in vivo. Lastly, DREADDs can be overexpressed leading to non-physiologically relevant findings. It is suggested when using DREADDs to quantify expression levels in order to determine whether measured results can translate physiologically [27].

5.5 Nanoparticles

The use of gold nanoparticles to excite neurons is another approach that can be implemented to selectively excite a population of neurons. This method is a newly developed tool with an advantage over optogenetics in that genes do not need to be inserted into the neuron. This technique uses 20 nm wide gold particles, which attach to three different ion channels (voltage-gated sodium channels, TRPV1 ion channel, and $P2X_3$ receptor ion channel) on the neuron's surface. By flashing a millisecond pulse of near infrared light, the particles are heated inducing neuronal firing [29].

6 Summary

Each approach used to isolate synaptic currents has advantages for dissecting synaptic function. For example, mPSCs can be isolated in order to measure quantal parameters of synaptic currents, while qPSCs can perform this similar task, but with the added advantage of being able to record from quantal events from specific synaptic inputs. Furthermore, evoked currents can be generated electrically or if precise neurocircuit analysis is required, optical stimulation can be used.

Accurate measurements of each approach require attention to detail such as checking access resistance during recordings, minimizing baseline noise, and preventing stimulus artifacts from contaminating the PSCs. With practice, all of these approaches can be mastered and provide reliable, reproducible data for years to come.

References

1. Kerchner GA, Nicoll RA (2008) Silent synapses and the emergence of a postsynaptic mechanism for LTP. Nat Rev Neurosci 9(11): 813–825
2. De Koninck Y, Mody I (1994) Noise analysis of miniature IPSCs in adult rat brain slices: properties and modulation of synaptic GABAA receptor channels. J Neurophysiol 71(4): 1318–1335
3. Kavalali ET, Chung C, Khvotchev M, Leitz J, Nosyreva E, Raingo J, Ramirez DM (2011) Spontaneous neurotransmission: an independent pathway for neuronal signaling? Physiology (Bethesda) 26(1):45–53
4. Ramirez DM, Kavalali ET (2011) Differential regulation of spontaneous and evoked neurotransmitter release at central synapses. Curr Opin Neurobiol 21(2):275–282
5. Xu-Friedman MA, Regehr WG (1999) Presynaptic strontium dynamics and synaptic transmission. Biophys J 76(4):2029–2042
6. Thomas MJ, Beurrier C, Bonci A, Malenka RC (2001) Long-term depression in the nucleus accumbens: a neural correlate of behavioral sensitization to cocaine. Nat Neurosci 4(12): 1217–1223
7. Bender KJ, Allen CB, Bender VA, Feldman DE (2006) Synaptic basis for whisker deprivation-induced synaptic depression in rat somatosensory cortex. J Neurosci 26(16): 4155–4165
8. Bezanilla F (2008) How membrane proteins sense voltage. Nat Rev Mol Cell Biol 9(4): 323–332
9. Boyden ES, Zhang F, Bamberg E, Nagel G, Deisseroth K (2005) Millisecond-timescale, genetically targeted optical control of neural activity. Nat Neurosci 8(9):1263–1268
10. Li X, Gutierrez DV, Hanson MG, Han J, Mark MD, Chiel H, Hegemann P, Landmesser LT, Herlitze S (2005) Fast noninvasive activation and inhibition of neural and network activity

by vertebrate rhodopsin and green algae channelrhodopsin. Proc Natl Acad Sci U S A 102(49):17816–17821

11. Nagel G, Brauner M, Liewald JF, Adeishvili N, Bamberg E, Gottschalk A (2005) Light activation of channelrhodopsin-2 in excitable cells of Caenorhabditis elegans triggers rapid behavioral responses. Curr Biol 15(24): 2279–2284
12. Bamann C, Kirsch T, Nagel G, Bamberg E (2008) Spectral characteristics of the photocycle of channelrhodopsin-2 and its implication for channel function. J Mol Biol 375(3):686–694
13. Feldbauer K, Zimmermann D, Pintschovius V, Spitz J, Bamann C, Bamberg E (2009) Channelrhodopsin-2 is a leaky proton pump. Proc Natl Acad Sci U S A 106(30): 12317–12322
14. Nagel G, Szellas T, Huhn W, Kateriya S, Adeishvili N, Berthold P, Ollig D, Hegemann P, Bamberg E (2003) Channelrhodopsin-2, a directly light-gated cation-selective membrane channel. Proc Natl Acad Sci U S A 100(24): 13940–13945
15. Lin JY, Lin MZ, Steinbach P, Tsien RY (2009) Characterization of engineered channelrhodopsin variants with improved properties and kinetics. Biophys J 96(5):1803–1814
16. Lin JY (2011) A user's guide to channelrhodopsin variants: features, limitations and future developments. Exp Physiol 96(1):19–25
17. Zhang F, Gradinaru V, Adamantidis AR, Durand R, Airan RD, de Lecea L, Deisseroth K (2010) Optogenetic interrogation of neural circuits: technology for probing mammalian brain structures. Nat Protoc 5(3):439–456
18. Zincarelli C, Soltys S, Rengo G, Rabinowitz JE (2008) Analysis of AAV serotypes 1-9 mediated gene expression and tropism in mice after systemic injection. Mol Ther 16(6): 1073–1080
19. Aponte Y, Atasoy D, Sternson SM (2011) AGRP neurons are sufficient to orchestrate feeding behavior rapidly and without training. Nat Neurosci 14(3):351–355
20. Atasoy D, Aponte Y, Su HH, Sternson SM (2008) A FLEX switch targets Channelrhodopsin-2 to multiple cell types for imaging and long-range circuit mapping. J Neurosci 28(28):7025–7030
21. Cao ZF, Burdakov D, Sarnyai Z (2011) Optogenetics: potentials for addiction research. Addict Biol 16(4):519–531
22. Tye KM, Deisseroth K (2012) Optogenetic investigation of neural circuits underlying brain disease in animal models. Nat Rev Neurosci 13(4):251–266
23. Suska A, Lee BR, Huang YH, Dong Y, Schluter OM (2013) Selective presynaptic enhancement of the prefrontal cortex to nucleus accumbens pathway by cocaine. Proc Natl Acad Sci U S A 110(2):713–718
24. Alexander GM, Rogan SC, Abbas AI, Armbruster BN, Pei Y, Allen JA, Nonneman RJ, Hartmann J, Moy SS, Nicolelis MA, McNamara JO, Roth BL (2009) Remote control of neuronal activity in transgenic mice expressing evolved G protein-coupled receptors. Neuron 63(1):27–39
25. Zhu H, Roth BL (2014) Silencing synapses with DREADDs. Neuron 82(4):723–725
26. Ferguson SM, Eskenazi D, Ishikawa M, Wanat MJ, Phillips PE, Dong Y, Roth BL, Neumaier JF (2011) Transient neuronal inhibition reveals opposing roles of indirect and direct pathways in sensitization. Nat Neurosci 14(1):22–24
27. Urban DJ, Roth BL (2015) DREADDs (designer receptors exclusively activated by designer drugs): chemogenetic tools with therapeutic utility. Annu Rev Pharmacol Toxicol 55:399–417
28. Guettier JM, Gautam D, Scarselli M, Ruiz de Azua I, Li JH, Rosemond E, Ma X, Gonzalez FJ, Armbruster BN, Lu H, Roth BL, Wess J (2009) A chemical-genetic approach to study G protein regulation of beta cell function in vivo. Proc Natl Acad Sci U S A 106(45): 19197–19202
29. Carvalho-de-Souza JL, Treger JS, Dang B, Kent SB, Pepperberg DR, Bezanilla F (2015) Photosensitivity of neurons enabled by cell-targeted gold nanoparticles. Neuron 86(1):207–217

Chapter 9

Fast and Slow Synaptic Currents

Nicholas Graziane and Yan Dong

Abstract

There are different types of synapses and receptors that regulate fast and slow synaptic currents. This chapter discusses two classes of synapses (e.g., chemical and electrical) and the receptors that populate these synapses including ionotropic receptors, metabotropic receptors, and gap junctions. We discuss the speed with which these receptors mediate or regulate ionic currents with the purpose of supplying the reader with a general idea of current kinetics. In addition, we include technical considerations when measuring fast and slow synaptic currents as well as, in some cases, the physiological relevance of current kinetics (see Chap. 17 for more details regarding ionotropic receptor kinetics).

Key words Chemical synapses, Excitatory ionotropic currents, Inhibitory ionotropic currents, Metabotropic receptors, Electrical synapses

1 Introduction

There are different types of synapses and receptors that regulate fast and slow synaptic currents. This chapter discusses two classes of synapses (e.g., chemical and electrical) and the receptors that populate these synapses including ionotropic receptors, metabotropic receptors, and gap junctions. We discuss the speed with which these receptors mediate or regulate ionic currents with the purpose of supplying the reader with a general idea of current kinetics. In addition, we include technical considerations when measuring fast and slow synaptic currents as well as, in some cases, the physiological relevance of current kinetics (see Chap. 17 for more details regarding ionotropic receptor kinetics).

2 Chemical Synapses

Chemical synapses are the most commonly studied and predominantly expressed synapses in the mammalian central nervous system (this is not the case for cold-blooded animals where electrical

Nicholas Graziane and Yan Dong, *Electrophysiological Analysis of Synaptic Transmission*, Neuromethods, vol. 112,
DOI 10.1007/978-1-4939-3274-0_9, © Springer Science+Business Media New York 2016

synapses predominate). The term chemical synapse refers to synapses that rely on neurotransmitters capable of relaying messages between synaptically connected neurons. The neurotransmitters are released presynaptically and diffuse through the synaptic cleft to the postsynaptic receptors, which become activated and induce changes in membrane potentials. The membrane potential can become depolarized or hyperpolarized depending upon the type of receptor that is activated (e.g., excitatory ionotropic receptors, inhibitory ionotropic receptors, or metabotropic receptors). Within each of these classes of receptors brings about characteristic differences in the speed of currents, which are modulated by the gating properties of the receptor (e.g., ionotropic receptors) or by the intracellular signaling cascades that are activated by the receptor (e.g., metabotropic receptors).

2.1 Excitatory Synaptic Ionotropic-Mediated Currents

Two commonly studied excitatory ionotropic synaptic currents are those generated by AMPA (α-amino-3-hydroxy-5-methyl-4-isoxazolepropionic acid) or NMDA (*n-methyl-d-aspartate*) receptors. They are often colocalized postsynaptically[1] and are activated by presynaptically released glutamate. This colocalization suggests that postsynaptic AMPARs and NMDARs are both exposed to the same concentration and duration of neurotransmitter [1, 2]. Despite this, their synaptic currents are easily differentiated. AMPAR-mediated currents are quite fast, lasting ≤8 ms under 'ideal' conditions [3–6], while NMDAR currents are slow, lasting as long as 200 ms (Fig. 1) [7–11]. Note that NMDARs possess much higher affinity for glutamate than AMPARs, which consequently affects NMDAR gating [5, 10] (see Chap. 17 for a more in-depth discussion of AMPAR and NMDAR kinetics).

AMPAR- or NMDAR-mediated currents can be conveniently isolated by their selective antagonists. A commonly used NMDAR-selective antagonist is APV (also called AP-5). NMDARs can be selectively inhibited by DNQX, NBQX, or CNQX, which have different binding affinities for AMPARs and NMDARs, so caution is needed to determine which blocker and what concentration are best for the proposed experiments. If experimental conditions do not permit using these receptor blockers, an alternative approach can be used by taking advantage of the differential decay kinetics of AMPAR- and NMDAR-mediated currents. First, measure evoked excitatory currents while voltage-clamping the cell at −70 mV. Then, depolarize the cell to +50 mV and notice the much slower current. At the holding potential of −70 mV, AMPAR current predominates

[1] In addition to their postsynaptic expression, AMPARs, NMDARs, 5-HT3Rs, nAChRs, $GABA_A$Rs, and GlyRs are also likely expressed on the presynaptic terminal to regulate neurotransmitter release. Since presynaptic ionotropic receptor's function was not discussed in this chapter, see the following references for more information [51–53].

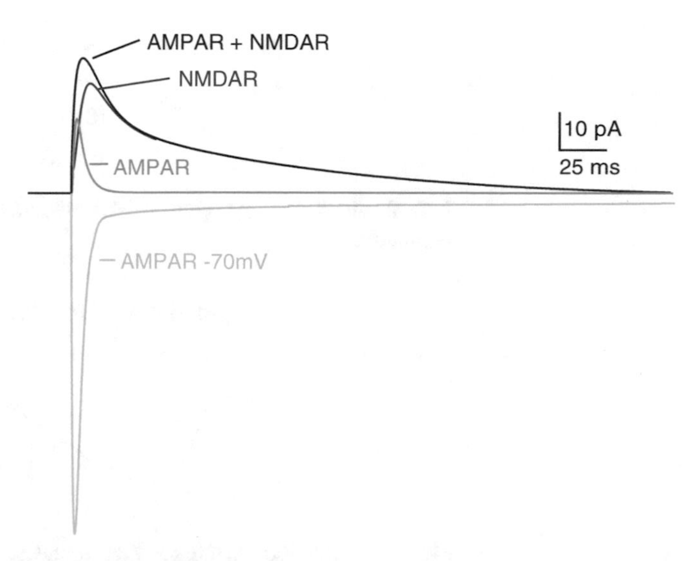

Fig. 1 Hypothetical AMPAR and NMDAR mediated currents. At holding potentials of −70 mV, NMDAR-mediated currents are negligible due to being blocked by Mg^{2+}, but AMPAR-mediated currents are present and consist of fast channel kinetics. At positive holding potentials, NMDAR-mediated currents are present and are noticeably slower relative to AMPAR-mediated currents at positive holding potentials (+50 mV). At +50 mV NMDAR-mediated currents were isolated using NBQX (2 μM) and AMPAR-mediated currents were isolated by subtracting dual component traces from isolated NMDAR-mediated currents

as NMDARs are blocked by Mg^{2+} and therefore produce negligible currents. However, at positive holding potentials, the Mg^{2+} block of NMDARs is relieved, resulting in a dual-component current mediated by both AMPARs and NMDARs. Because AMPAR-mediated currents decay fast, at +50 mV, the current is predominantly NMDAR mediated about 35-40 ms after the peak amplitude at −70 mV [12]. Thus, by measuring the current amplitudes at different time points, AMPAR- and NMDAR-mediated currents can be quantified (Fig. 1).

The significance of the fast and slow synaptic currents at excitatory ionotropic synapses is evident from the temporal precision required to elicit physiologically relevant plasticity such as long-term potentiation and depression (see Chaps. 6 and 12 for a description of spike-timing dependent plasticity and long-term plasticity, respectively).

Two other commonly studied excitatory ionotropic receptors are 5-hydroxytryptamine-3 receptors (5-HT3Rs) and nicotinic acetylcholine receptors (nAChRs) (see Chap. 17 for more details regarding 5-HT3Rs and nAChRs). Both 5-HT3Rs and nAChRs have similar properties including their molecular structure [13] and agonist-induced activation and desensitization of ionic currents [14–17]. They generate fast currents, with 5-HT3Rs having fast and slow mean-open channel time constants equaling ~0.38

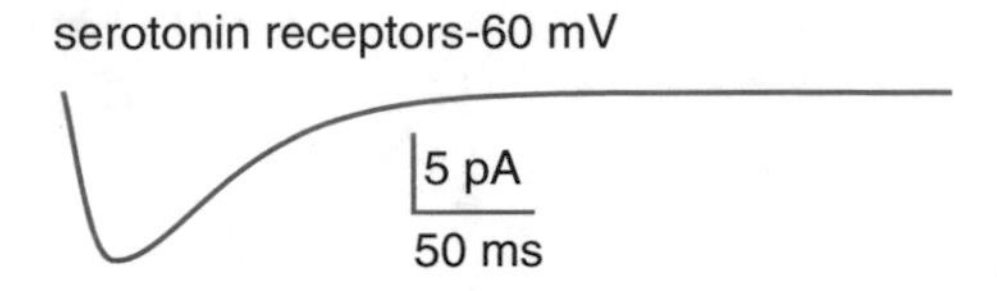

Fig. 2 Simulated synaptic 5-HT3-mediated currents evoked by electrical stimulation

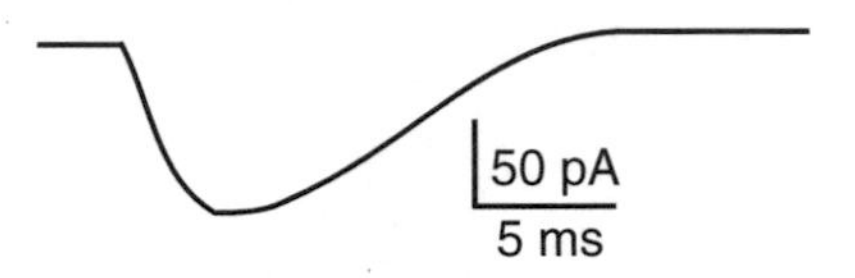

Fig. 3 Simulated synaptic P2x-mediated currents evoked by electrical stimulation

and ~5.0 ms, respectively (exogenously applied agonist-induced currents) [18] and nAChR producing a single mean-open channel time constant of ~1 ms (exogenously applied agonist-induced currents) [19–21]. It should be pointed out that these values are subject to change depending upon receptor subunit composition, exposure time to agonist, or exposure to exogenous compounds capable of altering receptor gating [22]. However, if interested in seeing the kinetics for electrically evoked 5-HT3R- and nAChR-mediated currents, see Roerig et al. [23] (5-HT3) and Skok et al. [24] (nAChR) [23, 24] (Fig. 2).

Lastly, P2X receptors are another cationic channel activated by ATP and mediate fast synaptic currents with rise times that can be <1 ms, and last <2 ms. P2X receptors also possess a fast decay time with a constant of ~17 ms (Fig. 3) [25] (see Chap. 17 for more details about P2X receptors).

2.2 Inhibitory Synaptic Ionotropic-Mediated Currents

Fast acting inhibitory currents in the central nervous system rely on either γ-aminobutyric $acid_A$ receptors ($GABA_ARs$) or glycine receptors (GlyRs) (see Chap. 17 for kinetic properties of both $GABA_ARs$ and GlyRs). For $GABA_ARs$, the subunit composition dictates the speed of the receptor's kinetics. For instance, receptors containing α1 subunits have characteristically faster currents (6–7 times faster) than those containing the α3 subunits [26–28]. Interestingly, α1 subunit expression is low during early developmental periods, but is predominately expressed in adulthood [29, 30]. Likewise, for GlyRs, the subunit expression dictates the speed of synaptic currents. During early development glycinergic IPSCs in the spinal cord are relatively slow compared to later in development, which is likely caused by the molecular switching from the

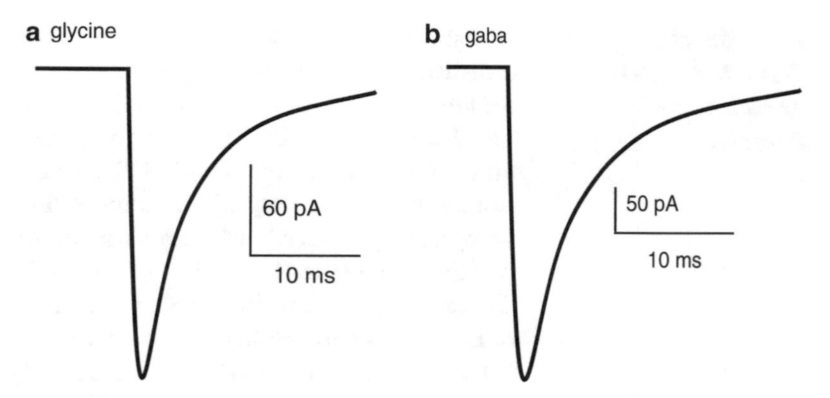

Fig. 4 Simulated synaptic (**a**) glycine- and (**b**) GABAR-mediated currents evoked by electrical stimulation

α2 to the α1 subunit at the postsynaptic membrane [31] (see Fig. 4 for representative synaptic IPSCs).

It is still unclear why the developmental switch between subunits is necessary or why the speed in inhibitory currents changes from early development to adulthood. However, it is speculated that the inhibitory current kinetics are important for regulating the neural synchrony of excitatory neurons leading to synchronized oscillations of brain frequencies. The slower inhibitory currents may decrease the oscillation frequency by prolonging the hyperpolarized state of the targeted neuron [32].

2.3 Technical Considerations When Measuring Synaptic Current Kinetics

When measuring receptor kinetics there are some technical issues to consider. First, pH can change receptor kinetics [26]. Therefore, it is necessary to always check the pH of both the external and internal solutions before every day use or at least after the solutions have been prepared. Second, if recording in the whole-cell configuration, poor cellular access may artificially distort the recorded synaptic currents, resulting in erroneous estimation of current kinetics. Recordings that have low access resistance (~10–20 MΩ) decrease the risk of erroneous kinetic measurements (as the access resistance increases, the current kinetics will become slower). Preferably, the access resistance should be checked before and after the measurement (or monitored throughout the recording) to ensure that it is within the acceptable range. Third, different brain regions can express contrasting receptor subunit combinations. Consequently, brain region A may have mismatched synaptic currents compared to brain region B. For example, the majority of $GABA_AR$ subunit combinations are α1/β2/γ2 or α2/β3/γ2, but in the cerebellum, α3, α4, α5, α6 are expressed in combination with β and γ subunits. Since the α5 subunit has slower kinetics than the α1 subunit, inhibitory currents in the cerebellum may be slower than those recorded from another brain region. Therefore, attention should be paid to the anatomical region being measured.

2.4 Synaptic Currents Mediated by Metabotropic Receptors

Metabotropic receptors (named for their metabolic processing) constitute an abundant class of receptors, including metabotropic glutamate receptors (mGluRs), $GABA_B$Rs, 5-HT (not to be confused with 5-HT3 discussed above), muscarinic AChR (mAChR), endocannabinoid receptors (eCB), catecholamine receptors (i.e., dopamine, norepinephrine, epinephrine receptors), adrenergic receptors (activated by catecholamines), histamine receptors, purinergic receptors (P2Y receptors), opioid receptors, and others. These receptors are incapable of forming ionic-permeable channels, but instead mediate slow synaptic currents through indirect activation of ionic receptors/channels. By binding to an agonist, metabotropic receptors couple to a receptor/channel via an intermediate regulatory protein (e.g., G-protein) or couple through the diffusible second messenger signaling cascade (e.g., G-protein activation of effector proteins). Subsequently, the downstream receptor/channel undergoes conformational changes due to receptor subunit phosphorylation or activation by cyclic nucleotides. Metabotropic receptor generated currents or effects on synaptic currents can last from hundreds of milliseconds to minutes or even longer, which contrasts the fast currents generated by ionotropic receptors. In addition, unlike ionotropic receptors, whose effects are relatively localized, activation of metabotropic receptors can lead to a more widespread effect. This is due to the diffusion of signaling molecules through the cell sometimes resulting in activation of distally located membrane channels or altering gene expression (second messenger diffusion to the cell nucleus).

3 Electrical Synapses

Electrical synapses are synapses formed between adjacent cells via specialized connections known as gap junctions; although rare, electrical synapses can also be formed independent of gap junctions [33, 34]. Vertebrate gap junctions are composed of connexins (Fig. 5a), which are four transmembrane proteins arranged in a hemichannel (Fig. 5b). They are expressed in neurons, astrocytes, oligodendrocytes, and microglia [35]. When one connexin attaches to a connexin from an adjacent cell, a connexon is formed (Fig. 5b, c) enabling two neurons to directly share cytosolic components. Due to the large diameters of connexons (~1.2 nm), the rapid transfer of ions, small metabolites, and second messengers is facilitated (for a thorough review on mammalian gap junctions see [35]).

The direct transfer of ions suggests that the electrical synapses are faster than chemical synapses at transferring current/voltage. To further support this claim, the presynaptic and postsynaptic membrane separation at electrical synapses is 2–4 nm versus 20–40 nm separation at chemical synapses. Strikingly though, electrical synapses in some occasions can have slightly longer synaptic

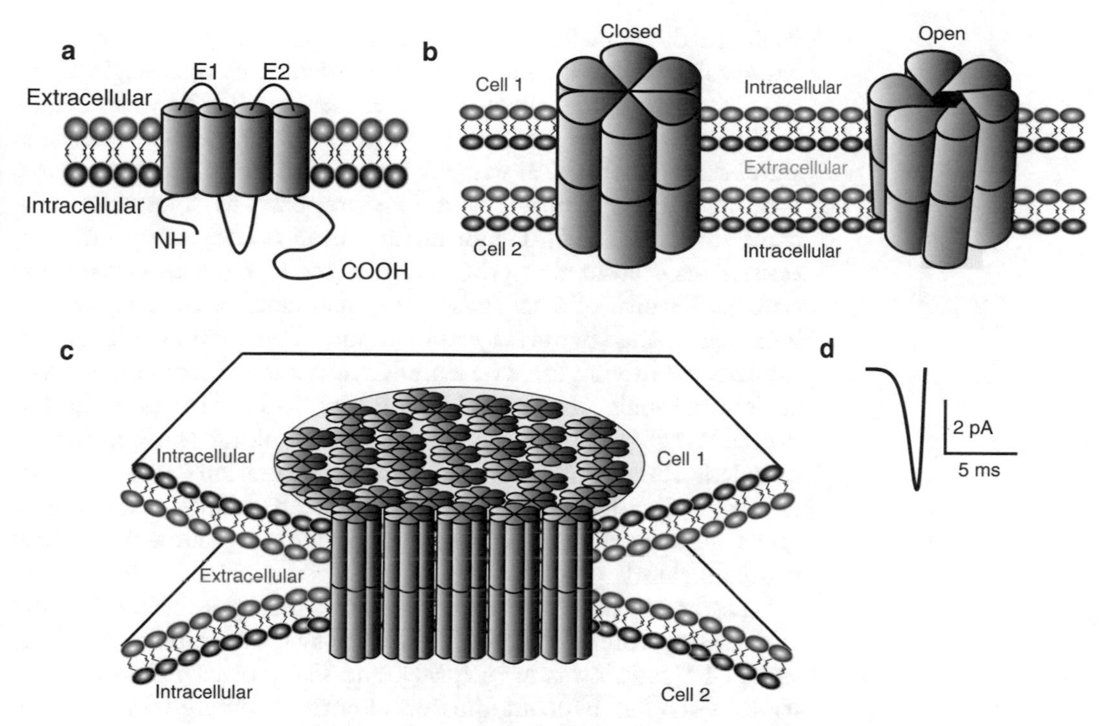

Fig. 5 Illustration of assembled connexons into gap junctions between neighboring cells. (**a**) The structure of a connexin molecule consisting of four transmembrane domains, two extracellular domains (E1 and E2), and an N and C-terminus located in the intracellular compartment. (**b**) The arrangement of six connexins between two cells forming a connexon in both the open and closed states. (**c**) Illustration of a gap junction between neighboring cells. Typically, gap junctions are arranged with multiple connexons clustered together around a central locus. (**d**) Simulated trace of an excitatory current produced at a gap junction

delays than chemical synapses, thus exceeding the ~150 μs delays typical of chemical synapses in warm blooded mammals [36]. Electrical synapses are slowed by the conduction impulses spreading through the presynaptic fiber as well as by the charging of the membrane capacitance postsynaptically [37, 38]. However, the speed of current/voltage transfer at electrical synapses is cell-type specific as fast transfer (faster than chemical synapses) is seen in the neocortical interneurons, cerebellar interneurons, neurons in the thalamic reticular nucleus, and in the hippocampus [35, 39]. This fast coupling between neurons permits synchronized firing, which is critically important for generating brain oscillations [40–42]. In addition, fast coupling is useful for hormone secretion. When a group of neurons fires action potentials simultaneously, a burst of hormones is released [43].

Electrical synapses are frequently seen in dendrodendritic, dendrosomatic, and somatosomatic synaptic arrangements most often between inhibitory neurons and rarely between excitatory neurons. An example of the functional properties of gap junctions at the dendrodendrtic connection is discussed here. Gap junctions facilitate the transfer of excitatory synaptic charges generated at chemical synapses of one cell into the dendrites of a neighboring cell. In other words, chemical synaptic integration between neighboring cells can be achieved at electrical synapses. The functional significance of chemical synaptic integration along gap junctions is proposed to be twofold. First, it is proposed to counteract dendritic saturation [44]. Dendritic saturation occurs because of axial resistances present along the dendrite (see Chap. 6 and axial resistance). Because of axial resistances, depolarizing currents remain locally near the chemical synaptic inputs. This local depolarization reduces the driving force on simultaneous and subsequent postsynaptic potentials/currents (PSPs/PSCs) [45]. However, in the presence of gap junctions, the resistance along the dendrite is reduced, thus counteracting dendritic saturation [44]. Second, dendritic electrical synapses enable localized synaptic excitation to affect a larger fraction of neurons, thus driving network activity more efficiently [44].

Lastly, electrical synapses are modulated by neuromodulators such as dopamine [46], noradrenaline [47], serotonin [48], histamine [49], nitric oxide [49, 50], and likely other neuromodulators. In general, neuromodulators alter the coupling/conductance of gap junctions via second messengers such as protein kinase A (PKA), guanylate cyclase (cGMP), or protein kinase C (PKC). These data suggest that, although not directly, electrical synapses are dynamic structures capable of undergoing activity-dependent modification.

4 Summary

At chemical synapses, there are two largely types of receptors expressed at the synapse, ionotropic (ion permeable) and metabotropic receptors. Ionotropic receptors produce characteristically fast currents lasting<tens of milliseconds in most cases (but see NMDARs), while metabotropic receptors produce relatively longer lasting currents in the range of milliseconds to minutes. Ionotropic receptors constitute receptors that either depolarize (excitatory ionotropic receptors) or hyperpolarize (inhibitory ionotropic receptors) membrane potentials directly via channel opening following agonist-receptor coupling. Metabotropic receptors, on the other hand, indirectly alter membrane potentials via G-coupled protein second messenger pathway activation of downstream receptors/channels.

Electrical synapses are formed by gap junctions between adjacent membranes allowing for the direct diffusion of ions and bioactive molecules. Electrical synapses typically generate fast currents and are posited to enhance neural synchrony within a neural ensemble, which is a group of neurons characterized by similar afferent/efferent relationships as well as closely related functions in behavior, neuroendocrine regulation, and sensorimotor gating (cite).

References

1. Clements JD (1996) Transmitter timecourse in the synaptic cleft: its role in central synaptic function. Trends Neurosci 19(5):163–171
2. Clements JD, Feltz A, Sahara Y, Westbrook GL (1998) Activation kinetics of AMPA receptor channels reveal the number of functional agonist binding sites. J Neurosci 18(1):119–127
3. Kiskin NI, Krishtal OA, Tsyndrenko A (1986) Excitatory amino acid receptors in hippocampal neurons: kainate fails to desensitize them. Neurosci Lett 63(3):225–230
4. Nelson PG, Pun RY, Westbrook GL (1986) Synaptic excitation in cultures of mouse spinal cord neurones: receptor pharmacology and behaviour of synaptic currents. J Physiol 372:169–190
5. Patneau DK, Mayer ML (1990) Structure-activity relationships for amino acid transmitter candidates acting at N-methyl-D-aspartate and quisqualate receptors. J Neurosci 10(7): 2385–2399
6. Trussell LO, Fischbach GD (1989) Glutamate receptor desensitization and its role in synaptic transmission. Neuron 3(2):209–218
7. Hestrin S, Sah P, Nicoll RA (1990) Mechanisms generating the time course of dual component excitatory synaptic currents recorded in hippocampal slices. Neuron 5(3):247–253
8. Jahr CE (1992) High probability opening of NMDA receptor channels by L-glutamate. Science 255(5043):470–472
9. Lester RA, Clements JD, Westbrook GL, Jahr CE (1990) Channel kinetics determine the time course of NMDA receptor-mediated synaptic currents. Nature 346(6284):565–567
10. Lester RA, Jahr CE (1992) NMDA channel behavior depends on agonist affinity. J Neurosci 12(2):635–643
11. Sah P, Hestrin S, Nicoll RA (1990) Properties of excitatory postsynaptic currents recorded in vitro from rat hippocampal interneurones. J Physiol 430:605–616
12. Huang YH, Lin Y, Mu P, Lee BR, Brown TE, Wayman G, Marie H, Liu W, Yan Z, Sorg BA, Schluter OM, Zukin RS, Dong Y (2009) In vivo cocaine experience generates silent synapses. Neuron 63(1):40–47
13. Maricq AV, Peterson AS, Brake AJ, Myers RM, Julius D (1991) Primary structure and functional expression of the 5HT3 receptor, a serotonin-gated ion channel. Science 254(5030):432–437
14. Eisele JL, Bertrand S, Galzi JL, Devillers-Thiery A, Changeux JP, Bertrand D (1993) Chimaeric nicotinic-serotonergic receptor combines distinct ligand binding and channel specificities. Nature 366(6454):479–483
15. Neijt HC, te Duits IJ, Vijverberg HP (1988) Pharmacological characterization of serotonin 5-HT3 receptor-mediated electrical response in cultured mouse neuroblastoma cells. Neuropharmacology 27(3):301–307
16. van Hooft JA, Vijverberg HP (1996) Selection of distinct conformational states of the 5-HT3 receptor by full and partial agonists. Br J Pharmacol 117(5):839–846
17. Yakel JL, Lagrutta A, Adelman JP, North RA (1993) Single amino acid substitution affects desensitization of the 5-hydroxytryptamine type 3 receptor expressed in Xenopus oocytes. Proc Natl Acad Sci U S A 90(11):5030–5033
18. Derkach V, Surprenant A, North RA (1989) 5-HT3 receptors are membrane ion channels. Nature 339(6227):706–709
19. Cuevas J, Adams DJ (1994) Local anaesthetic blockade of neuronal nicotinic ACh receptor-channels in rat parasympathetic ganglion cells. Br J Pharmacol 111(3):663–672
20. Mathie A, Cull-Candy SG, Colquhoun D (1991) Conductance and kinetic properties of single nicotinic acetylcholine receptor channels in rat sympathetic neurones. J Physiol 439:717–750
21. Valles AS, Garbus I, Barrantes FJ (2007) Lamotrigine is an open-channel blocker of the nicotinic acetylcholine receptor. Neuroreport 18(1):45–50
22. Feinberg-Zadek PL, Davies PA (2010) Ethanol stabilizes the open state of single 5 hydroxytryptamine(3A)(QDA) receptors. J Pharmacol Exp Ther 333(3):896–902
23. Roerig B, Nelson DA, Katz LC (1997) Fast synaptic signaling by nicotinic acetylcholine and serotonin 5-HT3 receptors in developing visual cortex. J Neurosci 17(21):8353–8362
24. Skok VI, Voitenko SV, Bobryshev AY, Voitenko LP, Skok MV (1998) Heterogeneity of neuronal nicotinic acetylcholine receptors: structural and functional aspects. Neurophysiology 30(4-5):200–202
25. Edwards FA, Gibb AJ, Colquhoun D (1992) ATP receptor-mediated synaptic currents in the central nervous system. Nature 359(6391):144–147
26. Barberis A, Mozrzymas JW, Ortinski PI, Vicini S (2007) Desensitization and binding properties determine distinct alpha1beta2gamma2 and alpha3beta2gamma2 GABA(A) receptor-channel kinetic behavior. Eur J Neurosci 25(9):2726–2740

27. Gingrich KJ, Roberts WA, Kass RS (1995) Dependence of the GABAA receptor gating kinetics on the alpha-subunit isoform: implications for structure-function relations and synaptic transmission. J Physiol 489(Pt 2):529–543
28. Mozrzymas JW, Barberis A, Mercik K, Zarnowska ED (2003) Binding sites, singly bound states, and conformation coupling shape GABA-evoked currents. J Neurophysiol 89(2):871–883
29. Dunning DD, Hoover CL, Soltesz I, Smith MA, O'Dowd DK (1999) GABA(A) receptor-mediated miniature postsynaptic currents and alpha-subunit expression in developing cortical neurons. J Neurophysiol 82(6):3286–3297
30. Fritschy JM, Paysan J, Enna A, Mohler H (1994) Switch in the expression of rat GABAA-receptor subtypes during postnatal development: an immunohistochemical study. J Neurosci 14(9): 5302–5324
31. Takahashi T, Momiyama A, Hirai K, Hishinuma F, Akagi H (1992) Functional correlation of fetal and adult forms of glycine receptors with developmental changes in inhibitory synaptic receptor channels. Neuron 9(6):1155–1161
32. Gonzalez-Burgos G, Lewis DA (2008) GABA neurons and the mechanisms of network oscillations: implications for understanding cortical dysfunction in schizophrenia. Schizophr Bull 34(5):944–961
33. Furukawa T, Furshpan EJ (1963) Two inhibitory mechanisms in the Mauthner neurons of goldfish. J Neurophysiol 26:140–176
34. Korn H, Axelrad H (1980) Electrical inhibition of Purkinje cells in the cerebellum of the rat. Proc Natl Acad Sci U S A 77(10):6244–6247
35. Bennett MV, Zukin RS (2004) Electrical coupling and neuronal synchronization in the mammalian brain. Neuron 41(4):495–511
36. Sabatini BL, Regehr WG (1996) Timing of neurotransmission at fast synapses in the mammalian brain. Nature 384(6605):170–172
37. Bennett MV (1997) Gap junctions as electrical synapses. J Neurocytol 26(6):349–366
38. Bennett MV (2000) Seeing is relieving: electrical synapses between visualized neurons. Nat Neurosci 3(1):7–9
39. Connors BW, Long MA (2004) Electrical synapses in the mammalian brain. Annu Rev Neurosci 27:393–418
40. Buzsaki G, Wang XJ (2012) Mechanisms of gamma oscillations. Annu Rev Neurosci 35: 203–225
41. Hormuzdi SG, Pais I, LeBeau FE, Towers SK, Rozov A, Buhl EH, Whittington MA, Monyer H (2001) Impaired electrical signaling disrupts gamma frequency oscillations in connexin 36-deficient mice. Neuron 31(3): 487–495
42. Nakazawa K, Zsiros V, Jiang Z, Nakao K, Kolata S, Zhang S, Belforte JE (2012) GABAergic interneuron origin of schizophrenia pathophysiology. Neuropharmacology 62(3):1574–1583
43. Purves D, Fitzpatrick D, Katz LC, Lamantia AS, McNamara JO, Williams SM, Augustine GJ (2001) Neuroscience. Sinauer Associates, Sunderland, MA
44. Vervaeke K, Lőrincz A, Nusser Z, Silver RA (2012) Gap junctions compensate for sublinear dendritic integration in an inhibitory network. Science 335(6076):1624–1628
45. Bush PC, Sejnowski TJ (1994) Effects of inhibition and dendritic saturation in simulated neocortical pyramidal cells. J Neurophysiol 71(6):2183–2193
46. Ribelayga C, Cao Y, Mangel SC (2008) The circadian clock in the retina controls rod-cone coupling. Neuron 59(5):790–801
47. Zsiros V, Maccaferri G (2008) Noradrenergic modulation of electrical coupling in GABAergic networks of the hippocampus. J Neurosci 28(8):1804–1815
48. Rorig B, Sutor B (1996) Serotonin regulates gap junction coupling in the developing rat somatosensory cortex. Eur J Neurosci 8(8): 1685–1695
49. Hatton GI, Yang QZ (1996) Synaptically released histamine increases dye coupling among vasopressinergic neurons of the supraoptic nucleus: mediation by H1 receptors and cyclic nucleotides. J Neurosci 16(1):123–129
50. Rorig B, Sutor B (1996) Nitric oxide-stimulated increase in intracellular cGMP modulates gap junction coupling in rat neocortex. Neuroreport 7(2):569–572
51. Jeong H-J, Jang I-S, Moorhouse AJ, Akaike N (2003) Activation of presynaptic glycine receptors facilitates glycine release from presynaptic terminals synapsing onto rat spinal sacral dorsal commissural nucleus neurons. J Physiol 550(2):373–383
52. MacDermott AB, Role LW, Siegelbaum SA (1999) Presynaptic ionotropic receptors and the control of transmitter release. Annu Rev Neurosci 22:443–485
53. Xiong W, Chen SR, He L, Cheng K, Zhao YL, Chen H, Li DP, Homanics GE, Peever J, Rice KC, Wu LG, Pan HL, Zhang L (2014) Presynaptic glycine receptors as a potential therapeutic target for hyperekplexia disease. Nat Neurosci 17(2):232–239

Chapter 10

Measuring Kinetics of Synaptic Current

Nicholas Graziane and Yan Dong

Abstract

Obtaining an accurate measurement of receptor kinetics is best done using single-channel recordings. However, single channel recordings require activation of channels directly under the patch pipette using either cell-attached patches or inside-out patches, which is difficult when sampling synaptic receptors. In addition, single-channel recordings require receptor activation via exogenously applied agonists. Consequently, receptor activation is elicited by exogenous agonists instead of endogenous neurotransmitters. In order to determine the kinetics of synaptic receptors under physiological environment, the synaptic current can be isolated using the methods described in Chap. 8 and the kinetics can be measured using the approaches described in this chapter.

Key words Current rise time, Current decay time, Spectral analysis

1 Introduction

Obtaining an accurate measurement of receptor kinetics is best done using single-channel recordings. However, single channel recordings require activation of channels directly under the patch pipette using either cell attached patches or inside-out patches, which is difficult when sampling synaptic receptors. In addition, single-channel recordings require receptor activation via exogenously applied agonists. Consequently, receptor activation is elicited by exogenous agonists instead of endogenous neurotransmitters. In order to determine the kinetics of synaptic receptors under physiological environment, the synaptic current can be isolated using the methods described in Chap. 8 and the kinetics can be measured using the approaches described in this chapter.

After reading this chapter the experimenter will understand the different approaches that can be used to measure kinetics of synaptic currents. We detail several approaches for data collection as well as how to handle the data once they are collected. By implementing these approaches, beginning electrophysiologists can enhance the field's current knowledge regarding synaptic input's temporal

Nicholas Graziane and Yan Dong, *Electrophysiological Analysis of Synaptic Transmission*, Neuromethods, vol. 112,
DOI 10.1007/978-1-4939-3274-0_10,

properties, which has thus far provided valuable information about the properties of neuronal connections.

2 Measuring Synaptic Current Kinetics

2.1 Synaptic Current Rise Time

In the absence of agonists, postsynaptic receptors are in a closed state. Upon neurotransmitter release and binding, the postsynaptic receptors enter into a closed, bound state, which then leads to a bound, open state. Once open, ions are able to penetrate the cellular membrane causing changes in intracellular ionic composition. In electrophysiological measurements of synaptic currents, the changes in intracellular ionic compositions are seen as either net inward or outward currents, depending on the ionic charge that moves to the intracellular compartment (Fig. 1). Time elapses from the onset of this measured current to the peak amplitude of the current is the rise time, which represents how quickly the activated receptors go from low to high ionic permeability. In neuroscience, the rise time is often measured as the time it takes for the average current (the average current is an average of ~30–100 traces) to go from 10 to 90 % of its final peak value. For example, if the average peak current amplitude is 200 pA, the rise time would be calculated as the time it takes for the current to go from 20 to 180 pA (Fig. 2a). Now-a-days, software is available that quickly calculates the 10–90 % rise time of synaptic currents. This is done by fitting a best-fit line using simple linear regression analysis of a straight line:

$$y = mx + b \tag{1}$$

where m is the slope of a line, b is the y-intercept, and x and y refer to points on the horizontal and vertical axis, respectively. Calculating the slope of the line (Fig. 2b), is simply calculated with the equation:

$$m = r\left(S_y / S_x\right) \tag{2}$$

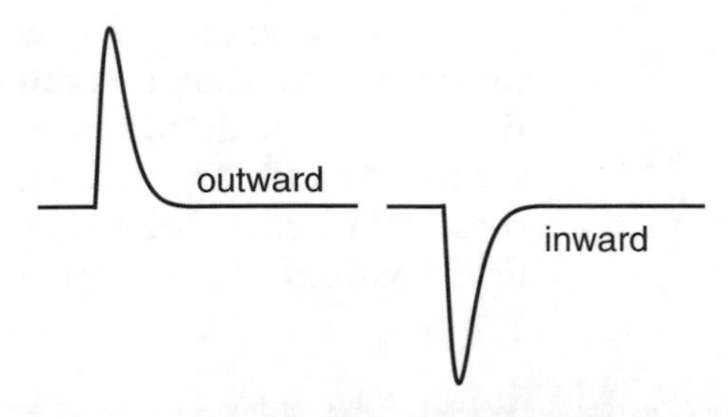

Fig. 1 Example traces of outward (*left*) and inward (*right*) currents

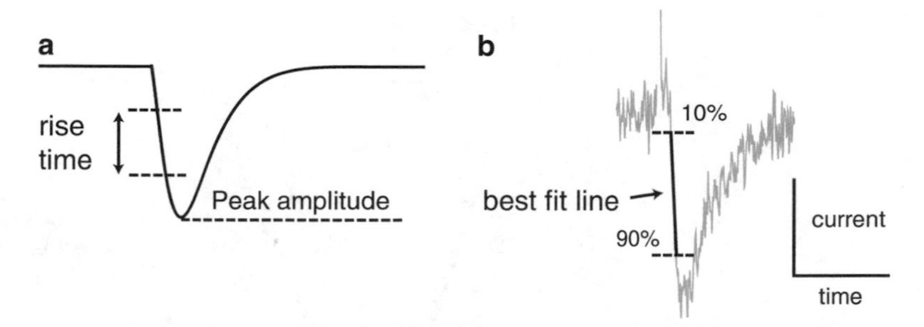

Fig. 2 Rise time calculation from a current trace. (**a**) A current trace showing which segment of the trace is used to calculate the rise time. This segment falls between 10 and 90 % of the peak amplitude. (**b**) An example trace generated by synaptically evoked currents showing a best fit line, which fits the 10–90 % rise time. The best fit line can then be used to calculate the rise time using the Eqs. (1) and (2) described in the text

where *m* equals the slope, *r* is the correlation between *x* and *y*, and *Sy* and *Sx* are the standard deviations of the *y*-values and *x*-values, respectively. Once the best-fit line is determined, the 10–90 % rise time can be calculated.

The above-described approach is the most common way to calculate rise times because the regression analysis provides an acceptably accurate model, making it easier and more practical to use than exponential functions (see Sect. 3.1).

2.2 Technical Considerations for Calculating Rise Times

For consistent comparisons in rise times, the capacitance of the microelectrode should be compensated such that the −3 decibel (dB) bandwidth of the microelectrode is the same as the −3 dB bandwidth of the measurement circuit (see Fig. 14 of Chap. 3 and Fig. 6b *for cutoff frequency*). The −3 dB bandwidth refers to the half power or cutoff frequency defined by

$$P_{out} / P_{in} = 0.5 \tag{3}$$

where P_{out} refers to the output power and P_{in} refers to the input power. Power ratios are often given in dB and can be calculated using the following equation:

$$\text{dB} = 10\log_{10} P_{out} / P_{in} \tag{4}$$

therefore when the power ratio is 0.5, the half power frequency is −3 dB.

The best time to correct for the microelectrode capacitance is during cell-attached patch. Both the fast and slow capacitance transients can be adjusted until the capacitive currents are no longer visible (Fig. 3b) leaving only a straight line. Modern electrophysiology equipment/software comes with an auto capacitance

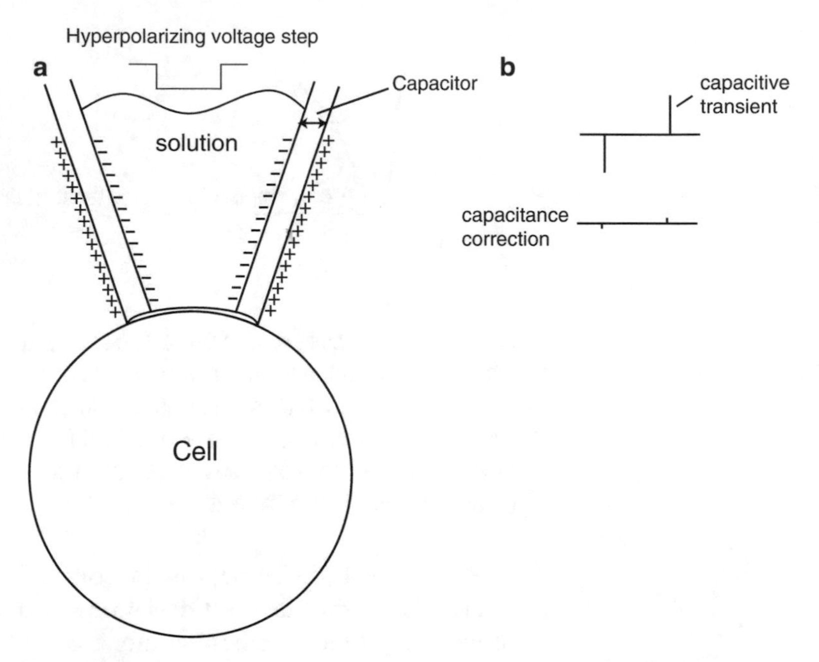

Fig. 3 Microelectrode capacitance correction. (**a**) A hyperpolarizing voltage step generates a negative charge on the inside of the microelectrode resulting in positive charges accumulating on the outer wall of the micropipette. This separation of charges is a capacitor, which causes the capacitive transients seen in (**b**). (**b**) In a cell-attached patch, the micropipette forms a tight, high-resistance seal on the cellular membrane. Passing current results in capacitive transients, which are caused by the charge separation shown in (**a**)

neutralization feature. With a simple click of a button in the cell-attached mode, both the fast and slow capacitive transients can be corrected.

Another technical feature to consider is what type of filter to use during electrophysiological measurements. A practical filtering strategy during rise time measurements is the Bessel or Gaussian filter, because they add less than 1 % overshoot to pulses (Fig. 4). Using these filters prevents an overestimation of rise times. Filters that should be avoided are Butterworth filters and sharp cutoff filters. Butterworth filters can add >10 % overshoot for a fourth-order filter and >15 % overshoot for an eight-order filter [1] (Fig. 4). Like Butterworth filters, sharp cutoff filters ($H(f) = 0.01$, −40 dB) can produce significant overshoot causing an overestimation of rise times. That is why sharp cutoff filters should be used only for frequency domain measurements and should be avoided when measuring time domains.

If the experimenter notices that a large overshoot is still present despite making the above mentioned corrections, they can try to reduce the length of the signal or ground wires. As the length of the signal or ground wire increases, the capacitance increases sometimes causing increases in the overshoot.

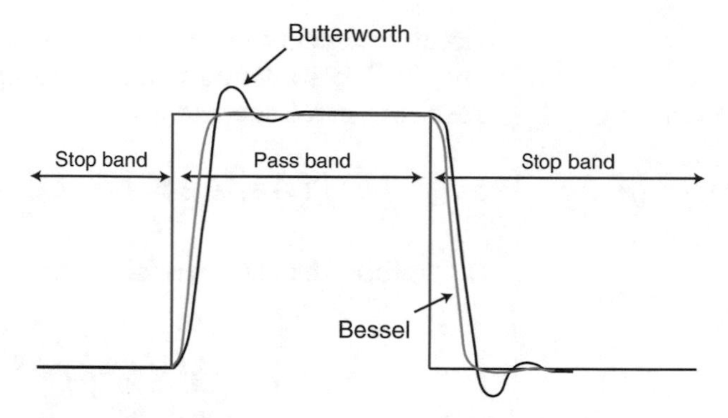

Fig. 4 Comparison of Bessel and Butterworth filters. A perfect filter is illustrated in red demonstrating the stop bands and pass band. The Bessel filter (*gray*) adds little overshoot in the passband. However, using a Butterworth filter causes a noticeable overshoot in the passband

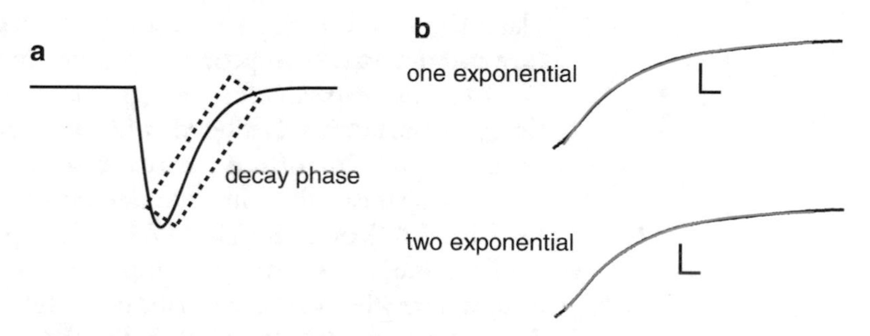

Fig. 5 Decay phase of a synaptically evoked current. (**a**) The decay phase of an inward postsynaptic current (PSC) is isolated just after the peak amplitude and before the PSC returns to baseline. (**b**) A decay phase of a PSC fit with one (*top*) or two (*bottom*) exponentials. In this case, increasing the number of exponentials produced a better fit line

2.3 Synaptic Current Decay Time

Following receptor activation is receptor deactivation in which the ion channel pore closes, thus decreasing ionic movement from the synaptic cleft into the intracellular compartment. Receptor deactivation is associated with the current decay phase that is noticeable in electrophysiological synaptic measurements (Fig. 5a, b). The time constant for the decay phase can be found by fitting the postsynaptic current (PSC) curve with an exponential function:

$$f(t) = A_i e^{-t/\tau} + C \quad (5)$$

where Ai is the amplitude, C is the y-axis offset (the baseline is not 0), and $e^{-t/\tau}$ is the exponential term. Often the literature gives decay time constants as decay fast and decay slow. This is because the PSC is sometimes best fit using a two-exponential function resulting in a fast τ and a slow τ:

$$f(t)=\left[A_i e^{-t/\tau(\text{fast})}+C\right]+\left[A_i e^{-t/\tau(\text{slow})}+C\right] \tag{6}$$

Therefore, the exponential function becomes

$$f(t)=\sum_{i=1}^{n} A_i e^{-t/\tau}+C \tag{7}$$

as n stands for the number of exponential terms needed to fit the PSC curve.

To calculate the decay time constant, it is best to isolate the part of the PSC curve that is associated with the decay phase. Therefore, the start of the measured decay curve should be just after the peak amplitude in order to avoid contamination with activation kinetics. In addition, end of the measured curve should take place right before the PSC returns to baseline, thus preventing unwanted inactive receptor components in the measurement.

In many experiments, the goal is to determine whether the decay of the current is altered, while the values of decay time constants are not important. In this case, experimenters can take a shortcut by measuring the time elapsed from the peak amplitude to ½ (Time1/2) or 1/3 (Time1/3) of the peak amplitude current. While Time1/2 is a rough estimate of the decay speed, Time1/3 indeed provide more information. In Eq. (5) above, when $t_1=\tau$, $f(t_1)=Ae^{-1}=\sim A/(2.7)=\sim 1/3A$. So Time1/3 is also a rough estimate of the decay τ.

Measuring the decay time constant can also be performed by dividing the area under the PSC curve from the peak amplitude. Typically, this is done using the average of 30 current traces and then measuring the decay time. This approach avoids the need to estimate the number of exponentials present.

3 Spectral Analysis (Noise Analysis)

In addition to calculating the rise and decay times of synaptic current, one can also measure the mean channel open time. This can be calculated using noise analysis (provided that the receptor's transition rates between open and closed states are constant), which can decompose the electrical signal revealing varying frequencies along a spectrum. This in turn allows the experimenter to estimate the kinetic parameters of a channel. Noise analysis can also be used to measure unitary channel conductance see [2–4].

How do we go from electrical noise to measureable channel kinetics? Electrical noise is composed of a combination of sinusoids

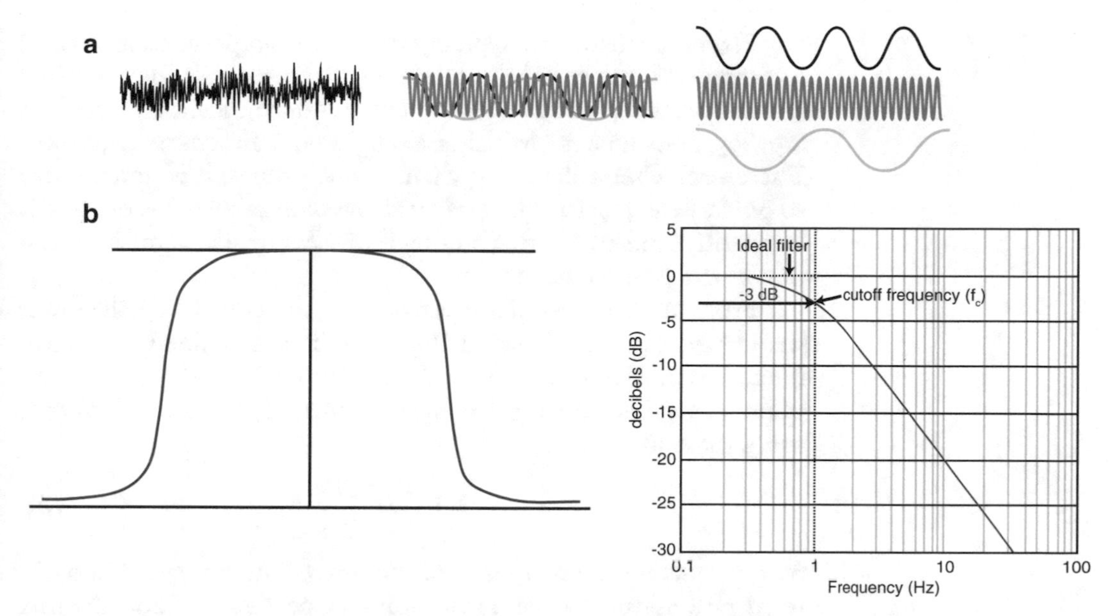

Fig. 6 Sinusoids that make up electrical noise. (**a**) Electrical noise (*left*) is made up of various sine and cosine waveforms at varying frequencies (*middle*) that can be separated (*right*) using the Fourier transform. (**b**) A Lorentzian curve (*left*) is a bell shaped curve, but with wider tails than a Gaussian curve. A power spectrum (*right*) shows the right side of a Lorentzian curve. Within the power spectrum is the cutoff frequency (f_c), which occurs at the −3 dB bandwidth. Once f_c is calculated, τ can be solved for using Eq. (8) described in the text

(sine and cosine waveforms) at varying frequencies (Fig. 6a). Much like the visible light characterized along a spectrum, electrical signals are characterized by a frequency spectrum. To put it in neuroscience/electrophysiological terms, during a whole-cell recording, neurotransmitters are released into the synaptic cleft. Postsynaptic receptors bind to neurotransmitters causing perturbations that change the waveforms making up the recorded electrical signal. This in turn changes the frequency of the sine and cosine waveforms making up the signal. The goal of the analysis is to find the best-fit curve that accurately depicts the combination of frequencies that make up the signal. Once the best-fit curve is calculated, the time constant can be determined giving the channel open time, which is explained below (for a more in depth explanation see van Dronglelen [5] or Mallot [6]).

Luckily, for neuroscientists who are not comfortable with advanced mathematics, noise analysis can be completed using computer software. The easiest approach is to compute a power spectrum of the electrical signal in log scale. Since power is the current^2 flowing in a 1 Ω resistor, the power spectrum has the units $A^2Hz^{-1} = A^2s$ [3] and is plotted against frequency (Hz). What the power spectrum basically tells is what frequency or frequencies are dominating the waveform.

Power spectrums are typically graphed as single-sided spectrums meaning only the positive values along the x-axis are shown. Why? Power spectrums are Lorentzian curves (since they are plotted on log/log coordinates), which are single-peak functions (Fig. 6b). Therefore, when calculating channel open times, it is unnecessary to portray the negative frequency values along the x-axis because it is repetitive information. Meaning it tells exactly the same information as the positive values.

From the single-sided spectrum, f_c (the cutoff or half-power frequency) can be calculated (Fig. 6b). This is defined as the frequency that the power carried has dropped to ½ of the zero frequency value. Once f_c has been calculated, τ can be calculated using the equation:

$$\tau = 1/2pf_c \tag{8}$$

with τ equaling the reciprocal of the sum of the forward and backward rate constants (beta and alpha, respectively, Model A) that define channel gating. In other words τ is equal to the mean open lifetime.

A single τ signifies a simple model predicting a channel opening with a single open and closed state.

$$C \overset{\beta}{\rightarrow} O \qquad C \underset{\alpha}{\leftarrow} O \qquad \text{((Model A))}$$

However, in some cases, more than one τ can be calculated signifying a more complex model of channel opening (potentially with multiple closed states bound to agonist). In general, the number of Lorentzian functions needed to fit the power spectrum = the total number of channel states (open and closed) − 1.

Up to this point, we have described noise analysis with limited mathematical equations. However, it is important to provide a mathematical description in order to answer some important questions. For example, electrophysiological recordings are measured in amplitude against a function of time. How can we go from a function of time to the power spectrum, which is defined by a spectral density function? To do this, we need to use the Fourier transform. The Fourier transform takes the function of time and converts it into a frequency component.

The original electrophysiological recording, represented graphically as current against time, can be used to calculate the covariance function $C(\Delta t)$. The covariance function describes the correlation between signal values at different times (t and $t + \Delta t$). Typically, there is a high correlation between signal values when the time interval Δt is small. This is because the probability that channels are open or closed at time t remains so at $t + \Delta t$. In contrast, there is a low correlation between signal values as the time interval increases. This is because multiple channel openings become increasingly likely to occur within the larger time interval [4]. The covariance function $C(\Delta t)$ for a channel with one open and one closed state is an exponential term given by

$$C(\Delta t) = Ni^2 p e^{-\Delta t/\tau} \tag{9}$$

Where N is equal to the number of receptors, i is equal to the current passed through the channel and p refers to the probability that the channel is open. The covariance function can also be written as

$$C(\Delta t) = I\gamma(V - V_{eq})e^{-\Delta t/\tau} \tag{10}$$

where I is the average current ($I = NiP$), γ is the single channel conductance ($I = \gamma(V - V_{eq})$), and ($V - V_{eq}$) is the electrochemical driving potential. The problem with using the covariance function is that it is difficult to handle the data. Therefore, an easier approach using the spectral density function has been developed, which relies on the help of the Fourier transform (for a complete review of the Fourier transform as it applies to biological transients see Harris [7]). The Fourier transform of $e^{-\Delta t/\tau}$ is

$$\frac{2}{1+(2\pi f\tau)^{\wedge}2} \tag{11}$$

Therefore, the spectral density function ($G(f)$) can be defined as

$$G(f) = 4I\gamma(V - V_{eq})\frac{1}{1+(2\pi f\tau)^{\wedge}2} \tag{12}$$

where f is the frequency. Since $\tau = 1/2\pi f_c$, the spectral density function becomes

$$G(f) = 4I\gamma(V - V_{eq})\frac{1}{1+(f/fc)^{\wedge}2} \tag{13}$$

which is further simplified to

$$G(f) = G(0)/\left[1+(f/f_c)\right] \tag{14}$$

where $G(0)$ is the asymptotic spectral density at zero frequency. Equation (11) describes a Lorentzian function for a two-state channel (one closed and one open state). For more complex models of channel opening, Eq. (11) becomes

$$G(f) = G(0)/\left[1+(f/f_c)\right] + G(0)/\left[1+(f/f_c)\right] + \cdots \tag{15}$$

which represents more than one Lorentzian function. The number of Lorentzian functions used depends upon what fits the power spectrum most accurately. As mentioned above, the number of Lorentzian functions needed to fit the power spectrum = the total number of channel states (open and closed) − 1. An example is available showing the power spectrum and mean channel open times for $GABA_A$ receptors recorded from mIPSCs by De Koninck and Mody [2].

A clear advantage of noise analysis is that it allows for an estimation of *synaptic* channel kinetics, which is not possible using single channel analysis. However, when using noise analysis to analyze synaptic currents, there are important assumptions. For example, it is assumed that all receptors function independently, possess the same kinetic properties, have one nonzero conductance level, and have transition states that constitute a Markov process. A Markov process refers to predictions for the future that are conditional on the present state of the system and are independent of the future and past events. If this is hard to fathom, think of a Markov process like gambling at the roulette table. Each roll on the wheel produces the same odds of winning no matter where the ball landed previously or where the ball will land in the future. In this case, the synaptic current kinetics (e.g., rise times, decay times) during each trial is not affected by the previous trial or future trials [2, 8–11]. In addition to these assumptions, there are also technical considerations when performing noise analysis. First, noise analysis constitutes all noise in an electrical signal. Therefore, it is important to filter out unwanted signals so that signals-of-interest can be analyzed. Using low and high pass filters during recordings can prevent unwanted signals from contaminating the analysis [3]. Second, the analysis may not always provide the most accurate results. This is because low frequency points can be contaminated with slow changes in mean current. Variances associated with some Lorentzian components may be too low to be detected. In addition, rapid gating kinetics may not be resolved [4].

4 Summary

The kinetics of synaptic current can be measured by looking at the rise time, the decay time, and the mean open channel time. The rise time is commonly measured as the time it takes for the current to go from 10–90 % its peak value. This is computed by finding the best-fit line using simple linear regression analysis. The decay time can be measured using two approaches. The first approach calculates the decay time constant by curve fitting using an exponential function, while the second approach determines the decay time by dividing the area under the curve by the peak current amplitude. Finally, the mean open channel time can be calculated by using spectral (noise) analysis.

Each approach allows the experimenter to dissect different components of receptor kinetics, which can provide clues as to what types of changes may occur at the synapse following a specific manipulation. For example, changes in receptor kinetics can be caused by changes in receptor subunit expression, in receptor regulation by postsynaptic signaling molecules, and by changes in neurotransmitter release and re-uptake.

Measuring synaptic receptor kinetics can provide valuable information, which, when combined with other techniques, can lead to a more complete picture of synaptic dynamics.

References

1. Axon Instruments I (1993) The Axon guide for electrophysiology & biophysics laboratory techniques. Axon Instruments, Foster City
2. De Koninck Y, Mody I (1994) Noise analysis of miniature IPSCs in adult rat brain slices: properties and modulation of synaptic GABAA receptor channels. J Neurophysiol 71(4): 1318–1335
3. Ogden D (1994) Microelectrode techniques: the Plymouth Workshop handbook. Company of Biologists Limited, Cambridge
4. Traynelis SF, Jaramillo F (1998) Getting the most out of noise in the central nervous system. Trends Neurosci 21(4):137–145
5. Mallot HA (2013) Computational neuroscience: a first course. Springer, New York
6. van Drongelen W (2006) Signal processing for neuroscientists: an introduction to the analysis of physiological signals. Elsevier Science, Amsterdam
7. Harris CM (1998) The Fourier analysis of biological transients. J Neurosci Methods 83(1):15–34
8. Colquhoun D, Hawkes AG (1977) Relaxation and fluctuations of membrane currents that flow through drug-operated channels. Proc R Soc Lond B Biol Sci 199(1135):231–262
9. Sigworth FJ (1980) The conductance of sodium channels under conditions of reduced current at the node of Ranvier. J Physiol 307:131–142
10. Sigworth FJ (1981) Covariance of nonstationary sodium current fluctuations at the node of Ranvier. Biophys J 34(1):111–133
11. Sigworth FJ (1981) Interpreting power spectra from nonstationary membrane current fluctuations. Biophys J 35(2):289–300

Chapter 11

Measuring Presynaptic Release Probability

Nicholas Graziane and Yan Dong

Abstract

Chemical synapses make up the majority of synaptic connections in warm-blooded animals. Their primary purpose is to depolarize or hyperpolarize downstream neurons and this process is mediated by presynaptic neurotransmitters and postsynaptic receptors. These neurotransmitters bind to postsynaptic receptors causing pore opening and ion influx/outflux. The flow of ions alters the postsynaptic membrane potential increasing or decreasing the likelihood of action potential firing; a depolarizing potential can increase the probability that an action potential is triggered, while a hyperpolarizing potential can hold the neuron at a membrane potential below threshold (Fig. 1).

Key words Presynaptic release machinery, Probability of release, Variance-mean analysis, Paired-pulse ratio

1 Introduction

Chemical synapses make up the majority of synaptic connections in warm-blooded animals. Their primary purpose is to depolarize or hyperpolarize downstream neurons and this process is mediated by presynaptic neurotransmitters and postsynaptic receptors. These neurotransmitters bind to postsynaptic receptors causing pore opening and ion influx/outflux. The flow of ions alters the postsynaptic membrane potential increasing or decreasing the likelihood of action potential firing; a depolarizing potential can increase the probability that an action potential is triggered, while a hyperpolarizing potential can hold the neuron at a membrane potential below threshold (Fig. 1).

Neurotransmitters are released from presynaptic terminals. These neurotransmitters are stored in vesicles, which recycle to and from the plasma membrane. Neurotransmitter release relies on many molecules and their interactions for vesicle membrane fusion and exocytosis, processes that are regulated by intracellular messengers and extracellular modulators that are constantly being altered during synaptic use. Thus, neurotransmitter release

Nicholas Graziane and Yan Dong, *Electrophysiological Analysis of Synaptic Transmission*, Neuromethods, vol. 112,
DOI 10.1007/978-1-4939-3274-0_11, © Springer Science+Business Media New York 2016

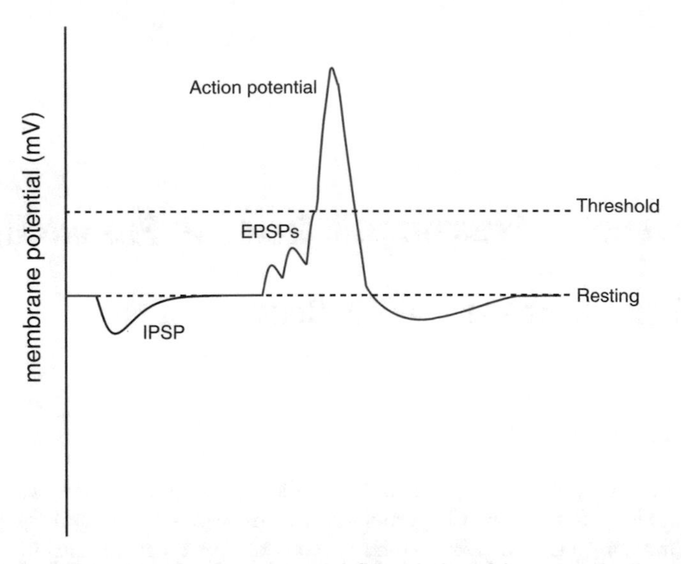

Fig. 1 An illustration of action potential firing induced by synaptically evoked potentials. At the resting membrane potential, IPSPs prevent cellular depolarization, thus keeping the cell at a hyperpolarized potential. However, accumulating EPSPs can drive the membrane to the action potential threshold triggering action potential generation

continuously changes in response to intra- and extracellular signals [1]. Given this, the probability of neurotransmitter release is not always identical at different release sites or between trials. For example, at certain pyramidal-to-interneuron synapses in the neocortex the probability of release is low (<0.3) [2]. At hippocampal CA1 pyramidal-to-pyramidal cell connections the probability of release is between 0.3 and 0.6 [3]. At the calyx of Held the probability of release can be as high as 0.9 [4]. Therefore, with different brain regions and cellular connections displaying different release probabilities, it is often necessary to measure release probability at specific synapses. This chapter discusses strategies for performing release probability measurement/analysis, including additional information about the molecules that regulate neurotransmitter release for data interpretation purposes.

2 Presynaptic Release Machinery

The presynaptic terminal contains active zones (Fig. 2), where neurotransmitters are released. At the active zone, synaptic vesicles are docked and primed through a series of protein–protein interactions. At the presynaptic membrane, at least five key proteins RIM, Munc13, RIM-BP, α-liprin, and ELKS form the active zone core (Fig. 2). These core proteins form a large protein complex, which recruits Ca^{2+} channels to the presynaptic terminal, tethers vesicles and Ca^{2+} channels to cell-adhesion molecules, docks, and primes

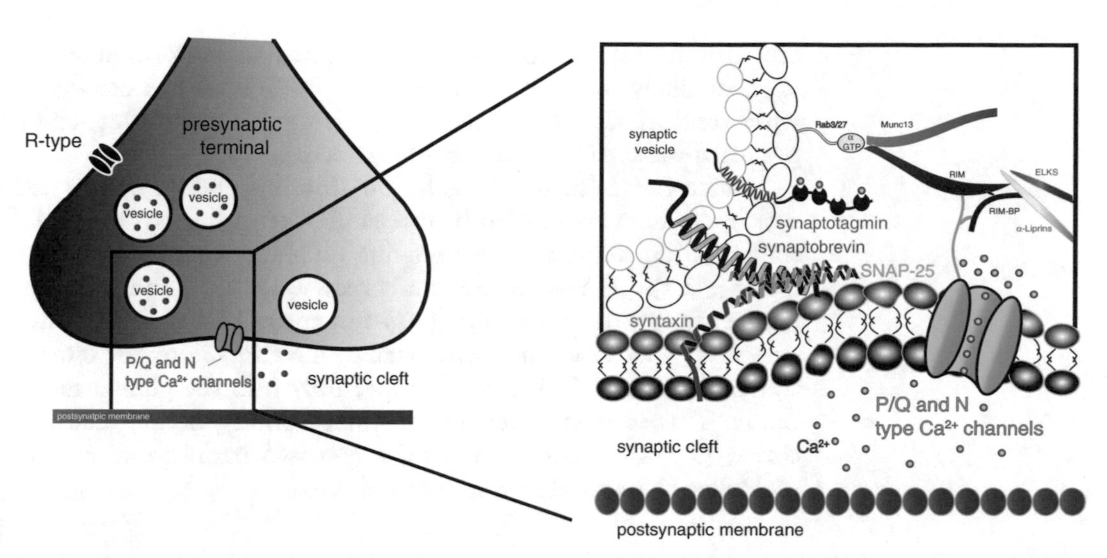

Fig. 2 Presynaptic release machinery. A presynaptic terminal (*left*) contains neurotransmitter-containing vesicles, which bind to the presynaptic membrane releasing neurotransmitters into the synaptic cleft. This process relies on Ca^{2+} influx through P/Q- or N-type channels and rarely through R-type channels. Once Ca^{2+} enters the presynaptic terminal (*right*), it binds to synaptotagmins, which are located on presynaptic vesicles. This binding of Ca^{2+} to synaptotagmins regulates synchronous and asynchronous neurotransmitter release. Vesicles are docked to the presynaptic membrane via protein–protein interactions. Synaptobrevin attached to the synaptic vesicle binds to syntaxin located on the presynaptic membrane and also binds to SNAP-25. Rab3/27 forms a complex with Munc13 and RIM. RIM and RIM-BP form complexes with Ca^{2+} channels, thus recruiting Ca^{2+} channels and docking them nearby synaptotagmins. RIM and RIM-BP are tethered near the postsynaptic membrane by α-Liprins and ELKS, which link to membrane docking proteins (not pictured) [Adapted from [5]]

vesicles [5, 6]. The vesicles containing neurotransmitters express key proteins necessary for vesicle docking and priming such as Rab3/27, synaptotagmin, and synaptobrevin. Rab3/27 binds to Munc13 and RIM (two of the five key proteins mentioned above). Synaptotagmin (with four Ca^{2+} binding sites) binds to Ca^{2+} entering the cytosol from nearby Ca^{2+} channels, and synaptobrevin binds to the SNARE proteins syntaxin and SNAP-25 located on the presynaptic membrane (for a full review on protein–protein interactions and their functions see [6]).

The catalyst for setting the release machinery in motion is Ca^{2+}. During an action potential, voltage-gated Ca^{2+} channels open, causing an influx of Ca^{2+}. This Ca^{2+} influx is typically through P/Q- ($Ca_v2.1$) or N-type ($Ca_v2.2$) Ca^{2+} channels and rarely through R ($Ca_v2.3$) or l-type Ca^{2+} channels [7, 8]. Once Ca^{2+} enters the cell it can trigger two mechanistically distinct types of release: A rapidly generated synchronous component facilitated by a low-affinity Ca^{2+} sensor [9] and an asynchronous component that can last for more than 1 s, which is facilitated by a high-affinity Ca^{2+} sensor [10–13]. How Ca^{2+} mediates synchronous and asynchronous release is posited to be through synaptotagmins (a vesicular protein—see above). Synaptotagmins with high Ca^{2+} affinity (synaptotagmin 3, 6, and 7) are posited to stimulate asynchronous release as this form of release is most likely induced by lower Ca^{2+} concentrations associated with residual Ca^{2+} (Ca^{2+} remaining in the terminal after action potential-induced neurotransmitter release).

Once Ca^{2+} initiates vesicle fusion, priming, and exocytosis, the vesicle can then be recycled by three different pathways: (1) vesicles remain in the active zone for refilling (kiss-and-stay). This process is fast (≤1 s). (2) Vesicles are locally recycled in a clathrin-independent endocytotic mechanism (kiss-and-run). This process occurs very fast (≤0.1 s) and happens distal to the release site just outside the active zone. (3) Vesicles collapse fully into the plasma membrane. This process is dependent upon clathrin, occurs relatively slowly (10–20 s) compared to the first two pathways discussed, and takes place distal to the release site.

3 Probability of Release

The probability of neurotransmitter release depends upon the coordinated activity of many molecules in the active zone. For example, synaptotagmins and their respective Ca^{2+} binding affinities can alter Pr as a twofold change in the Ca^{2+} sensor association time constant is predicted to result in a 16-fold change in the probability of exocytosis [14, 15]. Cysteine string proteins (proteins that promote Ca^{2+} channel opening) can have variable expression at release sites, thus altering Pr [16]. Furthermore, kinases, such as protein kinase C (PKC), can increase the number of available vesicles for fusion as well as decrease the Ca^{2+} threshold for exocytosis [17]. PKC elicits this response by phosphorylating proteins involved in transmitter release (e.g., synaptotagmins and Munc13). All in all, Ca^{2+} affinity or buffering at the presynaptic terminal is critically involved in controlling Pr.

Another factor that contributes to the probability of release is the number of functional release sites (N). These functional release sites include presynaptic terminals containing unrestrained, fusion-competent vesicles. Upon firing of a single action potential, these functional terminals release neurotransmitters based on their intrinsic Pr. In contrast to functional release sites, there are nonfunctional or refractory release sites [18]. Following an action potential and neurotransmitter release, these terminals are in a refractory period lasting ~10 ms [19]. During this time, refractory release sites have a zero Pr. Therefore, when measuring Pr, it is important to define N so that either all release sites can be measured (nonfunctional/refractory + functional) or only functional release sites can be measured.

Lastly, the size of the readily releasable pool of vesicles (docked and primed synaptic vesicles) can also dictate Pr [20, 21]; as the size of the readily releasable pool increases so too does Pr.

4 Measuring and Analyzing Release Probability

4.1 Paired-Pulse Protocol (Qualitative Measurement)

A paired-pulse protocol in electrophysiological measurements refers to two evoked pulses/stimuli with interpulse intervals typically separated by 50 ms. However, this interpulse interval can be adjusted depending upon the measurement. For example, with currents or potentials that have slow desensitization kinetics (e.g., IPSCs/IPSPs), separating the pulses by 100 ms may be beneficial. This longer duration allows the receptors to restore after inactivation or deactivation. Therefore, changes in the paired-pulse ratio (peak current/potential of the second stimulus divided by the peak current/potential of the first stimulus) can likely be attributed to presynaptic mechanisms. In addition, varying the interpulse intervals (50–400 ms) can illustrate the recovery of paired-pulse facilitation or depression.

Paired-pulse facilitation refers to the second pulse generating a larger postsynaptic current (PSC)/potential (PSP) than the first pulse (Fig. 3a). Facilitation occurs when the presynaptic cell has low release probability causing a small postsynaptic response that occurs after the first pulse. However, this first pulse generates a build-up of Ca^{2+} in the presynaptic terminal. Because of Ca^{2+} build-up, much of the release machinery is occupied by Ca^{2+} (e.g., Ca^{2+} bound to 1–3 of the four synaptotagmin Ca^{2+} binding sites), thus priming release sites. Therefore, following the second pulse, the release probability is greater causing a larger postsynaptic response [22–26].

Paired-pulse depression refers to the second pulse generating a smaller PSC/PSP than the first pulse (Fig. 3b). This can occur in one of three ways: (1) a change in Pr whereby the presynaptic terminal has a high release probability. Therefore, following the first pulse, the available readily releasable vesicles are much decreased causing fewer vesicles to be released following the second pulse,

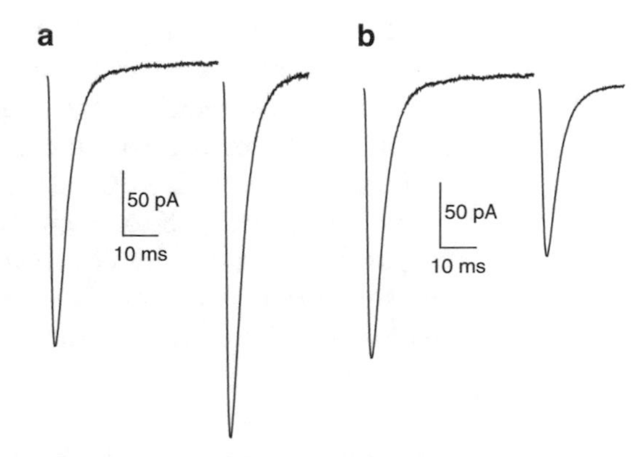

Fig. 3 A paired electrical pulse given at a 50 ms interval elicited postsynaptic currents, which underwent paired-pulse facilitation (**a**) or depression (**b**)

(2) a change in N caused by neurotransmitter depletion after the first pulse. With a depletion of neurotransmitters, the once functional synapse becomes nonfunctional [18, 25], thus changing N, and (3) a change in the postsynaptic response due to the accumulation of desensitized receptors that were activated following the first pulse [27].

In general, paired-pulse facilitation reflects a low Pr whereas paired-pulse depression is associated with a high Pr, a decrease in N, or postsynaptic receptor desensitization. Therefore, using a paired-pulse protocol gives a qualitative idea of Pr at a specific synapse, but further investigation (especially for paired-pulse depression) or quantitative analysis may be necessary.

4.2 Multiple-Probability Fluctuation Analysis or Variance-Mean Analysis (Quantitative Measurement)

Variance-mean analysis (V–M analysis) is a useful method for quantifying the probability of release [28]. This method was developed to overcome shortcomings associated with existing techniques. For example, using V–M analysis enables the researcher to calculate the variance about the mean at multiple release probability settings versus just one, thus providing a more complete analysis of synaptic function. In addition, by relating the analysis to nonstationary noise analysis of ion-channel function [29], the V–M analysis can be applied to isolated single-fiber synaptic inputs as well as multiple inputs from several presynaptic axons. Therefore, if activating multiple presynaptic axons with electrically or optically evoked currents, V–M analysis can provide quantitative measurements of Pr.

In order to implement V–M analysis, Pr needs to be experimentally changed. This can be accomplished by altering Ca^{2+} or Mg^{2+} concentrations in the bath solution or by using a stimulus pulse train. The disadvantage of changing Ca^{2+} or Mg^{2+} concentrations is that it requires long recording times due to the wash-in of new bath solution. Because of this, a multi-pulse train is used as it efficiently changes the probability of release simply through electrical stimulation.

Measuring Pr using V–M analysis can be performed as follows: After 30–100 postsynaptic currents (PSCs) have been recorded from each Pr condition, the peak amplitude of each PSC can be subtracted from the baseline and averaged. Following this, the variance of the peak PSCs are calculated and plotted against the mean amplitude PCS for each condition. Once plotted, the quantal parameters are estimated from a best fit curve to the V–M relationship [30]. The best fit curve has a parabolic shape defined mathematically using the equation of a parabola:

$$x = -ax^2 + bx + c \quad (1)$$

where the negative sign in front of a defines an inverted parabola as shown in Fig. 4, a is inversely proportional to N ($N = 1/a$), and b can be used to solve for the quantal size, Q ($Q = b/1 + \mathrm{CV}^2$). Q

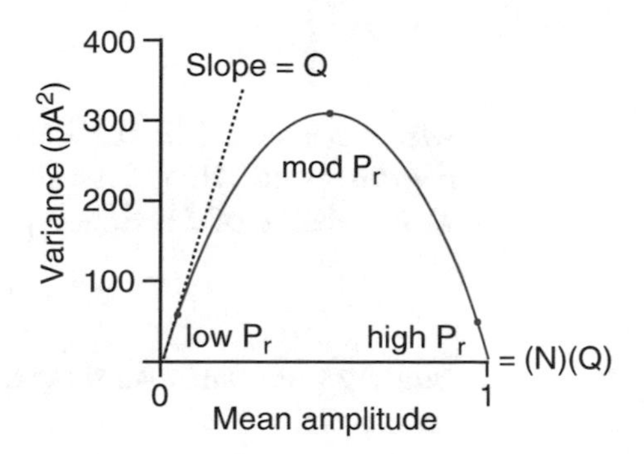

Fig. 4 Measuring Pr at synapses. After plotting the variance of the peak amplitude over the mean amplitude for each Pr condition, a best fit curve with a parabolic shape is graphed. In this simulated curve, synapses with low, moderate, or high release probability are plotted demonstrating their locations on the best fit curve. *Q* is equal to the slope of the best fit line in a synapse with low Pr and the product of *N* and *Q* is equal to the point where the parabola crosses the *x*-axis as the probability of release approaches 100 %

becomes even easier to solve for at low release probabilities as *Q* directly equals the initial slope of the V–M relationship (Fig. 4). Also, one can solve for the product of *N* and *Q* [i.e., (*N*)(*Q*)] since this is equal to the point where the parabola crosses the *x*-axis as the probability of release approaches 100 % (Fig. 4). Lastly, *c* is typically removed from the above quadratic equation because when performing V–M analysis one of the two *x* values from the quadratic equation typically equals 0. Removing *c* from the parabolic equation can be justified mathematically using the quadratic equation to solve for *x*:

$$x = \frac{-b \pm \sqrt{b^2 - 4ac}}{2a} \tag{2}$$

with *c* equal to 0, $-b$ is added or subtracted to the square root of b^2. Depending on the value of *b*, one of the two fractions will contain a 0 making $x = 0$.

At this point, *N* and *Q* can be solved for using software that can fit the parabolic curve generated by plotting the variance of the mean amplitude over the mean amplitude at different Prs. Finding the best fit parabolic curve provides numerical estimates of *a* and *b*, which can then be used to calculate *N* and *Q*. After calculating *N* and *Q*, Pr can then be calculated.

When calculating Pr, it is important that the experimenter choose the formula that best describes the synapse being investigated. The important questions to consider are, (1) Is synaptic release synchronous or asynchronous? (2) Is quantal size uniform or nonuniform? and (3) Is the probability of release uniform or nonuniform? The simplest model used to calculate Pr is the binomial model [31]:

$$I = N \Pr Q \tag{3}$$

where I is equal to the mean amplitude of the PSC (if using the 5-pulse train, then I would be calculated from the first pulse). The variance of I is then equal to:

$$\sigma_I^2 = NQ^2 \Pr(1 - \Pr) \tag{4}$$

Making the relationship between the mean amplitude and the variance equal to:

$$\sigma^2 = QI - I^2 / N \tag{5}$$

However, by using this model to solve for Pr, the experimenter makes some underlying assumptions such as: (1) release sites operate independently of one another, (2) neurotransmitter release is synchronous, (3) Pr is uniform across release sites, and (4) Q is uniform at individual sites as well as across all release sites. If it is known that any of these assumptions are not true, there are other equations that can be used and have been elegantly reviewed and explained by Silver [32]. For example, Pr can be calculated from a synapse with asynchronous transmitter release (Fig. 5) with uniform quantal parameters using the equation:

$$\sigma^2_{\text{total latency}} = N \Pr \sigma^2_{\text{quantal latency}} \tag{6}$$

where $\sigma^2_{\text{quantal latency}}$ refers to the quantal size variance at the time of the peak PSC at one release site [32]. At a synapse with nonuniform quantal size, Pr can be solved for using the following equation, but is only valid when quantal behavior approaches the mean quantal size at the time of the PSC peak (Q_P) [32]:

$$\sigma_I^2 = NQ_P^2 \Pr(1 - \Pr)\left(1 + CV^2_{QII}\right) + NQ_P^2 \Pr CV^2_{QI} \tag{7}$$

where CV^2_{QI} is coefficient of variance calculated from the quantal variability and CV^2_{QII} refers to the coefficient of variation of quantal variability across all release sites. At a synapse with nonuniform

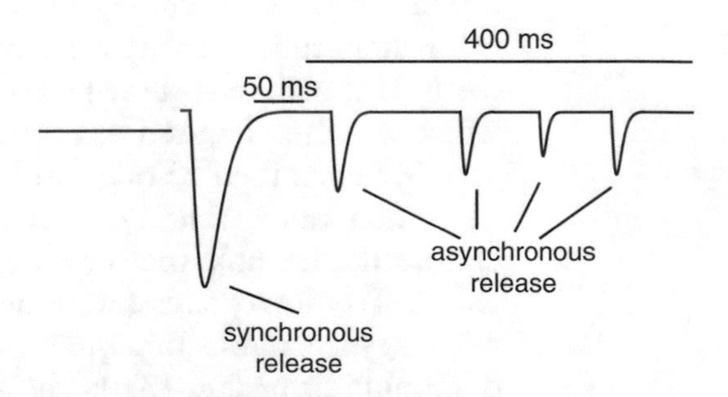

Fig. 5 Asynchronous release at a synapse following electrically evoked synchronous release

quantal size and nonuniform release probability, Pr can be calculated using this equation:

$$\sigma_I^{\ 2} = NQ_P^{\ 2}\,\mathrm{Pr}\left[1-\mathrm{Pr}\left(1+\mathrm{CV}^2{}_P\right)\right]\left(1+\mathrm{CV}^2{}_{QII}\right)+NQ_P^{\ 2}\,\mathrm{Pr}\,\mathrm{CV}^2{}_{QI} \qquad (8)$$

where CV_P is the coefficient of variation of release probability across all release sites [32].

4.3 Technical Considerations for V–M Analysis

Like with most electrophysiological recordings, there are artifacts that can be introduced causing inaccurate estimations of Pr. To avoid inaccurate measurements and overestimations of synaptic variance [28], we have supplied a list of components to consider:

1. Avoid analysis of synaptic currents when there is a change in series resistance. This can be avoided by monitoring the series resistance throughout the recording. If the experimenter does not know what percent change in series resistance leads to changes in synaptic current, it can be tested empirically. Another option is to dismiss recordings that have undergone >20 % change in series resistance.
2. Avoid analysis of currents that rundown. Checking synaptic current stability can be tested statistically by using the Spearman rank correlation test. The Spearman correlation measures the relationship between two variables from a scale of 1 to –1. A value of 1 means that both variables have a perfect increasing relationship (Fig. 6), while a value of –1 means that both variables have a perfect decreasing relationship (Fig. 6). An electrophysiologist expects to see a Spearman coefficient of 0 or close to 0, which means that there is no tendency for the amplitude (variable *y*) to increase or decrease as time (variable *x*) increases.
3. Avoid active dendritic conductances. By including in the pipette solution cesium ions to block K^+ channels or QX314 to block Na^+ channels, active conductances can be minimized.

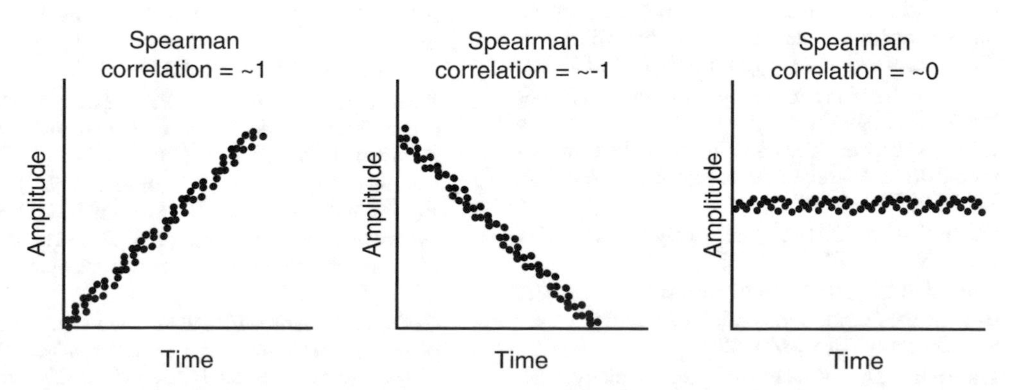

Fig. 6 Measured amplitudes from a PSC graphed over time have a Spearman correlation of 1 (*left*), –1 (*middle*), or 0 (*right*)

5 Summary

Measuring Pr can provide valuable information about synapses in identified brain regions. By using the paired-pulse protocol, the experimenter can qualitatively measure Pr at synapses through interpretation of paired-pulse facilitation or depression. If further characterization of Pr is necessary, V–M analysis can be used for quantitative measure. With these measurements and interpretations in hand, the experimenter can then further explore molecular substrates that control Pr at the presynaptic active zone. These important molecular components include RIM, Munc13, RIM-BP, α-liprin, ELKS, and synaptotagmin. In addition, Ca^{2+} and protein kinases are key modulators of Pr through their regulation of vesicle fusion/release at the active zones.

Taking into consideration our current knowledge of presynaptic release machinery and adding Pr analysis at specific presynaptic and postsynaptic terminals can tell us a lot about how specific neurocircuits operate. This type of information extended to models of neurological disease is worth testing as the information can provide valuable insights as to how neurocircuits are altered by external stimuli or internal developmental impairments. We hope that after reading this chapter, experimenters will be able to measure release probability accurately and confidently.

References

1. Sudhof TC (2004) The synaptic vesicle cycle. Annu Rev Neurosci 27:509–547
2. Thomson AM, West DC, Deuchars J (1995) Properties of single axon excitatory postsynaptic potentials elicited in spiny interneurons by action potentials in pyramidal neurons in slices of rat neocortex. Neuroscience 69(3): 727–738
3. Thomson AM, Bannister AP (1999) Release-independent depression at pyramidal inputs onto specific cell targets: dual recordings in slices of rat cortex. J Physiol 519(Pt 1):57–70
4. von Gersdorff H, Schneggenburger R, Weis S, Neher E (1997) Presynaptic depression at a calyx synapse: the small contribution of metabotropic glutamate receptors. J Neurosci 17(21):8137–8146
5. Sudhof TC (2012) The presynaptic active zone. Neuron 75(1):11–25
6. Sudhof TC (2013) Neurotransmitter release: the last millisecond in the life of a synaptic vesicle. Neuron 80(3):675–690
7. Dietrich D, Kirschstein T, Kukley M, Pereverzev A, von der Brelie C, Schneider T, Beck H (2003) Functional specialization of presynaptic Cav2.3 Ca2+ channels. Neuron 39(3):483–496
8. Wu LG, Borst JG, Sakmann B (1998) R-type Ca2+ currents evoke transmitter release at a rat central synapse. Proc Natl Acad Sci U S A 95(8):4720–4725
9. Sabatini BL, Regehr WG (1996) Timing of neurotransmission at fast synapses in the mammalian brain. Nature 384(6605):170–172
10. Atluri PP, Regehr WG (1998) Delayed release of neurotransmitter from cerebellar granule cells. J Neurosci 18(20):8214–8227
11. Barrett EF, Stevens CF (1972) The kinetics of transmitter release at the frog neuromuscular junction. J Physiol 227(3):691–708
12. Geppert M, Goda Y, Hammer RE, Li C, Rosahl TW, Stevens CF, Sudhof TC (1994) Synaptotagmin I: a major Ca2+ sensor for transmitter release at a central synapse. Cell 79(4):717–727
13. Goda Y, Stevens CF (1994) Two components of transmitter release at a central synapse. Proc Natl Acad Sci U S A 91(26):12942–12946
14. Kasai H (1999) Comparative biology of Ca2+-dependent exocytosis: implications of kinetic

diversity for secretory function. Trends Neurosci 22(2):88–93

15. Thomson AM (2000) Facilitation, augmentation and potentiation at central synapses. Trends Neurosci 23(7):305–312
16. Seagar M, Leveque C, Charvin N, Marqueze B, Martin-Moutot N, Boudier JA, Boudier JL, Shoji-Kasai Y, Sato K, Takahashi M (1999) Interactions between proteins implicated in exocytosis and voltage-gated calcium channels. Philos Trans R Soc Lond B Biol Sci 354(1381):289–297
17. Hilfiker S, Augustine GJ (1999) Regulation of synaptic vesicle fusion by protein kinase C. J Physiol 515(Pt 1):1
18. Betz WJ (1970) Depression of transmitter release at the neuromuscular junction of the frog. J Physiol 206(3):629–644
19. Stevens CF, Wang Y (1995) Facilitation and depression at single central synapses. Neuron 14(4):795–802
20. Rosenmund C, Stevens CF (1996) Definition of the readily releasable pool of vesicles at hippocampal synapses. Neuron 16(6):1197–1207
21. Zucker RS (1973) Changes in the statistics of transmitter release during facilitation. J Physiol 229(3):787–810
22. Balnave RJ, Gage PW (1974) On facilitation of transmitter release at the toad neuromuscular junction. J Physiol 239(3):657–675
23. Katz B, Miledi R (1968) The role of calcium in neuromuscular facilitation. J Physiol 195(2): 481–492
24. Katz B, Miledi R (1970) Further study of the role of calcium in synaptic transmission. J Physiol 207(3):789–801
25. Mallart A, Martin AR (1968) The relation between quantum content and facilitation at the neuromuscular junction of the frog. J Physiol 196(3):593–604
26. McLachlan EM (1978) The statistics of transmitter release at chemical synapses. Int Rev Physiol 17:49–117
27. Chen C, Blitz DM, Regehr WG (2002) Contributions of receptor desensitization and saturation to plasticity at the retinogeniculate synapse. Neuron 33(5):779–788
28. Clements JD, Silver RA (2000) Unveiling synaptic plasticity: a new graphical and analytical approach. Trends Neurosci 23(3):105–113
29. Traynelis SF, Jaramillo F (1998) Getting the most out of noise in the central nervous system. Trends Neurosci 21(4):137–145
30. Suska A, Lee BR, Huang YH, Dong Y, Schluter OM (2013) Selective presynaptic enhancement of the prefrontal cortex to nucleus accumbens pathway by cocaine. Proc Natl Acad Sci U S A 110(2):713–718
31. Kuno M (1964) Quantal components of excitatory synaptic potentials in spinal motoneurones. J Physiol 175:81–99
32. Silver RA (2003) Estimation of nonuniform quantal parameters with multiple-probability fluctuation analysis: theory, application and limitations. J Neurosci Methods 130(2): 127–141

Chapter 12

Long-Term Measurements

Nicholas Graziane and Yan Dong

Abstract

Long-term measurements in slice electrophysiology typically constitute long-term potentiation (LTP) and long-term depression (LTD). These measurements can last 1–2 h, requiring minimal baseline noise and easy detection of activated synaptic receptors. This can be relatively easy to achieve in field recordings, but can be challenging in whole cell configurations. In this chapter, we discuss the concepts behind LTP and LTD as well as the approaches that the beginning electrophysiologist can use in order to complete these long-term measurements. We finish with potential technical issues to consider during experimentation as well as suggestions that may increase the likelihood of successful recordings.

Key words Long-term potentiation, Long-term depression

1 Introduction

Long-term measurements in slice electrophysiology typically constitute long-term potentiation (LTP) and long-term depression (LTD). These measurements can last 1–2 h, requiring minimal baseline noise and easy detection of activated synaptic receptors. This can be relatively easy to achieve in field recordings, but can be challenging in whole-cell configurations. In this chapter, we discuss the concepts behind LTP and LTD as well as the approaches that the beginning electrophysiologist can use in order to complete these long-term measurements. We finish with potential technical issues to consider during experimentation as well as suggestions that may increase the likelihood of successful recordings.

2 Long-Term Measurements

2.1 Long-Term Potentiation (LTP)

LTP refers to the persistent potentiation of postsynaptic responses (e.g., potentials or currents) to a presynaptic action potential (Fig. 1a). The physiological relevance of LTP is posited to encode learning and memory as well as activity-dependent brain

Nicholas Graziane and Yan Dong, *Electrophysiological Analysis of Synaptic Transmission*, Neuromethods, vol. 112, DOI 10.1007/978-1-4939-3274-0_12, © Springer Science+Business Media New York 2016

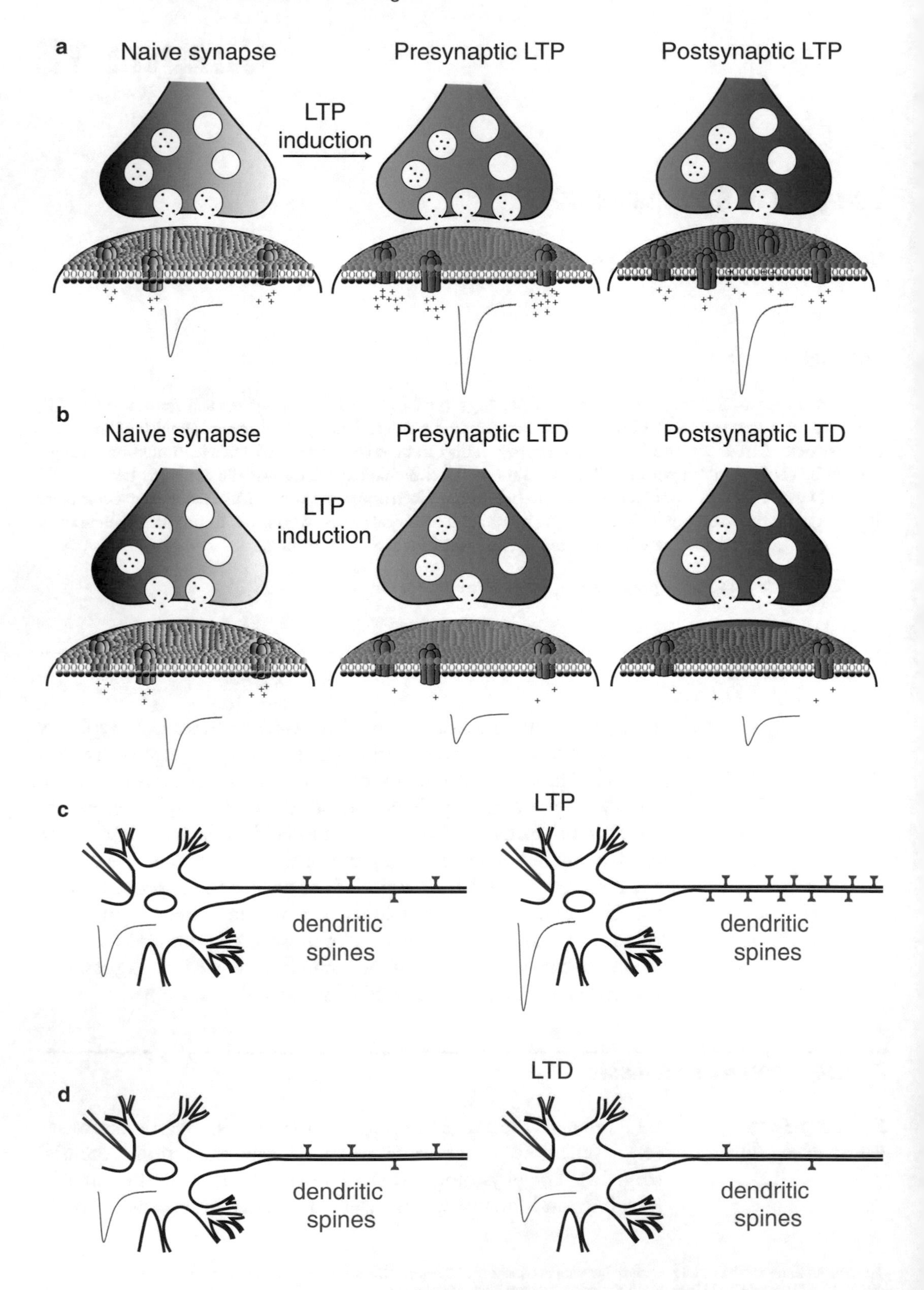
a
Naive synapse
LTP
induction
Presynaptic LTP
Postsynaptic LTP
b
Naive synapse
LTP
induction
Presynaptic LTD
Postsynaptic LTD
c
LTP
dendritic
spines
dendritic
spines
LTD
d
dendritic
spines
dendritic
spines

development [1, 2]. Since the discovery of LTP at mammalian hippocampal synapse in vivo [3], and the proposed physiological relevance, an enormous amount of work has gone into identifying cellular mechanisms mediating LTP throughout the central nervous system (CNS). This work has resulted in the use of in vitro experimental approaches designed to understand the presynaptic or postsynaptic modifications necessary for LTP induction. For LTP in the hippocampal CA1 region, a clear postsynaptic mechanism has been characterized that relies on NMDAR activation, which is thus called NMDAR-dependent LTP. Postsynaptic changes mediating LTP constitute increases in the number of functional AMPARs in the postsynaptic membrane, caused by activity-dependent changes in AMPAR trafficking, or biophysical and biochemical modifications AMPARs via phosphorylation [1, 2, 4–13]. These processes are initiated by activation of NMDARs, which induces calcium entry, and activation of downstream calcium-dependent signaling proteins such as CaMKII, triggering the cellular/molecular actions needed for LTP. This and other forms of LTP can persist for hours, days, and even months under certain conditions. Therefore, LTP must not only be expressed but also be maintained at the synapse. In the CA1 region, one way that LTP is maintained is through activation of signaling molecules linking to the nucleus of the cell. These signaling molecules include PKA, CaMKIV, and MAPK. Once activated, these signaling molecules turn on transcription factors such as CREB and immediate early genes such as zif268 [14–16].

Presynaptic expression of LTP is also common in the CNS. For example, at the hippocampal CA3 region a presynaptic form of plasticity predominates at mossy fiber synapses (synapses between the axons of the dentate gyrus and granule cells). This form of plasticity is calcium dependent. However, it does not rely on NMDARs, but instead is dependent upon presynaptic calcium channels such as the P/Q-, N-, and R-type calcium channels (R-type necessary for expression of LTP only) [17–20]. Enhanced neurotransmitter release at these synapses depends upon increases in calcium, which then leads to the downstream activation of cyclic AMP (cAMP) and PKA [21–23]. Once activated, cAMP/PKA

Fig. 1 Mechanisms mediating LTP and LTD. (**a**) LTP is triggered at an excitatory synapse leading to increased positive ion influx into the postsynaptic membrane. The increase in ion influx can be due to increased neurotransmitter release presynaptically (*middle*) or due to increases in postsynaptic receptor number (*right*). (**b**) LTD is triggered at an excitatory synapse leading to decreased positive ion influx into the postsynaptic membrane. The decrease in ion influx can be due to either decreased neurotransmitter release presynaptically (*middle*) or due to decreased numbers of postsynaptic receptors (*right*). (**c**) LTP is triggered in a neuron as the expression of functional synapses is increased. (**d**) LTD is induced in a neuron as the expression of functional synapses is decreased

modulate the molecular machinery in the presynaptic active zone (see Chap. 11) such as Rab3A (a GTP-binding protein thought to reversibly attach to synaptic vesicle membranes) and Rim1α, leading to enhanced neurotransmitter release [24–26].

LTP is not restricted to the excitatory synapses, but is also expressed at inhibitory synapses (LTP_{GABA} or iLTP for inhibitory LTP). For example, in the ventral tegmental area (VTA) (a brain region involved in reward and motivation), LTP at GABAergic synapses onto VTA dopamine neurons exists and has been shown to be critically involved in stress-induced reinstatement to drugs of abuse [27, 28]. The mechanism for LTP induction at these synapses requires multiple steps whereby calcium enters the postsynaptic cell through NMDARs. This increase in calcium leads to the postsynaptic release of nitric oxide, which then retrogradely traffics to the presynaptic neuron. Presynaptically, nitric oxide activates PKA and/or PKG, thus causing increased release of GABA [28, 29].

2.2 Long-Term Depression (LTD)

LTD refers to the persistent depression of postsynaptic responses to a presynaptic action potential (Fig. 1b). The physiological significance of LTD is that it is essential for the fine-tuning of neural circuits including synaptic weakening, which potentially leads to synaptic elimination. Much like LTP, the mechanisms mediating LTD have been extensively studied throughout the CNS with LTD being initiated and maintained by presynaptic or postsynaptic modifications at excitatory or inhibitory synapses. For example, two postsynaptic forms of LTD include NMDAR-dependent LTD and metabotropic glutamate receptor-dependent LTD (mGluR LTD). Unlike NMDAR-dependent LTP where strong activation elicits a large rise in intracellular calcium, LTD is induced by modest activation of NMDARs causing a small rise in postsynaptic calcium. This small rise in calcium triggers signaling molecules such as serine/threonine phosphatases, which can then dephosphorylate AMPAR subunits [2, 30]. The dephosphorylation of AMPARs then signals endocytotic machinery to either reduce the function of AMPARs or remove AMPARs from the postsynaptic membrane. AMPAR internalization is achieved by dynamin- and clathrin-mediated endocytosis [31]. The mechanism for mGluR-dependent LTD also is mediated by clathrin-dependent endocytosis of synaptic AMPARs requiring mGluR activation of voltage-gated calcium channels or mGluR activation alone [30]. A well-studied presynaptic modification causing LTD is endocannabinoid (eCB) mediated LTD. eCB-LTD requires postsynaptically derived eCBs to travel retrogradely across the synapse activating cannabinoid-1 receptors (CB1 receptors) or presynaptic NMDARs. The activation of these receptors induces decrease in neurotransmitter release by altering PKA and RIM1α-dependent signaling [32, 33].

2.3 Approach for Eliciting LTP or LTD

2.3.1 LTP

A few approaches discussed here are typically used to elicit LTP in in vitro brain slices. A commonly used approach to induce LTP is to use electrically evoked high frequency stimulation (e.g., 100 Hz). Many studies use optogenetic approaches to stimulate specific pathways, using the high-frequency tetanic stimulation as long as the channelrhodopsin does not desensitize [34, 35]. In addition to using high frequency protocols, LTP can be induced chemically via bath perfusion of compounds such as forskolin (activates cAMP pathways), glycine, and tetraethylammonium, to name a few [36–38]. Lastly, a proposed physiologically relevant induction of LTP is called spike-timing dependent plasticity (STDP) or otherwise known as Hebbian plasticity (see Chap. 6 for a description of Hebbian/STDP). This form of plasticity is considered physiologically relevant in that it relies on the temporal firing of both presynaptic and postsynaptic neurons rather than a high frequency train of stimulation. Spike-timing dependent LTP can be induced at excitatory or inhibitory synapses by pairing a presynaptic action potential with a postsynaptic depolarization (pairing the presynaptic action potential typically occurs 40 ms after the postsynaptic depolarization) [39–44]. However, there are some cases where presynaptic action potentials precedes the postsynaptic depolarization causing LTP [45]. This form of plasticity is known as anti-Hebbian LTP.

2.3.2 LTD

To induce LTD in in vitro brain slices the following approaches are can be used. Unlike LTP protocols, which require high-frequency stimulation (HFS) protocols, LTD can be induced by weaker or low frequency stimuli (LFS) (0.5–10 Hz stimulation). This relatively weak stimulation can be evoked using electrical or optogenetic tools and can induce LTD either with short or long durations [46–50].

In addition to using LFS approaches, LTD can also be induced using chemical means (chemLTD) such as bath perfusion with the agonist NMDA [51, 52]. This is a different form of plasticity when compared to evoked NMDAR-dependent LTD in that chemLTD activates both extrasynaptic and synaptic NMDARs, while evoked currents are predominantly mediated by synaptic NMDARs. Therefore, finding differences in cellular mechanisms mediating the two should not be surprising as reported previously [52].

Lastly, LTD can be induced using anti-Hebbian STDP whereby the postsynaptic neuron depolarizes within milliseconds prior to a presynaptic action potential [39, 53].

2.4 Identifying Mechanisms of LTP or LTD Using Electrophysiology Methods

Once LTP or LTD has been induced, the next step is to systematically identify the cellular and molecular mechanisms involved. To begin, the experimenter can identify whether the plasticity is expressed by presynaptic and/or postsynaptic modifications. To do this, a very straightforward approach is to incorporate a measurement of paired-pulse ratio (see Chap. 11) before and after the expression of

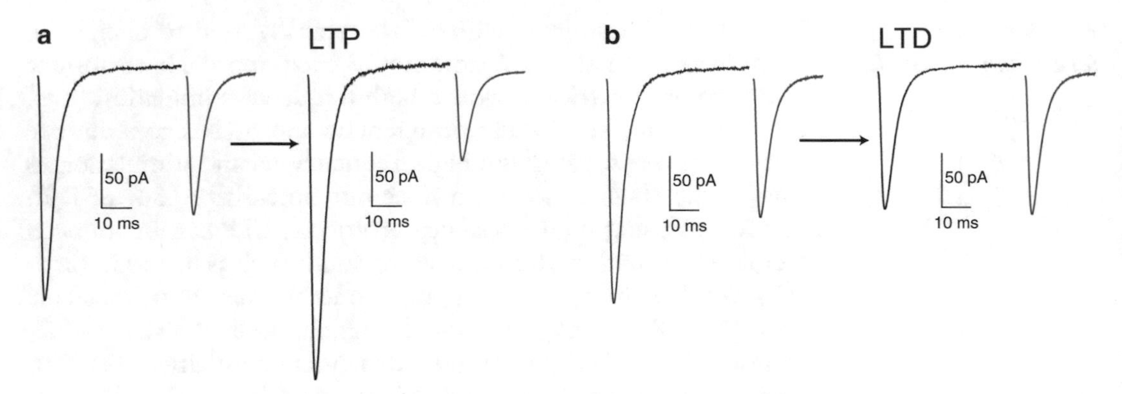

Fig. 2 Plasticity effects on paired-pulse currents. (**a**) LTP is triggered leading to a decreased paired-pulse ratio (pulse #2/pulse #1). (**b**) LTD is induced leading to increases in the paired-pulse ratio

the plasticity (Fig. 2). Typically, changes in paired-pulse ratios (PPRs) are associated with presynaptic changes, but this is not always the case and extra evidence is often desired. The extra evidence can come from the variance-mean analysis (see Chap. 11 for experimental approach needed for variance-mean analysis), in which postsynaptic modifications can be reflected, to certain degree, by the changes in quantal size (*Q*), while presynaptic modifications are associated with changes in the probability of release (Pr) or the number of release sites (*N*) [54, 55]. However, these associations are not always set in stone as it is debatable whether changes in *N* are mediated entirely by presynaptic mechanisms (see silent synapses Chap. 19). Other approaches include measuring miniature events or measuring the coefficient of variation (CV). Measuring miniature events is performed by bath applying tetrodotoxin or by replacing calcium with strontium during evoked stimulus recordings (see Chap. 8 for experimental approaches). In performing these experiments, the amplitude and frequency of quantal events can be measured before and after plasticity. Typically, but not always, changes in frequency are associated with presynaptic modifications, while changes in amplitude are associated with postsynaptic modifications. The CV, which depends strongly on presynaptic neurotransmitter release, is a measure of how much a single postsynaptic response varies from the overall mean postsynaptic response (see Chap. 15 a detailed explanation of the CV). The CV is inversely proportional to the quantal release of neurotransmitter, thus CV^{-2} directly proportional. Calculating a decrease in CV (inversely proportional to quantal release) or an increase in CV^{-2} (direct measure of quantal release) during LTP measurements (a potentiation of synaptic responses) indicates a presynaptic mechanism. However, the association between CV and presynaptic function is not concrete as changes in quantal content are also dependent upon *N*, which can increase or decrease depending upon presynaptic and/or postsynaptic unsilencing. Therefore, postsynaptic alterations can also contribute to changes in CV, implicating the necessity to use multiple measurement strategies to determine the mechanism of LTP or LTD.

For excitatory synapses, another convenient measurement is to compare AMPAR/NMDAR ratios before and after the plasticity induction. Because the AMPAR/NMDAR ratio is relatively independent of the number of synapses, presynaptic release probability, and other presynaptic factors, a change in this ratio should reflect changes in postsynaptic AMPARs or NMDARs. In most experimental conditions, changes in NMDARs during plasticity induction are minimal compared to changes in AMPARs, so, when supported with additional data, an increase or decrease in AMPAR/NMDAR ratio is often interpreted as an increase or decrease in AMPARs. The AMPAR/NMDAR ratio is best performed in whole-cell configuration by evoking synaptic currents while voltage clamping the cell at positive membrane potentials to remove the magnesium block on NMDARs (e.g., +50 mV). Once a stable baseline is achieved using a stimulus protocol of 0.1 Hz (an evoked stimulus every 10 s), NMDARs can be blocked using an antagonist (e.g., APV) via bath perfusion to isolate AMPARs (another option is to isolate NMDARs by blocking AMPARs using an AMPAR antagonist). Once a stable baseline is again achieved the experiment can be completed and analyzed as follows: First, take an average of ~30 sweeps from the AMPAR + NMDAR-mediated currents (the last 30 sweeps before blocker was applied) and from the isolated AMPAR currents (the last 30 sweeps of the experiment). Then subtract the averaged dual current by the AMPAR-mediated only current, resulting in the NMDAR current. Finally, divide the peak amplitude of the AMPAR-mediated current by the peak amplitude of the NMDAR-mediated current to obtain the ratio. To interpret these results, it is important to first check whether the NMDAR component has changed [6]. If not, any changes in the ratio can be interpreted as changes in AMPAR expression/function. Lastly, postsynaptic mechanisms mediating the induction or expression of synaptic plasticity can be measured in whole-cell configurations by filling the internal solution with antagonists or antibodies to block intracellular signaling molecules, proteins, or ions. Using this approach allows the experimenter to isolate specific cellular mechanisms involved in the plasticity-induced postsynaptic modification and has led to important cellular/molecular mechanisms that mediate LTP and LTD [2, 52].

2.5 Analysis/Interpretation

In LTP/LTD experiments, some experienced electrophysiologists may use a paired-pulse protocol throughout the experiment, but measure the peak amplitude of the first current for analyzing LTP/LTD and incorporate the second current for PPR analysis (to see how to measure amplitudes see Chap. 14). This analysis includes the peak amplitude generated during a stable baseline

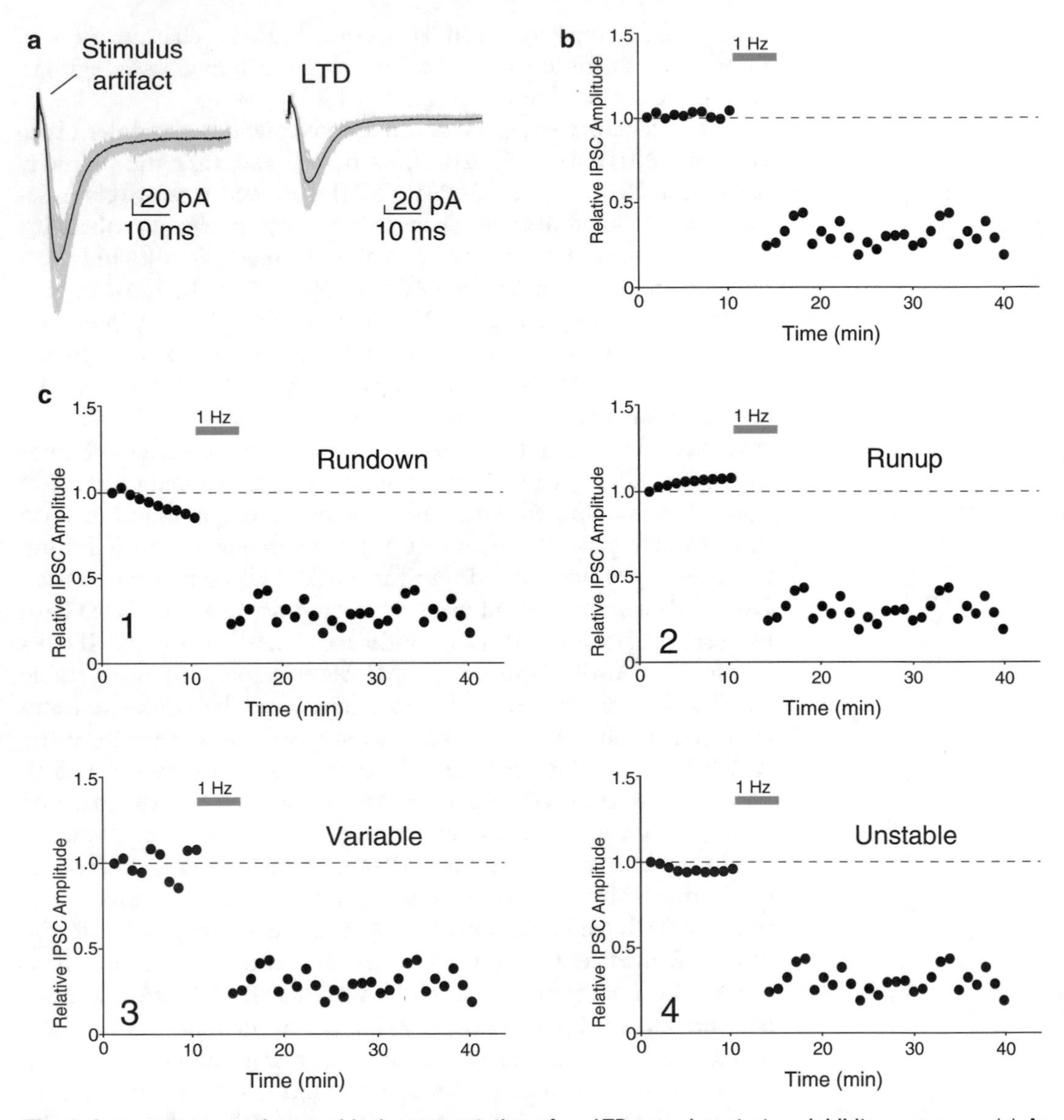

Fig. 3 Current traces and a graphical representation of an LTD experiment at an inhibitory synapse. (**a**) An acceptable LTD experiment showing representative IPSC traces taken before and after LTD induction. (**b**) An ideal time course showing a stable 10 min baseline followed by a depression in synaptic currents whose peak amplitude is <10 % of the baseline IPSCs. (**c**) Examples of unacceptable 10 min baselines before LTD induction: (*1*) run-down, (*2*) run-up, (*3*) variable, (*4*) run-down followed by stable baseline

period (10–20 min) and after plasticity induction (Fig. 3). The peak amplitudes of the synaptic currents after plasticity induction are then normalized to those during the baseline period. The results can then be plotted versus time. What should be interpreted as LTP/LTD? Statistics can be used to test significance (e.g., Student's *t*-test comparing the averaged traces 2 min before LTP/LTD induction with the averaged traces 2 min after induction; anywhere in the time frame of ~15–60 min after). Due to the long-term recording and intrinsic variability of synaptic transmission, a relatively high variability is often present across recorded cells. Many experienced electrophysiologists often use 10 % change as a cutoff to determine whether the observed phenotype is worth pursuing.

Often, the in vitro test for plasticity may follow an in vivo behavioral manipulation. For example, one animal is given a saline injection while another is injected with an addictive drug. Tissue slices are prepared and LTP is elicited using HFS. The results may show the presence of LTP in saline-treated animals, but not in the drug-treated animals. The absence of LTP in drug-treated animals can be interpreted in two ways: either the induction of plasticity is disabled or blocked (cannot be induced by HFS) or it is occluded by prior in vivo experience (LTP has already occurred in vivo, and due to the ceiling effect, further potentiation is occluded) in the drug-treated animals. To determine whether the plasticity is blocked or occluded, knowing where the LTP is expressed (e.g., presynaptic or postsynaptic) is helpful, using the approaches discussed in the "Identifying mechanisms of LTP and LTD" section. For example, if following LTP induction the PPR is changed, the PPR of the drug-treated animal can be compared to the PPR before and after LTP induction. If the PPR of the drug-treated animal is similar to the PPR after LTP induction, occlusion may have already occurred. If, however, the PPR in the drug-treated animal is similar to the PPR before LTP induction, the induction mechanisms of LTP may have been disrupted. The scenario proposed in this example can be applied to any of the above approaches used for identifying presynaptic or postsynaptic mechanisms (e.g., quantal analysis of amplitudes/frequencies, V–M analysis of *N*, *Q*, and Pr).

When analyzing LTP data, the issue of associativity should also be included in the interpretation. Associativity refers to LTP that occurs at a strongly activated pathway and a neighboring weakly activated pathway on the same cell. Normally, the weakly activated pathway would not undergo LTP. However, the conjoined activity of both pathways leads to LTP at both sets of synapses. Associativity should always be considered in the interpretation of optogenetic experiments whereby specific pathways are activated. For example, the absence of LTP from pathway x may not necessarily mean that LTP does not exist. Rather, pathway x may need conjoined activity from pathway y in order for LTP to occur. The beginning electrophysiologist needs to be aware of the complicated nature of synaptic communication in order to interpret the data wisely.

2.6 Technical Considerations

Long-term whole-cell recordings require healthy slices, well-made solutions/micropipettes, and expert patching techniques to achieve reliable measurements over the 1–2 h periods (see Chaps. 4 and 5 for a description of solution and micropipette preparation, respectively). In addition, the access resistance should be monitored throughout the recording as changes in the access resistance can lead to increases/decreases in the current amplitudes. Before LTP/LTD induction, stable baselines lasting 10–20 min are preferred as run-up or run-down of currents is common in long-term recordings of synaptic transmission (see Chap. 16). A stable baseline refers to stability in the variance of the peak amplitude (Fig. 3). The peak amplitude during the baseline period should be roughly 200 pA. If the current is too large, space clamp artifacts may be introduced into the recording (see Chap. 6 for space clamp description). If the current is too small during the baseline period, LTD may go undetected. By using baselines with similar amplitudes, consistency can be achieved from the beginning to the completion of a research project.

Filtering electrical signals during long-term measurements are Bessel filters, which prevent aliasing. Bandpass filters (low-pass (1 Hz) and high pass (5 kHz)) are also used (see Chaps. 3 and 10 for descriptions of aliasing as well as filters used for electrical signals).

3 Summary

Long-term measurements of synaptic currents can last 1–2 h and are commonly used to study long-term plasticity, such as LTP and LTD. Whereas LTP refers to synaptic potentiation, LTD is the depression of synaptic transmission. Both LTP and LTD can be mediated by presynaptic or postsynaptic modifications, with NMDAR-dependent or independent mechanisms.

There are varying forms of LTP or LTD in in vitro slice preparations including classic high/low frequency-evoked, chemical, spike-timing dependent, and others. Once plasticity has been detected, identifying presynaptic or postsynaptic modifications can be performed using measurements such as PPR, variance-mean analysis, quantal analysis, or measurements of AMPAR/NMDAR ratios.

Interpreting plasticity measurements include identifying block or occlusion of LTP/LTD as well as whether the plasticity requires associativity.

Lastly, by using caution with all these technical issues, long-term measurements can be made in a reliable manner, providing useful data to understand the ever adapting and plastic CNS.

References

1. Foeller E, Feldman DE (2004) Synaptic basis for developmental plasticity in somatosensory cortex. Curr Opin Neurobiol 14(1):89–95
2. Malenka RC, Bear MF (2004) LTP and LTD: an embarrassment of riches. Neuron 44(1):5–21
3. Bliss TV, Lomo T (1973) Long-lasting potentiation of synaptic transmission in the dentate area of the anaesthetized rabbit following stimulation of the perforant path. J Physiol 232(2):331–356
4. Benke TA, Luthi A, Isaac JT, Collingridge GL (1998) Modulation of AMPA receptor unitary conductance by synaptic activity. Nature 393(6687):793–797

5. Bredt DS, Nicoll RA (2003) AMPA receptor trafficking at excitatory synapses. Neuron 40(2):361–379
6. Kauer JA, Malenka RC, Nicoll RA (1988) NMDA application potentiates synaptic transmission in the hippocampus. Nature 334(6179):250–252
7. Kauer JA, Malenka RC, Nicoll RA (1988) A persistent postsynaptic modification mediates long-term potentiation in the hippocampus. Neuron 1(10):911–917
8. Lee HK, Barbarosie M, Kameyama K, Bear MF, Huganir RL (2000) Regulation of distinct AMPA receptor phosphorylation sites during bidirectional synaptic plasticity. Nature 405(6789):955–959
9. Malenka RC, Nicoll RA (1999) Long-term potentiation--a decade of progress? Science 285(5435):1870–1874
10. Malinow R, Malenka RC (2002) AMPA receptor trafficking and synaptic plasticity. Annu Rev Neurosci 25:103–126
11. Shi SH, Hayashi Y, Petralia RS, Zaman SH, Wenthold RJ, Svoboda K, Malinow R (1999) Rapid spine delivery and redistribution of AMPA receptors after synaptic NMDA receptor activation. Science 284(5421):1811–1816
12. Soderling TR, Derkach VA (2000) Postsynaptic protein phosphorylation and LTP. Trends Neurosci 23(2):75–80
13. Song I, Huganir RL (2002) Regulation of AMPA receptors during synaptic plasticity. Trends Neurosci 25(11):578–588
14. Abraham WC, Williams JM (2003) Properties and mechanisms of LTP maintenance. Neuroscientist 9(6):463–474
15. Lynch MA (2004) Long-term potentiation and memory. Physiol Rev 84(1):87–136
16. Pittenger C, Kandel ER (2003) In search of general mechanisms for long-lasting plasticity: aplysia and the hippocampus. Philos Trans R Soc Lond B Biol Sci 358(1432):757–763
17. Breustedt J, Vogt KE, Miller RJ, Nicoll RA, Schmitz D (2003) Alpha1E-containing Ca2+ channels are involved in synaptic plasticity. Proc Natl Acad Sci U S A 100(21):12450–12455
18. Castillo PE, Weisskopf MG, Nicoll RA (1994) The role of Ca2+ channels in hippocampal mossy fiber synaptic transmission and long-term potentiation. Neuron 12(2):261–269
19. Dietrich D, Kirschstein T, Kukley M, Pereverzev A, von der Brelie C, Schneider T, Beck H (2003) Functional specialization of presynaptic Cav2.3 Ca2+ channels. Neuron 39(3):483–496
20. Nicoll RA, Schmitz D (2005) Synaptic plasticity at hippocampal mossy fibre synapses. Nat Rev Neurosci 6(11):863–876
21. Huang YY, Kandel ER, Varshavsky L, Brandon EP, Qi M, Idzerda RL, McKnight GS, Bourtchouladze R (1995) A genetic test of the effects of mutations in PKA on mossy fiber LTP and its relation to spatial and contextual learning. Cell 83(7):1211–1222
22. Huang YY, Li XC, Kandel ER (1994) cAMP contributes to mossy fiber LTP by initiating both a covalently mediated early phase and macromolecular synthesis-dependent late phase. Cell 79(1):69–79
23. Weisskopf MG, Castillo PE, Zalutsky RA, Nicoll RA (1994) Mediation of hippocampal mossy fiber long-term potentiation by cyclic AMP. Science 265(5180):1878–1882
24. Castillo PE, Janz R, Sudhof TC, Tzounopoulos T, Malenka RC, Nicoll RA (1997) Rab3A is essential for mossy fibre long-term potentiation in the hippocampus. Nature 388(6642):590–593
25. Castillo PE, Schoch S, Schmitz F, Sudhof TC, Malenka RC (2002) RIM1alpha is required for presynaptic long-term potentiation. Nature 415(6869):327–330
26. Wang Y, Okamoto M, Schmitz F, Hofmann K, Sudhof TC (1997) Rim is a putative Rab3 effector in regulating synaptic-vesicle fusion. Nature 388(6642):593–598
27. Graziane NM, Polter AM, Briand LA, Pierce RC, Kauer JA (2013) Kappa opioid receptors regulate stress-induced cocaine seeking and synaptic plasticity. Neuron 77(5):942–954
28. Nugent FS, Penick EC, Kauer JA (2007) Opioids block long-term potentiation of inhibitory synapses. Nature 446(7139):1086–1090
29. Nugent FS, Niehaus JL, Kauer JA (2009) PKG and PKA signaling in LTP at GABAergic synapses. Neuropsychopharmacology 34(7): 1829–1842
30. Kauer JA, Malenka RC (2007) Synaptic plasticity and addiction. Nat Rev Neurosci 8(11):844–858
31. Carroll RC, Beattie EC, von Zastrow M, Malenka RC (2001) Role of AMPA receptor endocytosis in synaptic plasticity. Nat Rev Neurosci 2(5):315–324
32. Chevaleyre V, Heifets BD, Kaeser PS, Sudhof TC, Castillo PE (2007) Endocannabinoid-mediated long-term plasticity requires cAMP/PKA signaling and RIM1alpha. Neuron 54(5):801–812
33. Chevaleyre V, Takahashi KA, Castillo PE (2006) Endocannabinoid-mediated synaptic plasticity in the CNS. Annu Rev Neurosci 29:37–76
34. Lin JY (2011) A user's guide to channelrhodopsin variants: features, limitations and future developments. Exp Physiol 96(1):19–25

35. Lin JY, Lin MZ, Steinbach P, Tsien RY (2009) Characterization of engineered channelrhodopsin variants with improved properties and kinetics. Biophys J 96(5):1803–1814
36. Lu W, Man H, Ju W, Trimble WS, MacDonald JF, Wang YT (2001) Activation of synaptic NMDA receptors induces membrane insertion of new AMPA receptors and LTP in cultured hippocampal neurons. Neuron 29(1):243–254
37. Otmakhov N, Khibnik L, Otmakhova N, Carpenter S, Riahi S, Asrican B, Lisman J (2004) Forskolin-induced LTP in the CA1 hippocampal region is NMDA receptor dependent. J Neurophysiol 91:1955
38. Stewart MG, Medvedev NI, Popov VI, Schoepfer R, Davies HA, Murphy K, Dallerac GM, Kraev IV, Rodriguez JJ (2005) Chemically induced long-term potentiation increases the number of perforated and complex postsynaptic densities but does not alter dendritic spine volume in CA1 of adult mouse hippocampal slices. Eur J Neurosci 21(12):3368–3378
39. Bi GQ, Poo MM (1998) Synaptic modifications in cultured hippocampal neurons: dependence on spike timing, synaptic strength, and postsynaptic cell type. J Neurosci 18(24): 10464–10472
40. Gustafsson B, Wigstrom H, Abraham WC, Huang YY (1987) Long-term potentiation in the hippocampus using depolarizing current pulses as the conditioning stimulus to single volley synaptic potentials. J Neurosci 7(3): 774–780
41. Haas JS, Nowotny T, Abarbanel HD (2006) Spike-timing-dependent plasticity of inhibitory synapses in the entorhinal cortex. J Neurophysiol 96(6):3305–3313
42. Levy WB, Steward O (1983) Temporal contiguity requirements for long-term associative potentiation/depression in the hippocampus. Neuroscience 8(4):791–797
43. Markram H, Lubke J, Frotscher M, Sakmann B (1997) Regulation of synaptic efficacy by coincidence of postsynaptic APs and EPSPs. Science 275(5297):213–215
44. McNaughton BL, Douglas RM, Goddard GV (1978) Synaptic enhancement in fascia dentata: cooperativity among coactive afferents. Brain Res 157(2):277–293
45. Bell CC, Han VZ, Sugawara Y, Grant K (1997) Synaptic plasticity in a cerebellum-like structure depends on temporal order. Nature 387(6630):278–281
46. Chevaleyre V, Castillo PE (2004) Endocannabinoid-mediated metaplasticity in the hippocampus. Neuron 43(6):871–881
47. Gerdeman GL, Ronesi J, Lovinger DM (2002) Postsynaptic endocannabinoid release is critical to long-term depression in the striatum. Nat Neurosci 5(5):446–451
48. Kemp N, McQueen J, Faulkes S, Bashir ZI (2000) Different forms of LTD in the CA1 region of the hippocampus: role of age and stimulus protocol. Eur J Neurosci 12(1): 360–366
49. Lee HK, Takamiya K, Han JS, Man H, Kim CH, Rumbaugh G, Yu S, Ding L, He C, Petralia RS, Wenthold RJ, Gallagher M, Huganir RL (2003) Phosphorylation of the AMPA receptor GluR1 subunit is required for synaptic plasticity and retention of spatial memory. Cell 112(5):631–643
50. Thiels E, Xie X, Yeckel MF, Barrionuevo G, Berger TW (1996) NMDA receptor-dependent LTD in different subfields of hippocampus in vivo and in vitro. Hippocampus 6(1):43–51
51. Lee HK, Kameyama K, Huganir RL, Bear MF (1998) NMDA induces long-term synaptic depression and dephosphorylation of the GluR1 subunit of AMPA receptors in hippocampus. Neuron 21(5):1151–1162
52. Morishita W, Connor JH, Xia H, Quinlan EM, Shenolikar S, Malenka RC (2001) Regulation of synaptic strength by protein phosphatase 1. Neuron 32(6):1133–1148
53. Feldman DE (2012) The spike-timing dependence of plasticity. Neuron 75(4):556–571
54. Clements JD, Silver RA (2000) Unveiling synaptic plasticity: a new graphical and analytical approach. Trends Neurosci 23(3):105–113
55. Silver RA (2003) Estimation of nonuniform quantal parameters with multiple-probability fluctuation analysis: theory, application and limitations. J Neurosci Methods 130(2):127–141

Chapter 13

Measuring Reversal Potentials

Nicholas Graziane and Yan Dong

Abstract

The reversal potential refers to the membrane potential at which the current changes its flowing direction. For electrophysiological measurements, the change in direction can either mean inward currents becoming outward currents at a specific voltage or vice versa. Using reversal potential measurements allows the experimenter to determine ion specificity of a receptor/channel as well as the driving force of ions through the receptor/channel at a given voltage. For example, if the reversal potential measured for a given channel is close the reversal potential for K^+, it is likely that the receptor/channel is predominantly permeable to K^+. This chapter discusses the approaches that can be used to measure the reversal potentials of the isolated currents. In addition, we discuss technical considerations that can enhance the accuracy of the measurements.

Key words Current–voltage plot, Voltage step, Voltage ramp, Conductance

1 Introduction

The reversal potential refers to the membrane potential at which the current changes its flowing direction. For electrophysiological measurements, the change in direction can either mean inward currents becoming outward currents at a specific voltage or vice versa. Using reversal potential measurements allows the experimenter to determine ion specificity of a receptor/channel as well as the driving force of ions through the receptor/channel at a given voltage. For example, if the reversal potential measured for a given channel is close the reversal potential for K^+, it is likely that the receptor/channel is predominantly permeable to K^+. This chapter discusses the approaches that can be used to measure the reversal potentials of the isolated currents. In addition, we discuss technical considerations that can enhance the accuracy of the measurements.

Nicholas Graziane and Yan Dong, *Electrophysiological Analysis of Synaptic Transmission*, Neuromethods, vol. 112,
DOI 10.1007/978-1-4939-3274-0_13, © Springer Science+Business Media New York 2016

2 Measuring Reversal Potentials

Measuring the reversal potential for a channel that is selective for only one type of ion can be easily estimated using the Nernst equation discussed in Chap. 2. Briefly, the Nernst equation calculates the voltage at which there is no net movement of ions (i.e., no net current flowing in and out of the cell). The lack of ionic movement results from the ionic gradient reaching the equilibrium. Thus, the equilibrium potential is equal to the reversal potential. Therefore, as voltages become more positive than the equilibrium potential, the currents reverse direction (e.g., inward versus outward) relative to more negative voltages. The Nernst equation is a useful and fast way to determine the reversal potential of a given channel. However, it is often the case that the receptor/channel is permeable to more than one type of ion. If so, the Goldman–Hodgkin–Katz equation can be used to calculate the equilibrium/reversal potential by taking into account multiple ionic species (see Chap. 2 for the equation). Knowing which ions pass through a receptor/channel enables convenient calculations using the Nernst or Goldman–Hodgkin–Katz equations. The problem arises when the receptor/channel permeability is unknown. When this is the case, measuring the reversal potential can be accomplished by using a current–voltage plot (*I–V* plot) experimentally.

Identification of reversal potentials for both ionotropic receptors and voltage-gated channels is achieved using an *I–V* plot. For voltage-gated channels, *I–V* plots and therefore reversal potentials are measured using whole-cell patch configurations while voltage clamping the cell (the internal solution should be devoid of voltage-gated channel blockers such as Cs^{+} and QX314). Voltage steps with increments of 10 mV from −70 to +50 mV are applied and the peak amplitude of the current generated is measured. The peak amplitude of the current generated is then graphed versus voltage and the reversal potential is identified at the *x*-axis intercept (i.e., the point on the graph where the current of the activated channel equals 0 pA) (Fig. 1a). For identifying the reversal potentials of synaptic currents, a fixed presynaptic stimulation is applied while voltage clamping the postsynaptic cell in the whole-cell mode. The holding voltage is then increased by increments of 10 mV from −70 to +50 mV and at each new holding potential, synaptic currents are evoked and measured. The average current amplitude at each holding potential is then graphed as current versus voltage and the reversal potential can be identified (Fig. 1a). This approach is simple enough, but what if the measured currents do not have a defined peak at each voltage increment? If this is the case, the current amplitude can be calculated by measuring the steady-state values recorded at the end of the voltage step (Fig. 1b).

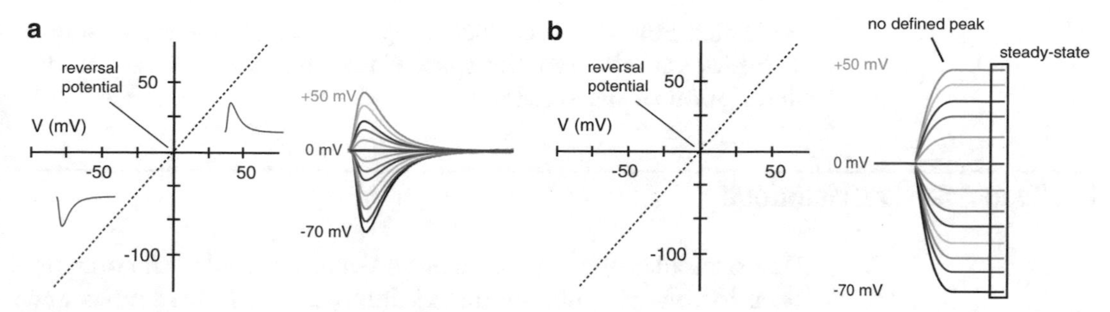

Fig. 1 Calculation of the reversal potential using an *I–V* plot. (**a**) In whole-cell, voltage-clamp configuration, a neuron is held at different holding potentials from −70 to +50 mV and excitatory currents are elicited. As the reversal potential is approached, the currents become smaller as the driving force is reduced until the equilibrium potential is reached at 0 mV causing no ionic flux across the postsynaptic membrane. When a defined current peak is absent (**b**), the peak amplitude of the current is measured at the steady-state (where the current has plateaued)

Another approach to measuring reversal potentials is to create an *I–V* curve using a voltage ramp. This approach is useful in calculating the reversal potential of transmitter responses assuming that the currents do not inactivate during the voltage ramp. For example, to measure the reversal potential of GABA-induced currents, a 200 mV and 400 ms voltage ramp can be applied in the absence and presence of GABA. By subtracting the current responses (before and after GABA application), the GABA-induced currents can be isolated and an *I–V* plot can be created at each voltage [1]. If the receptor is prone to desensitization/inactivation, the reversal potentials can still be roughly calculated via extrapolation. To do this, the cell can be maintained at −80 mV and a single +80 mV step for 50 ms is applies in the absence and presence of GABA. Peak current amplitudes at −80 and 0 mV are then subtracted and plotted versus the voltage. A line can then be drawn to fit the two data points providing a rough (i.e., not true) estimate of the reversal potential [1]. As implied, above, a linear *I–V* curve is assumed in this extrapolation strategy.

In addition to indicating the ionic specificity of a receptor/channel, measuring the reversal potential can also be used to calculate the conductance of a receptor/channel at a given voltage. This is done by simply dividing the current at each voltage by the driving force, which is the voltage minus the reversal potential ($V - V_{rev}$).[1] Furthermore, reversal potential calculations can also signify how well the cell is voltage clamped. Poorly clamped cells exhibit inconsistent reversal potentials over trials or possess reversal potentials that are inconsistent with the known equilibrium

[1] The idea that conductance can be calculated by dividing current by voltage should not be too much of a surprise. If we refer to Ohm's law ($V = IR$), the resistance is the reciprocal of conductance making the equation $V = I(1/C)$, where C is the conductance. Solving for C gives, $C = I/V$.

potential. Furthermore, calculating the reversal potential using an *I–V* plot can also test the space-clamp errors (see Chap. 6 for a description of space clamp).

3 Technical Considerations

When calculating the reversal potential using whole-cell configurations in voltage-clamp mode, it is important to keep in mind some technical considerations. First, the recording electrode needs to be coated with chloride, thus preventing current drift. With a properly coated electrode, the pipette offset can be accurately calculated, thus enabling true measures of currents and potentials. Another issue to be aware of is the liquid-junction potentials that form at the interface between the internal and external solutions. These liquid-junction potentials should be corrected in order to obtain an accurate offset of voltage between internal and external solutions (see Chap. 3 for junction potential corrections).

Sometimes it is difficult to measure the reversal potential because the currents are too small to be detected, which can be the case when the currents are dominated by background conductance or leak conductance (see Na^+ channel *I–V* plot). An elegant approach has been developed to overcome this problem by calculating the deactivation currents. For example, when a voltage step is terminated, tail currents are formed (Fig. 2 including voltage steps see Pusch and Sontheimer). Tail currents are currents that result from the deactivation of the channel and form as current continues to flow through the closing channel. The peak

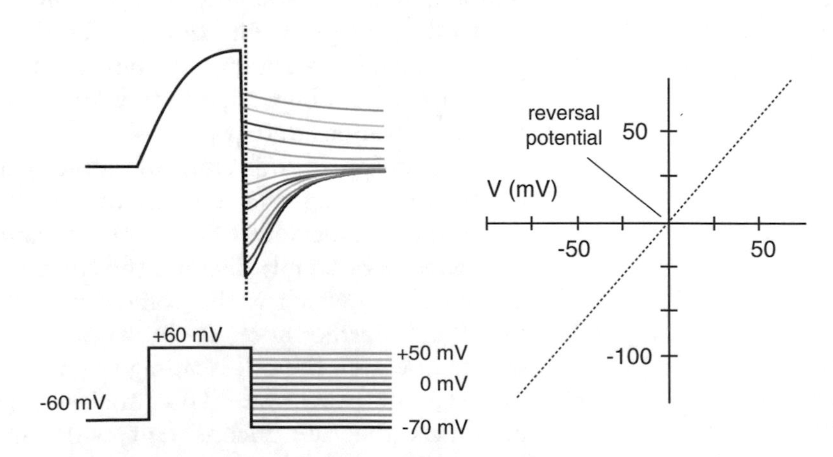

Fig. 2 Calculating reversal potentials via tail currents. In whole-cell, voltage-clamp configuration, a depolarizing pulse is applied to a neuron followed by hyperpolarizing or depolarizing pulses ranging from −70 to +50 mV resulting in tail currents. The current amplitude of the tail currents is measured at the *dotted line* illustrated in the current traces. The result is a linear *I–V* plot (*right*)

amplitudes of the tail currents can then be measured and plotted versus voltage giving a linear *I–V* curve (Fig. 2—see if there is one for sodium).

We mentioned above that measuring known reversal potentials using *I–V* relationships can indicate whether space-clamp errors are being introduced into the recording. However, we have not mentioned that space clamping can grossly distort the reversal potential measurement of unknown receptor/channel. Therefore, to avoid space clamping errors, the following approaches can be used [2]:

1. Avoid recording from cells with long processes. However, if this is unavoidable, cutting the processes to shorter lengths can reduce the impact of the problem.
2. If possible, close off attached axons.
3. Use the dendritic patch procedure (see Chap. 20) thereby improving the voltage clamp of nearby dendritic spines.

Lastly, it is important to remember that pharmacological blockers are important tools that can be used to isolate the currents of interest (see Chap. 4 for description of blockers that can be used in the external and internal solutions). For example, evoking currents in order to measure the reversal potential of an excitatory ionotropic receptor requires bath-applied blockers that prevent inhibitory ionotropic receptors (e.g., picrotoxin). This prevents the current measurements from being contaminated by inhibitory components. In addition, depending upon the level of depolarization, there is a possibility that voltage-gated channels could open contaminating the current of interest. To prevent contamination of voltage-gated channels, Na^+ and K^+ channel blockers (QX314 and Cs^+, respectively) are included in the intracellular solution and measurements should only begin once the internal solution has had time to perfuse the cell (e.g., after 15 min).

4 Summary

Calculating reversal potentials is a useful tool that can help identify receptor/channel's ion selectivity and the driving force of ions through the receptor/channel. Measuring the reversal potential can be performed using the Nernst equation for single ion permeabilities, the Goldman-Hodgkin-Katz equation for known ionic permeabilities of more than one ion, or an *I–V* plot can be constructed.

In order to calculate reversal potentials accurately, it is important to obtain sound measurements. By ensuring that the pipette offset is accurately adjusted and that space-clamp artefacts are not being introduced, the reversal potentials can be confidently calculated.

References

1. Walz W, Boulton AA, Baker GB (2002) Patch-clamp analysis: advanced techniques. Humana Press, Totowa, NJ
2. Axon Instruments (1993) The Axon guide for electrophysiology & biophysics laboratory techniques. Axon Instruments, Foster City, CA

Part III

Basic Experimentations of Synaptic Transmission

Chapter 14

Amplitude

Nicholas Graziane and Yan Dong

Abstract

Amplitudes in in vitro electrophysiological measurements refer to the value of currents or potentials recorded from one cell or a population of cells at a given time point. This chapter discusses amplitude measurements for in vitro slice electrophysiology and covers appropriate calculations for action potential amplitudes, field potential amplitudes, postsynaptic current/potential amplitudes, and steady-state current amplitudes. Like many electrophysiological measurements, technical considerations for precise calculations are important. These technical considerations are addressed at the end of this chapter.

Key words Action potentials, Local field potentials, Synaptic currents/potentials

1 Introduction

Amplitudes in in vitro electrophysiological measurements refer to the value of currents or potentials recorded from one cell or a population of cells at a given time point. This chapter discusses amplitude measurements for in vitro slice electrophysiology and covers appropriate calculations for action potential amplitudes, field potential amplitudes, postsynaptic current/potential amplitudes, and steady-state current amplitudes. Like many electrophysiological measurements, technical considerations for precise calculations are important. These technical considerations are addressed at the end of this chapter.

2 Amplitude Measurements

2.1 Action Potentials

The amplitude of an action potential relies entirely on the properties of the cellular membrane regardless of the input strength from excitatory presynaptic terminals. This is because an action potential is an "all-or-none" response elicited at the time that the potential threshold of the membrane is reached (Fig. 1). Once the potential threshold is reached, sodium channels open, causing the cell to

Nicholas Graziane and Yan Dong, *Electrophysiological Analysis of Synaptic Transmission*, Neuromethods, vol. 112, DOI 10.1007/978-1-4939-3274-0_14, © Springer Science+Business Media New York 2016

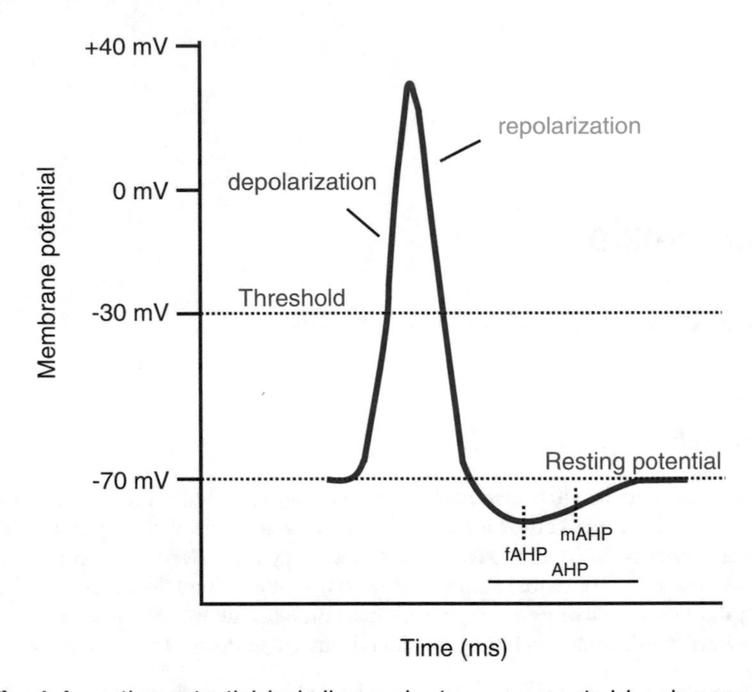

Fig. 1 An action potential including each stage represented by changes in membrane potential. An action potential is elicited when a depolarizing potential reaches the membrane's threshold (the minimum potential necessary to trigger an action potential). The depolarization phase of an action potential is triggered by the activation of voltage-gated Na^+ channels. The membrane repolarizes as voltage-gated Na^+ channels close and voltage-gated K^+ and Ca^{2+} channels are activated (the driving force for Ca^{2+} increases during the falling phase contributing to Ca^{2+}-activated K^+ channel currents). The afterhyperpolarization (AHP) is caused by voltage-gated K^+ channels, which remain open after the resting membrane potential is reached. Two K^+ channels contributing to the AHP are BK channels (big K^+ current) and SK channels (small K^+ current). BK channels generate hyperpolarizing currents occurring ~3 ms after the onset of the action potential (fAHP), while SK channels generate hyperpolarizing currents 10 ms after the onset of the action potential (mAHP)

depolarize until potassium channels open to bring the membrane back to its resting membrane potential. So why is it important to measure an action potential's peak amplitude if it never changes within a cell? The answer is because action potentials can be different between cells. The gating properties of K^+ channels in certain cell types may be faster than other cell types, narrowing action potentials or shortening refractory periods [1–8]. In addition, calcium currents, which typically make a larger contribution during the falling phase (during K^+ channel activation) rather than the rising phase, have a profound effect on the broadening of action potentials [9–13]. Because of differences in channel properties or channel expression between cell types, it is often necessary to measure action potential characteristics such as action potential width. To measure the action potential width the amplitude of the action

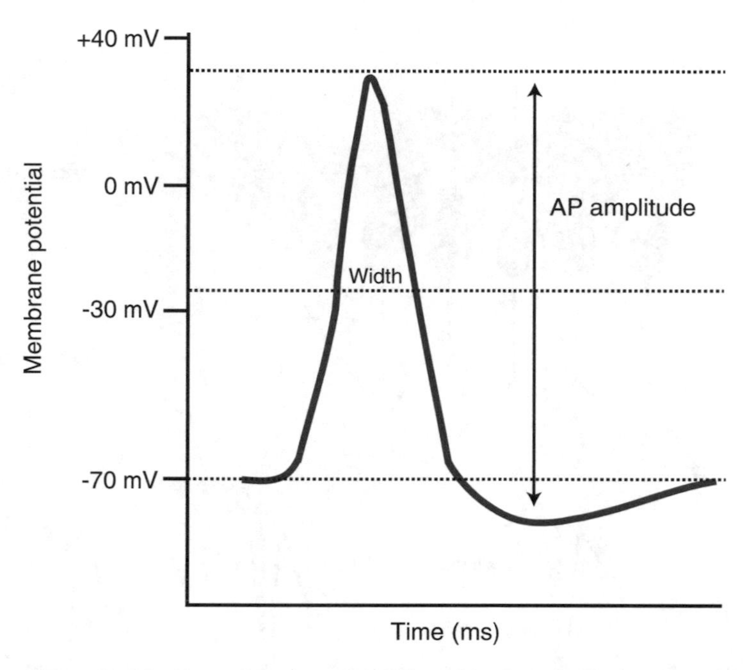

Fig. 2 To calculate the action potential (AP) width, the amplitude of the AP is calculated by measuring the difference in membrane potential from the peak to the most negative voltage reached after the spike

potential is often calculated to determine the width at half-maximal spike amplitude [2]. To measure the action potential spike amplitude, calculate the difference in membrane potential from the peak to the most negative voltage reached after the spike (Fig. 2).

2.2 Local Field Potentials (LFPs)

Local field potentials measure potential differences generated by a population of neurons and thus are a useful readout of the activity of multiple neurons in a particular brain region (see Chap. 2). In recording from a population of neurons, one can measure population spikes, population excitatory postsynaptic potentials (EPSPs), or population inhibitory postsynaptic potentials (IPSPs).

Population spikes (Fig. 3b) are summed synchronous action potentials from a pool of neurons. To measure population spikes, the recording electrode is positioned in a region populated with neuronal somas (e.g., CA1 stratum pyramidale layer of the hippocampus) (Fig. 3a). To measure the population spike amplitude, two methods can be used: the first approach uses the formula $[(x+y)/2]-z$, where x, y, and z are shown in Fig. 3b. The second approach is the tangent method, which calculates the population spike amplitude by taking the voltage difference between a tangent to the two positive peaks and the trough of the negative going wave (Fig. 3c). The tangent method can be subdivided into three varying approaches based on where the tangent line forms between the two positive peaks and the negative going trough. One, the

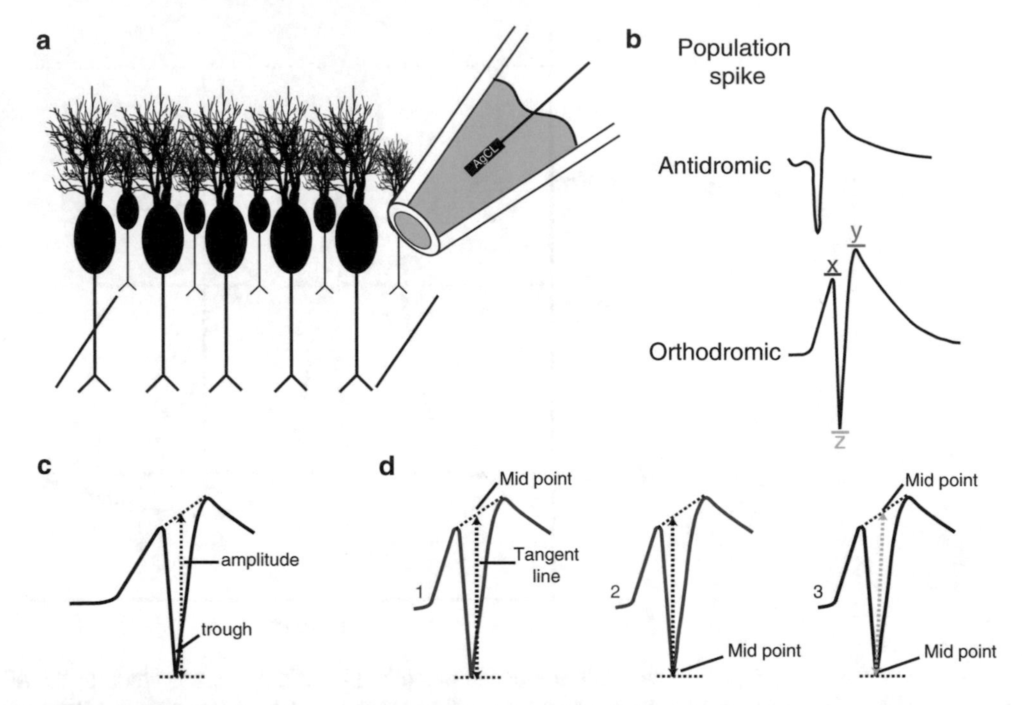

Fig. 3 Measuring amplitudes of population spikes. (**a**) An extracellular recording taken from an area populated by neuronal somas. (**b**) A simulated antidromic and orthodromic population spike trace recorded from the CA1 region of the hippocampus. The peak amplitude at points *x*, *y*, and *z* are used to calculate the population spike amplitude using the equation $[(x+y)/2]-z$. (**c**) An illustration of the tangent approach used to calculate the amplitude of a population spike. (**d**) The three version of the tangent method that are used to calculate the population spike amplitude. One, the tangent line is drawn from the middle of the line joining the two positive peaks straight down to the voltage where the negative going peak terminates (*left*). Two, the tangent line is drawn such that the line intersects the middle of the negative going peak and extends straight up to the line formed between the two positive peaks (*middle*). Three, the tangent line is drawn from the middle of the negative going peak to the middle of the line drawn between the two positive peaks (*right*)

tangent line is drawn from the middle of the line joining the two positive peaks straight down to the voltage where the negative going peak terminates (Fig. 3d). Two, the tangent line is drawn such that the line intersects the middle of the negative going peak and extends straight up to the line formed between the two positive peaks (Fig. 3d). Three, the tangent line is drawn from the middle of the negative going peak to the middle of the line drawn between the two positive peaks (Fig. 3d). What measurement approach works best? All of the above mentioned approaches are accepted as credible approaches for measuring population spike amplitudes. For the beginning electrophysiologist, it is advised to experiment a few cells with all three approaches and evaluate which one better reflects the data.

Population EPSPs and IPSPs are a measurement of excitatory and inhibitory synaptic activity, respectively, and can be measured in brain regions that contain predominantly dendrites (Fig. 4a). To generate EPSPs or IPSPs, an evoked stimulus is used to elicit a presynaptic action potential and subsequent neurotransmitter

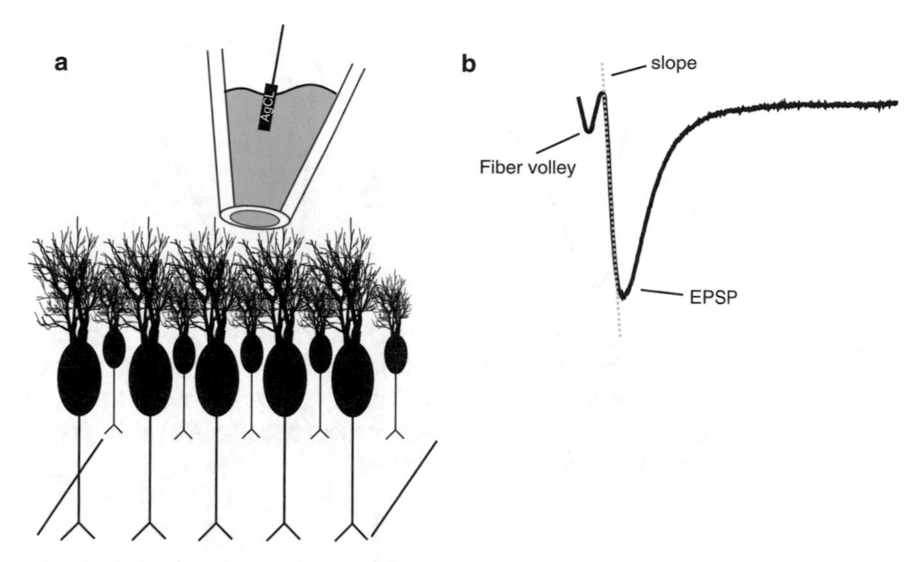

Fig. 4 Measuring amplitudes of population EPSPs. (**a**) Extracellular recording taken from a region populated by dendrites, thus currents generated are predominantly triggered by synaptic connections. (**b**) An example EPSP trace illustrating the fiber volley followed by the EPSP. The amplitude of the EPSP is calculated from the initial slope of the negative potential

release. The presynaptic action potential is detected in the recording as a fiber volley (this is a non-synaptic population spike) (Fig. 4b). A fiber volley is a negative, extracellularly recorded wave that is detected when an action potential in a population of axons passes the recording electrode [14]. The amplitude is proportional to the number of active presynaptic fibers, thus can be used to estimate the afferent input's strength. Changes in the fiber volley will result in changes in the amplitudes of EPSP or IPSP. If the investigation looks to identify changes in EPSP or IPSP amplitudes or slopes, it is essential that the fiber volley amplitude does not change during the recording.

To isolate and measure changes in EPSP amplitude or slopes, measurements are taken in the presence of an inhibitory receptor blocker (e.g., picrotoxin). To quantify EPSP, the initial slope of the negative potential is often used (Fig. 4b). This is in contrast to whole-cell recordings (discussed later) where the peak amplitudes is often used to measure potential strength. In EPSP *field recordings*, the initial slope is measured instead of the peak amplitude because the peak amplitude is often contaminated with population spikes (backpropagating action potentials in the dendrites), population IPSPs (if no inhibitory receptor blockers are used), and polysynaptic events[1] [15].

[1] Polysynaptic refers to two or more synaptic events that occur at different times following an evoked stimulus. Polysynaptic events can be mediated by chains of connected neurons whereby the initial synaptic event measured (PSP/PSC) is associated with the most direct pathway to the recorded neuron, while the synaptic events that occur at later time points are associated with indirect pathways (Fig. 5).

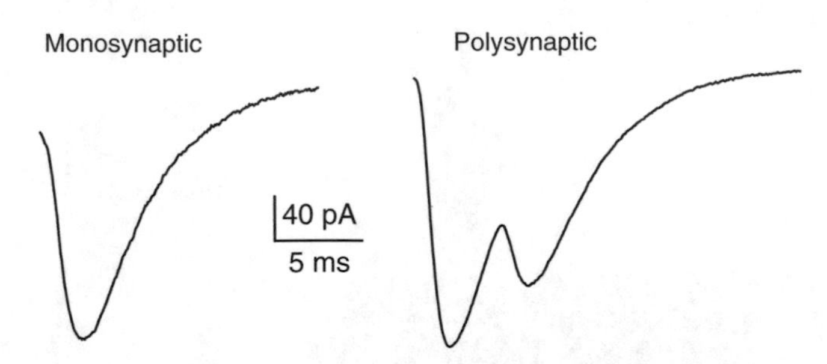

Fig. 5 A comparison between two EPSC traces triggered by electrical stimulation. (*Left*) A characteristic EPSC whereby the currents generated are produced from the simultaneous activation of postsynaptic receptors. (*Right*) A polysynaptic EPSC trace resulting from two or more synaptic events occurring at different times following an evoked stimulus

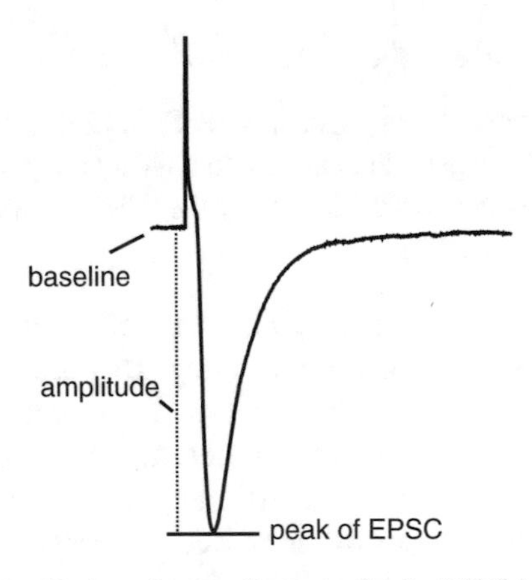

Fig. 6 Measuring amplitudes of synaptic currents. An EPSC trace illustrating that the amplitude of the current is measured from the baseline to the peak EPSC

To isolate and measure IPSP amplitude or slopes, measurements are taken in the presence of excitatory receptor blockers (e.g., NBQX and AP5 to block AMPARs and NMDARs, respectively). To quantify IPSPs, the initial slope of the outward current is calculated.

2.3 Current/Potential Amplitude (Intracellular Recordings)

Measuring the amplitudes of currents/potentials in intracellular recording configurations gives useful information about changes in synaptic strength. This is why measuring the size of currents/potentials following an experimental manipulation is common in electrophysiology. Measuring the amplitudes of currents or potentials can be performed by normalizing the peak amplitudes (positive or negative going) relative to their baseline values (Fig. 6). This analysis has been hinted at throughout this book. In measuring LTP and LTD (Chap. 12), measuring reversal potentials (Chap. 13), and run-up and run-down of synaptic currents (Chap. 16), the amplitudes of the currents or potentials are commonly used for intra- and inter-sample comparisons.

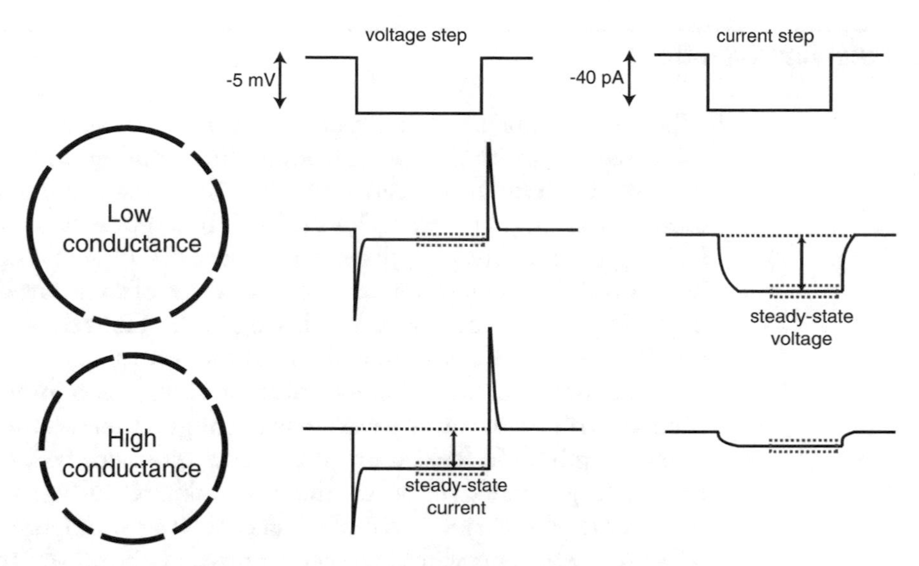

Fig. 7 Calculating the input resistance. The input resistance for two neurons is illustrated. One neuron has a low membrane conductance (high membrane resistance) (*top*), while the second neuron has a high membrane conductance (low membrane resistance) (*bottom*). Passing a hyperpolarizing voltage step in voltage-clamp mode produces a relatively smaller steady-state current (*dashed red box*) compared to the neuron with the higher membrane conductance. Passing a hyperpolarizing current step in current-clamp mode generates a relatively large membrane potential drop compared to the neuron with the higher membrane conductance

2.4 Input Resistance

The input resistance is monitored during intracellular recordings. It describes the conductance or resistance of the membrane upon current injection or membrane depolarization, mediated by ionic channels on cell surface. A low input resistance suggests that more channels are activated and the membrane conductance is high. Conversely, a low input resistance suggests that fewer channels are in the active state and the conductance level is low (Fig. 7). Measuring the input resistance is useful for determining what effects a compound has on receptors/channels expressed on the cell's surface. In order to calculate the input resistance, the amplitude of the voltage or current (depending on whether the cell is voltage- or current-clamped) needs to be determined. To calculate the input resistance using the voltage-clamp mode, a hyperpolarizing step (−0.5 to −5 mV) can be introduced through the recording micropipette to generate a current (Fig. 7). The amplitude of this current is calculated from the baseline to the steady-state amplitude (Fig. 7). Once the current is calculated, Ohm's law (voltage (V) = current (I) × resistance (R)) can be used to solve for the input resistance.

In the current-clamp mode, the input resistance is calculated by injecting a hyperpolarizing current into the cell and measuring the steady-state of the potentials that are generated (Fig. 7). Then Ohm's law can be used to calculate the input resistance ($V = IR$).

3 Technical Considerations

When measuring the amplitudes of currents or potentials, it is important to maintain constant ionic fluxes during a recording/experiment. Ionic fluxes can vary when there is a change in ionic composition (e.g., external/internal solutions) or a change in the holding potential (e.g., voltage clamping a cell in whole-cell configuration). Modifying the ionic compositions of solutions or altering holding potentials affects the driving force of ions subsequently, resizing the current or potential amplitude recorded.

In addition to maintaining constant ionic fluxes, monitoring the access resistance during whole-cell configurations is also important for amplitude measurements. Access resistance refers to the resistance generated between the micropipette and the cellular membrane (See Chap. 3). Ideally, micropipettes with large tips are preferred, which provides low access resistance. However, the negative pressure needed to obtain a whole-cell patch would ultimately cause the cell to be sucked up into the large pipette tip, ruining the experiment. Because the micropipette tip is limited by the cell size, access resistance exists and can resize the amplitude. As we know, a resistor impedes the flow of current in a circuit. That means that the access resistance, which lies in series with the recording pipette, causes the detected current signal (if in voltage-clamp mode) to be lower than the actual current passing through the cell. In addition, Ohm's law indicates that voltage is proportionally altered as resistance alters ($V = IR$). Therefore, a high access resistance introduces voltage errors affecting recorded potentials or holding potentials. Clearly, the access resistance can grossly distort the recorded signal. How do researchers remedy the access resistance issues? Electrophysiologists performing whole-cell measurements often select recordings with low access resistance (<20 MΩ) where the access resistance does not change by for example ≥ 20 % during the measurement.

4 Summary

This chapter covers three important amplitude measurements that are commonly used in electrophysiological calculations. First, amplitudes measured from action potential waveforms (e.g., action potential of a single neuron or a population of neurons) were discussed followed by amplitude measurements for current or potential waveforms. Finally, amplitude measurements for steady-state currents were explained. Each of these amplitude calculations are commonly seen in in vitro electrophysiological data because they can directly test the strength of neuronal connectedness including synaptic strengths or number of neurons synchronously firing (e.g., population spikes).

The beginning electrophysiologist will most likely need to implement amplitude measurements sooner or later, but it is advised to refrain from ignoring other key characteristics such as waveform kinetics, reversal potentials, and synchronous/asynchronous release. For example, noticing similar waveform kinetics during each experiment can ensure that the access resistance is consistent between cells and that the cells, receptors, or channels being recorded are from the same population (see Chaps. 10 and 17 for synaptic current kinetics). Measuring the reversal potentials of the cells being recorded can ensure that the driving force of ionic movement is similar between the cells sampled. Observing changes in synchronous/asynchronous release characteristics could potentially identify alterations in the release probability, which could explain amplitude changes. One of the most exciting aspects of electrophysiology is that the current traces measured withhold a wide array of useful data. It is up to the electrophysiologist to dissect all of the data appropriately so that accurate interpretations of the experiments can be deduced.

References

1. Baranauskas G, Tkatch T, Nagata K, Yeh JZ, Surmeier DJ (2003) Kv3.4 subunits enhance the repolarizing efficiency of Kv3.1 channels in fast-spiking neurons. Nat Neurosci 6(3):258–266
2. Bean BP (2007) The action potential in mammalian central neurons. Nat Rev Neurosci 8(6):451–465
3. Du J, Zhang L, Weiser M, Rudy B, McBain CJ (1996) Developmental expression and functional characterization of the potassium-channel subunit Kv3.1b in parvalbumin-containing interneurons of the rat hippocampus. J Neurosci 16(2):506–518
4. Erisir A, Lau D, Rudy B, Leonard CS (1999) Function of specific K(+) channels in sustained high-frequency firing of fast-spiking neocortical interneurons. J Neurophysiol 82(5):2476–2489
5. Lien CC, Martina M, Schultz JH, Ehmke H, Jonas P (2002) Gating, modulation and subunit composition of voltage-gated K(+) channels in dendritic inhibitory interneurones of rat hippocampus. J Physiol 538(Pt 2):405–419
6. Martina M, Schultz JH, Ehmke H, Monyer H, Jonas P (1998) Functional and molecular differences between voltage-gated K+ channels of fast-spiking interneurons and pyramidal neurons of rat hippocampus. J Neurosci 18(20):8111–8125
7. Massengill JL, Smith MA, Son DI, O'Dowd DK (1997) Differential expression of K4-AP currents and Kv3.1 potassium channel transcripts in cortical neurons that develop distinct firing phenotypes. J Neurosci 17(9):3136–3147
8. Rudy B, McBain CJ (2001) Kv3 channels: voltage-gated K+ channels designed for high-frequency repetitive firing. Trends Neurosci 24(9):517–526
9. Faber ES, Sah P (2002) Physiological role of calcium-activated potassium currents in the rat lateral amygdala. J Neurosci 22(5): 1618–1628
10. Sah P (1996) Ca(2+)-activated K+ currents in neurones: types, physiological roles and modulation. Trends Neurosci 19(4):150–154
11. Sah P, Faber ES (2002) Channels underlying neuronal calcium-activated potassium currents. Prog Neurobiol 66(5):345–353
12. Sah P, McLachlan EM (1992) Potassium currents contributing to action potential repolarization and the afterhyperpolarization in rat vagal motoneurons. J Neurophysiol 68(5): 1834–1841
13. Shao LR, Halvorsrud R, Borg-Graham L, Storm JF (1999) The role of BK-type Ca2+-dependent K+ channels in spike broadening during repetitive firing in rat hippocampal pyramidal cells. J Physiol 521(Pt 1):135–146
14. Nicoll RA, Schmitz D (2005) Synaptic plasticity at hippocampal mossy fibre synapses. Nat Rev Neurosci 6(11):863–876
15. Johnston D, Wu SMS (1995) Foundations of cellular neurophysiology. MIT Press, Cambridge, MA

Chapter 15

Pre vs. Post synaptic Effect

Nicholas Graziane and Yan Dong

Abstract

A commonly pursued goal in studying synaptic plasticity is to determine whether synaptic inputs undergo presynaptic and/or postsynaptic alterations following a stimulus and how these alterations affect the firing properties of the innervated neuron. In this chapter we continue to discuss approaches that can be implemented to test presynaptic mechanisms of plasticity. In addition, we explain modifications that can occur in the postsynaptic cell affecting the intrinsic membrane excitability. The intrinsic membrane excitability contributes to the firing properties of a neuron, and is mediated by voltage-gated channels as well as synaptic inputs. In this chapter, we discuss the voltage-gated channels that regulate the intrinsic membrane excitability on the dendrite explaining how these channels contribute to EPSP summation at the soma and back-propagating action potentials. We conclude this chapter with approaches that can be used to measure the intrinsic membrane excitability.

Key words Neurotransmitter release, Intrinsic membrane excitability, Sodium channels, Calcium channels, Potassium channels, Hyperpolarization-activated cation channels, EPSP-to-spike plasticity

1 Introduction

A commonly pursued goal in studying synaptic plasticity is to determine whether synaptic inputs undergo presynaptic and/or postsynaptic alterations following a stimulus and how these alterations affect the firing properties of the innervated neuron. In this chapter we continue to discuss approaches that can be implemented to test presynaptic mechanisms of plasticity. In addition, we explain modifications that can occur in the postsynaptic cell affecting the intrinsic membrane excitability. The intrinsic membrane excitability contributes to the firing properties of a neuron, and is mediated by voltage-gated channels as well as synaptic inputs. In this chapter, we discuss the voltage-gated channels that regulate the intrinsic membrane excitability on the dendrite explaining how these channels contribute to EPSP summation at the soma and back-propagating action potentials. We conclude this chapter with approaches that can be used to measure the intrinsic membrane excitability.

Nicholas Graziane and Yan Dong, *Electrophysiological Analysis of Synaptic Transmission*, Neuromethods, vol. 112,
DOI 10.1007/978-1-4939-3274-0_15, © Springer Science+Business Media New York 2016

2 Presynaptic

2.1 Neurotransmitter Release

There are useful analysis approaches already discussed in previous chapters that can indicate whether synaptic plasticity is presynaptic in origin. In Chap. 8 we discuss recording techniques designed to measure the quantal release of neurotransmitters (e.g., miniature postsynaptic current recordings and quantal postsynaptic current recordings) and in Chap. 12 we review that decreases or increases in quantal frequency, not amplitude, are associated with attenuation or augmentation of presynaptic release, respectively. In Chap. 11 we review paired-pulse ratios (PPRs) and in Chap. 12 discuss that changes in PPRs are associated with presynaptically mediated plasticity. In addition to measuring quantal content and PPRs, there are other electrophysiological analysis approaches that can be used to explore whether changes in synaptic strength are presynaptically mediated. The first method discussed below is implemented by Del Castillo and Katz [1].

As we know, changes in the probability of release (Pr), the number of transmitter release sites (*N*), and/or the postsynaptic quantal amplitude (*Q*), all contribute to changes in synaptic responses (see Chap. 11). In this case, Pr and *N* are presynaptic parameters and *Q* is often determined by postsynaptic properties. To determine whether plasticity is presynaptically mediated, the coefficient of variation (CV) can be assessed, which incorporates only the presynaptic components (Pr and *N*). To do this, the reciprocal of the square of the coefficient of variation (CV^{-2}) can be simply expressed using Eq. (1):

$$CV^{-2} = \frac{I^2}{\sigma^2} \tag{1}$$

where I is the mean synaptic current amplitude and σ refers to the variance of the mean. How then does CV^{-2} remain independent of postsynaptically mediated variables? This is best demonstrated mathematically, as the current and variance in Eq. (1) can be substituted as shown in Eq. (2):

$$I^2 / \sigma^2 = (N\,\mathrm{Pr}\,Q)^2 / N\,\mathrm{Pr}(1-\mathrm{Pr})(Q)^2 \tag{2}$$

This further simplifies to

$$I^2 / \sigma^2 = N\,\mathrm{Pr}/(1-\mathrm{Pr}) \tag{3}$$

Therefore, as shown in Eq. (3), measurements of CV^{-2} are independent of the postsynaptic response *Q*, but exclusively determined by Pr and N, thus providing a presynaptic readout. Data demonstrating the usefulness of using CV^{-2} measurements for analyzing presynaptic changes was exemplified in an early study [2], in which the authors varied the calcium concentration in order to test whether CV^{-2} measurements can accurately detect presynaptic changes. Since raising extracellular Ca^{2+} concentration increases presynaptic release, it is not surprising that increasing extracellular calcium from 2.5 to 3.5 mM increases the mean excitatory postsynaptic current (EPSC) amplitude. Furthermore, raising the calcium concentration also increases the CV^{-2}, which remains increased even following administration of an AMPAR antagonist. In other words, blocking the postsynaptic response had no effect on the increased CV^{-2} because the increase in EPSC amplitude if presynaptic in origin. This elegant experiment demonstrates the usefulness of the CV^{-2} measurement, but with one caveat.

Measurements of CV^{-2} rely on changes in Pr and N, which are both presynaptically mediated. However, these assumptions are not always accurate. Changes in N can also occur when AMPAR silent synapses become unsilenced and functional at hyperpolarized potentials (see Chap. 19 for silent synapses). A prescient warning in the interpretation of their results (testing whether LTP in the hippocampal CA1 region was presynaptic or postsynaptic; results using the CV^{-2} assay pointing toward a presynaptically mediated mechanism), Bekkers and Stevens pointed out that N could be a postsynaptic alteration masking as a presynaptic change [3]. Therefore, if measuring presynaptic alterations using the CV^{-2} measurement, it may be beneficial to also consider potential postsynaptic mechanisms.

2.2 Variance-Mean (V–M) Analysis-Revisited

In Chap. 11 we review V–M analysis as a way to calculate the probability of release at presynaptic terminals [4, 5]. We revisit V–M analysis here to discuss the equation as a useful tool that can be used to explore whether a measured alteration in synaptic currents is presynaptic in origin. The V–M analysis is not restricted to measurements containing only one synaptic input, but rather can be implemented when isolating compound inputs from multiple presynaptic axons. Therefore, the V–M analysis is useful when evoking currents, which typically activate a large number of presynaptic inputs. To determine whether synaptic plasticity occurs presynaptically, again N and Pr need to be calculated. By varying Pr and graphing the variance around the mean amplitude versus the mean amplitude, a parabolic curve can be formed. This parabolic curve can then be fit using software, which will assist the experimenter in solving for N, Q, and Pr (see Chap. 11 for a detailed description of the experimental approach and the analysis).

3 Postsynaptic

In the previous chapters we discuss experimental approaches designed to identify postsynaptically mediated plasticity. For example, changes in quantal amplitudes are associated with postsynaptic alterations (see Chap. 8 for mini PSC and quantal PSC descriptions). Using antagonists or antibodies in the internal solution to block synaptic plasticity or finding changes in Q using V–M analysis often indicates postsynaptically mediated plasticity mechanisms. In addition, Chap. 19 looks at silent synapse measurements, which can be used to support postsynaptic mechanisms of plasticity [6–8] (see Chap. 19). In this chapter, we discuss the intrinsic membrane excitability, which is another contributing factor to neuronal messaging in the central nervous system relying on voltage- or calcium-dependent ion channels. The intrinsic membrane excitability contributes to synaptic plasticity [9] by regulating EPSP summation at the soma. This in turn affects AP firing and subsequent back-propagating APs, which contribute to spike-timing dependent plasticity (STDP) and associative learning processing [10–12]. With the obvious link between the intrinsic membrane excitability, synaptic plasticity, and learning and memory, we provide an overview of the channels that control membrane excitability. We do this in order to provide the beginning electrophysiologist with the appreciation for the many varying components that embody electrical transfer in the CNS.

3.1 Intrinsic Membrane Excitability

Intrinsic membrane excitability refers to membrane's capacity of firing action potentials, which is determined by the intrinsic electrical properties of the cell membrane. The intrinsic membrane excitability sets the action potential (AP) threshold as well as determines how often APs fire [13, 14]. These APs are triggered most commonly in the axon hillock or axon. However, APs can also back propagate from the soma to the dendrites or they can be triggered in the dendrites [15–19]. The characteristic shape of dendritic APs is considerably broader and shorter than somatic APs and does not express a prominent afterhyperpolarization (AHP). APs triggered in the dendrites propagate forward to the soma, but do so unreliably [20]. The electrical properties that control AP shape at the dendrites as well as EPSPs are voltage-gated channels (discussed below).

3.2 Na^+ Channels

Triggering dendritic APs are Na^+ channels. These Na^+ channels, much like Na^+ channels expressed in the soma, have transient and persistent components most likely generating the biexponential inactivation time constants at ~5 and 50 ms (at −40 mV) [20–26]. The slow inactivation of the persistent Na^+ currents boosts the excitatory postsynaptic potential (EPSP) amplitude in the subthreshold range, while the transient Na^+ current underlies the upstroke of the action potential in the suprathreshold range [16, 18, 20].

3.3 Ca^{2+} Channels

Assisting in triggering sodium-mediated dendritic APs are dendritic calcium channels, which produce local calcium spikes. These calcium spikes remain in the dendrites and are prevented from actively propagating to the soma by K^+ channels (possibly D-type or Ca^{2+}-dependent K^+ channels) [25–27]. Although these cannot propagate to the soma, these calcium spikes are posited to activate low threshold Na^+ spikes in the dendrites [25, 26], potentially contributing to EPSP boosts in the subthreshold range. Low-voltage-activated Ca^{2+} channels are posited to contribute to EPSP boosts and contribute to the supralinear summation of EPSPs. Low-voltage-activated Ca^{2+} channels are expressed in dendrites, activate quickly around −70 mV (time-to-peak ~10–30 ms), and inactivate moderately quickly (time constant: ~30–50 ms) [20].

Another factor contributing to dendritic APs are Ca^{2+}-mediated plateau potentials [28–30]. Plateau potentials depolarize the membrane potential, thus keeping the cell in a stable up-state until hyperpolarizing stimuli [e.g., activation of inhibitory currents such as the activation of Ca^{2+}-dependent K^+ channels [31]] returns the cell to its resting membrane potential [32, 33]. Contributing to plateau potentials are L-type Ca^{2+} channels (high-voltage-activated Ca^{2+} channels expressed in proximal dendrites and to a lesser extent in distal dendrites), which can activate quickly at −25 mV (time-to-peak 10 ms) and do not inactivate [34]. The lack of inactivation is important for the stability of plateau potentials,

3.4 K^+ Channels

Unlike Na^+ and Ca^{2+} channels, which contribute to dendritic APs, most dendritically expressed K^+ channels oppose dendritic AP generation (but see GIRKs). Dendritic K^+ channels with properties similar to delayed rectifier K^+ channels [activation time constant ~30 ms at −40 mV with slow or no inactivation [20]] are active in the suprathreshold range and assist in repolarizing dendritic action potentials [35].

A-type K^+ currents also reduce cell excitability and do so by countering the effects of Na^+ and Ca^{2+} channels, by minimizing transient EPSP amplitudes, and by reducing the amplitudes of back propagating action potentials. In addition, A-type K^+ currents increase the threshold for AP initiation in dendrites, increase the threshold for activating Ca^{2+}-mediated bursts, and repolarize dendritic action potentials [20, 35–37]. A-type K^+ currents activate very fast with a time-to-peak of ~1 ms and inactivate with varying speeds depending upon the membrane potential (at −20 mV, 5–8 ms; at −70 mV, ~50 ms) [35, 37–39].

Lastly, G-protein activated inward rectifying K^+ channels (GIRKs) are another type of K^+ channel expressed in dendrites that can hyperpolarize resting membrane potentials as well as decrease the EPSP amplitude via shunting [20] (Fig. 1). These channels are regulated by neuromodulators (e.g., serotonin, adenosine, and $GABA_B$ receptors) potentially providing a means for neuromodulators to regulate neuronal excitability [20].

3.5 Hyperpolarization-Activated Cation Channels (I_H)

As discussed in Chap. 6, I_H channels are expressed in the dendrites. When activated they reduce the input resistance of the neuron [40, 41]. The reduced input resistance decreases the EPSP amplitude, lowering the probability of AP generation in the axon.

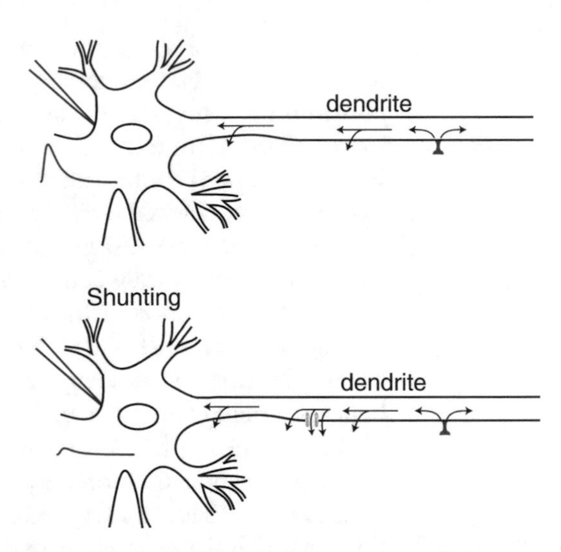

Fig. 1 GIRK channel activation and shunting of an EPSP. If a GIRK channel is activated, the input resistance is reduced locally causing the depolarizing current to leak out of the cell. Recordings from the soma show the reduced EPSP caused shunting

3.6 Intrinsic Membrane Excitability: EPSP-to-Spike (E-S) Plasticity

Population spikes refer to the APs produced by a neuronal population. E-S plasticity refers to changes in AP firing from a neuronal population to a given excitatory synaptic input. For example, E-S potentiation refers to an increase in the probability of AP firing triggered by an excitatory synaptic input [42], while E-S depression reduces this probability [43] (Fig. 2).

Data suggest that dendritic excitatory inputs can initiate three different types of firing patterns in the soma of specific cell types (e.g., cortical pyramidal neurons and motoneurons): repetitive firing, repetitive burst firing, and plateau potentials (Fig. 3). Plateau potentials elicit repetitive firing at the soma, but are different from repetitive firing in that plateau potentials rely on voltage-gated channels, while repetitive firing relies on synaptic currents. Potentially, all three different types of firing patterns are driven by different populations of presynaptic cells innervating the postsynaptic cell at different locations. For example, dendritic input preferentially triggers burst firing[1] likely due to EPSP summation with back-propagating APs [20]. Burst firing of neurons may contribute to the precise depolarization of the postsynaptic neuron, thereby contributing to plasticity either at innervated regions or at afferent inputs.

3.7 Intrinsic Membrane Excitability: Measurements

To measure the intrinsic membrane excitability, there are two approaches that can be used. The first approach uses current-steps to elicit APs in the whole-cell configuration (cell is current clamped).

[1] Cells can fire tonically or in bursts. These firing patterns are likely controlled by the interplay between inhibitory inputs, excitatory inputs, and voltage-gated channels.

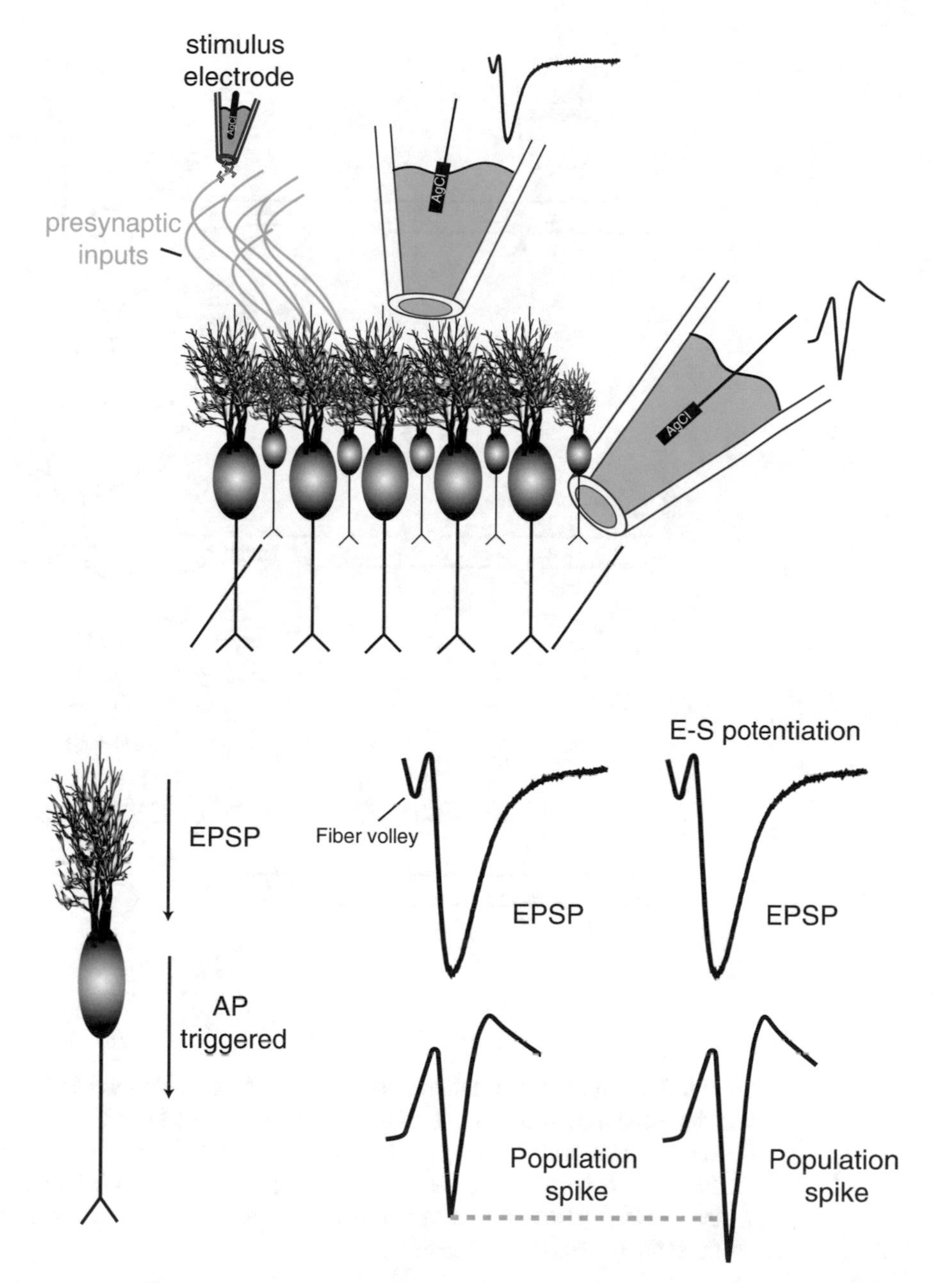

Fig. 2 EPSP-to-spike (E-S) potentiation. A stimulus electrode stimulates presynaptic inputs onto the dendrites of a population of postsynaptic neurons. The EPSPs are measured extracellularly. As the EPSP depolarizes the membrane, an action potential is triggered, which is detected as a population spike measured extracellularly in a region containing neuronal somas. E-S potentiation is seen (*lower right*) as an increase in the number of action potentials evoked by the same EPSP input

This is typically performed by injecting current-steps (e.g., with a 300-ms duration) from hyperpolarizing current (e.g., –400 pA) to a depolarizing current (e.g., +400 pA) at short (e.g., 50 pA) increments. Although APs can only be initiated at depolarizing current

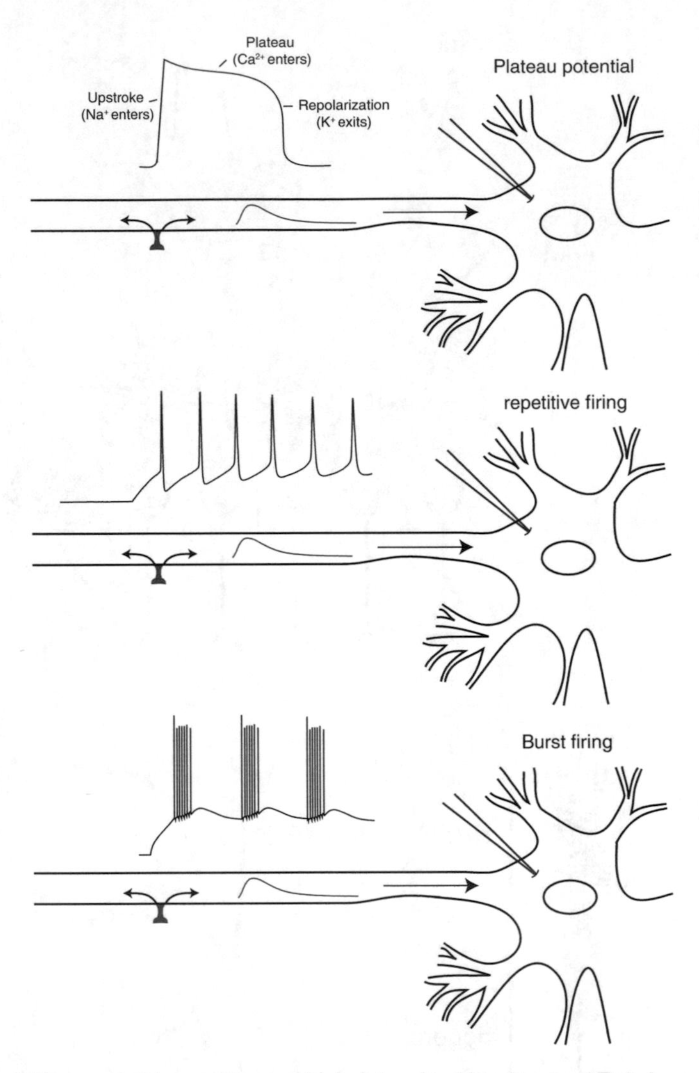

Fig. 3 Neuronal firing patterns initiated by dendritic inputs. (*Top*) A somatic whole-cell measurement of a plateau potential initiated by an EPSP. Ion channel activation for each component of the plateau potential is illustrated. (*Middle*) A somatic whole-cell measurement illustrating neuronal repetitive firing initiated by an EPSP. (*Bottom*) A somatic whole-cell measurement illustrating neuronal burst firing initiated by an EPSP

steps, the hyperpolarizing current steps are often applied so that voltage-gated channels such as sodium or T-type calcium channels can recover from inactivation induced at depolarizing potentials. Once these voltage-gated channels have recovered, they are available to regain permeability to depolarizing ions, thus contributing to neuronal firing. The hyperpolarizing currents generated can also be useful for demonstrating that the access to the cell has not changed during the recording (see bridge compensation in Chap. 3). The analysis takes the AP number at each voltage step, which is plotted against the current injected (Fig. 4a) [44].

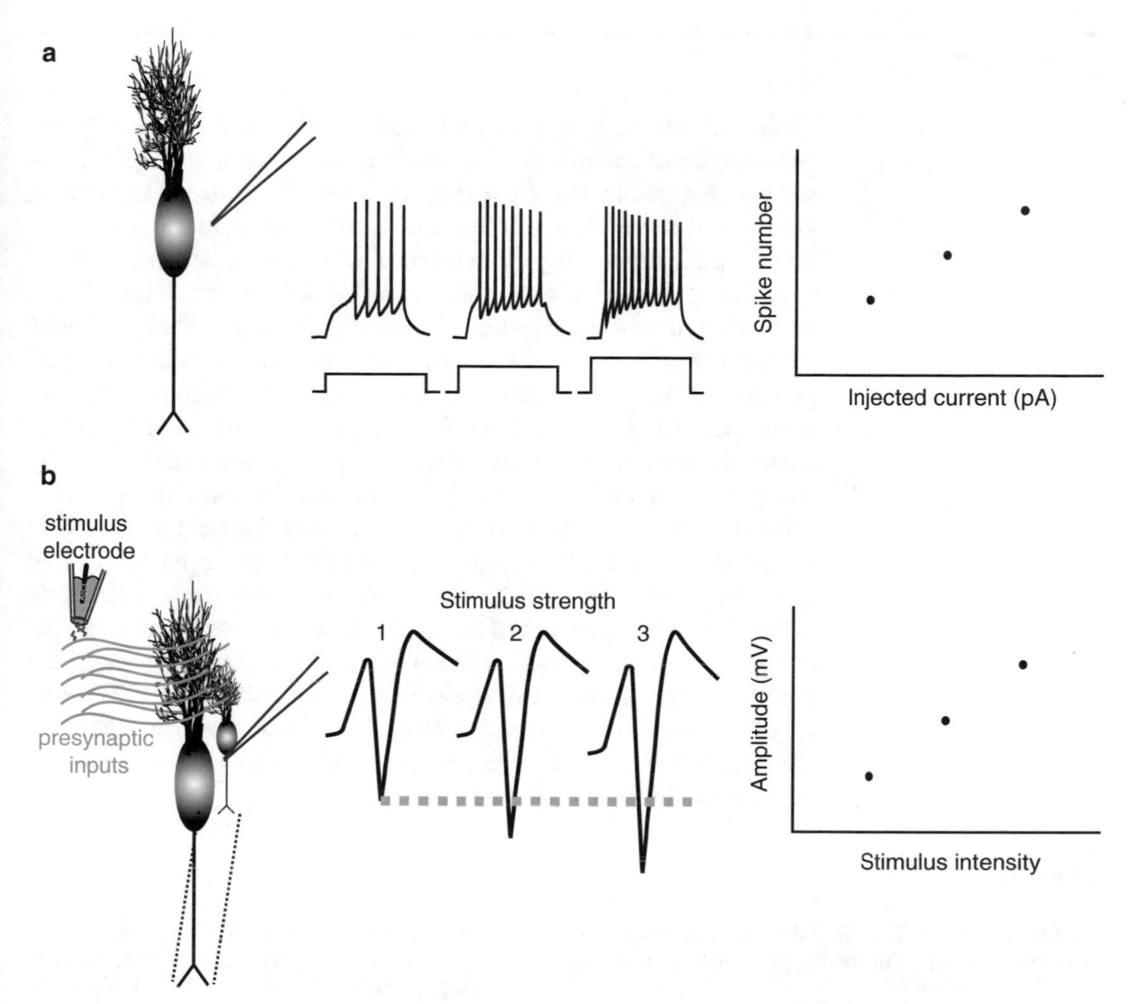

Fig. 4 Measuring intrinsic membrane excitability. (**a**) In a whole-cell configuration, depolarizing voltage steps trigger action potentials in a recorded neuron. The number of action potentials is then plotted against the injected current. (**b**) Increasing the evoked stimulus strength triggers increasing numbers of action potentials in a population of neurons. The increase in action potential output is measured via the amplitude of the population spike in an extracellular recording. The population amplitude is plotted against the stimulus intensity, which can be in units of amplitude or voltage

Statistical analysis is determined based on how many groups are compared and at how many time points [45]. In the simplest scenario a Student's test can be used when only one point on the abscissa is compared between two groups.

The second approach, like the first approach, is another form of an input/output assay, but instead of current-steps, the input is synaptic, initiated from presynaptic axons. This assay can be performed using extracellular field recordings, whereby the recording electrode is placed near the somas of a population of neurons and the population spike amplitude is measured at increasing stimulus intensities. The amplitude of the population spike is then plotted against the stimulus intensity (Fig. 4b).

4 Summary

There are multiple approaches that can be used to identify the presynaptic mechanisms of synaptic plasticity. For example, measuring changes in the frequency of quantal events or measuring changes in the PPR are frequently used measurement strategies. In addition, measuring the reciprocal of the squared coefficient of variation or using variance-mean analysis to solve for N and Pr are two other useful approaches. Combining presynaptically focused analyses with postsynaptic measurements can provide a detailed picture of the mechanisms contributing to synaptic changes. Postsynaptically mediated mechanisms can be measured by calculating changes in amplitudes of quantal postsynaptic currents or by using V–M analysis to solve for Q. In addition, analysis of silent synapse expression can also indicate whether the strengthening or weakening of synaptic inputs is mediated postsynaptically. All of these parameters discussed above are measurements of synaptic current, which is generated by synaptically expressed ionotropic or metabotropic receptors. However, postsynaptic voltage-gated channels may also affect synaptic responses via altering EPSP kinetics or propagation to the soma. In addition, voltage-gated channels can regulate back-propagating APs, which can alter the size of EPSPs generated at synapses.

References

1. Del Castillo J, Katz B (1954) Quantal components of the end-plate potential. J Physiol 124(3):560–573
2. Malinow R, Tsien RW (1990) Presynaptic enhancement shown by whole-cell recordings of long-term potentiation in hippocampal slices. Nature 346(6280):177–180
3. Bekkers JM, Stevens CF (1990) Presynaptic mechanism for long-term potentiation in the hippocampus. Nature 346(6286):724–729
4. Clements JD, Silver RA (2000) Unveiling synaptic plasticity: a new graphical and analytical approach. Trends Neurosci 23(3):105–113
5. Silver RA (2003) Estimation of nonuniform quantal parameters with multiple-probability fluctuation analysis: theory, application and limitations. J Neurosci Methods 130(2): 127–141
6. Durand GM, Kovalchuk Y, Konnerth A (1996) Long-term potentiation and functional synapse induction in developing hippocampus. Nature 381(6577):71–75
7. Isaac JT, Nicoll RA, Malenka RC (1995) Evidence for silent synapses: implications for the expression of LTP. Neuron 15(2):427–434
8. Liao D, Hessler NA, Malinow R (1995) Activation of postsynaptically silent synapses during pairing-induced LTP in CA1 region of hippocampal slice. Nature 375(6530): 400–404
9. Abraham WC (2008) Metaplasticity: tuning synapses and networks for plasticity. Nat Rev Neurosci 9(5):387
10. Alkon DL (1984) Calcium-mediated reduction of ionic currents: a biophysical memory trace. Science 226(4678):1037–1045
11. Alkon DL, Lederhendler I, Shoukimas JJ (1982) Primary changes of membrane currents during retention of associative learning. Science 215(4533):693–695
12. Zhao ML, Wu CF (1997) Alterations in frequency coding and activity dependence of excitability in cultured neurons of Drosophila memory mutants. J Neurosci 17(6): 2187–2199
13. Daoudal G, Debanne D (2003) Long-term plasticity of intrinsic excitability: learning rules and mechanisms. Learn Mem 10(6):456–465
14. Huang YH, Schluter OM, Dong Y (2011) Cocaine-induced homeostatic regulation and

dysregulation of nucleus accumbens neurons. Behav Brain Res 216(1):9–18
15. Chen WR, Midtgaard J, Shepherd GM (1997) Forward and backward propagation of dendritic impulses and their synaptic control in mitral cells. Science 278(5337):463–467
16. Golding NL, Spruston N (1998) Dendritic sodium spikes are variable triggers of axonal action potentials in hippocampal CA1 pyramidal neurons. Neuron 21(5):1189–1200
17. Martina M, Vida I, Jonas P (2000) Distal initiation and active propagation of action potentials in interneuron dendrites. Science 287(5451):295–300
18. Schwindt P, Crill W (1999) Mechanisms underlying burst and regular spiking evoked by dendritic depolarization in layer 5 cortical pyramidal neurons. J Neurophysiol 81(3):1341–1354
19. Stuart GJ, Sakmann B (1994) Active propagation of somatic action potentials into neocortical pyramidal cell dendrites. Nature 367(6458):69–72
20. Reyes A (2001) Influence of dendritic conductances on the input-output properties of neurons. Annu Rev Neurosci 24:653–675
21. Colbert CM, Magee JC, Hoffman DA, Johnston D (1997) Slow recovery from inactivation of Na+channels underlies the activity-dependent attenuation of dendritic action potentials in hippocampal CA1 pyramidal neurons. J Neurosci 17(17):6512–6521
22. Jung HY, Mickus T, Spruston N (1997) Prolonged sodium channel inactivation contributes to dendritic action potential attenuation in hippocampal pyramidal neurons. J Neurosci 17(17):6639–6646
23. Magee JC, Christofi G, Miyakawa H, Christie B, Lasser-Ross N, Johnston D (1995) Subthreshold synaptic activation of voltage-gated Ca2+ channels mediates a localized Ca2+ influx into the dendrites of hippocampal pyramidal neurons. J Neurophysiol 74(3):1335–1342
24. Mittmann T, Linton SM, Schwindt P, Crill W (1997) Evidence for persistent Na+current in apical dendrites of rat neocortical neurons from imaging of Na+-sensitive dye. J Neurophysiol 78(2):1188–1192
25. Schwindt PC, Crill WE (1997) Local and propagated dendritic action potentials evoked by glutamate iontophoresis on rat neocortical pyramidal neurons. J Neurophysiol 77(5): 2466–2483
26. Schwindt PC, Crill WE (1997) Modification of current transmitted from apical dendrite to soma by blockade of voltage- and Ca2+-dependent conductances in rat neocortical pyramidal neurons. J Neurophysiol 78(1):187–198
27. Golding NL, Jung HY, Mickus T, Spruston N (1999) Dendritic calcium spike initiation and repolarization are controlled by distinct potassium channel subtypes in CA1 pyramidal neurons. J Neurosci 19(20):8789–8798
28. Carlin KP, Jones KE, Jiang Z, Jordan LM, Brownstone RM (2000) Dendritic L-type calcium currents in mouse spinal motoneurons: implications for bistability. Eur J Neurosci 12(5):1635–1646
29. Hounsgaard J, Kiehn O (1993) Calcium spikes and calcium plateaux evoked by differential polarization in dendrites of turtle motoneurones in vitro. J Physiol 468:245–259
30. Lee RH, Heckman CJ (1996) Influence of voltage-sensitive dendritic conductances on bistable firing and effective synaptic current in cat spinal motoneurons in vivo. J Neurophysiol 76(3):2107–2110
31. Booth V, Rinzel J, Kiehn O (1997) Compartmental model of vertebrate motoneurons for Ca2+-dependent spiking and plateau potentials under pharmacological treatment. J Neurophysiol 78(6):3371–3385
32. Bennett MV, Hille B, Obara S (1970) Voltage threshold in excitable cells depends on stimulus form. J Neurophysiol 33(5):585–594
33. Kiehn O, Eken T (1998) Functional role of plateau potentials in vertebrate motor neurons. Curr Opin Neurobiol 8(6):746–752
34. Magee JC, Johnston D (1995) Synaptic activation of voltage-gated channels in the dendrites of hippocampal pyramidal neurons. Science 268(5208):301–304
35. Bekkers JM (2000) Distribution and activation of voltage-gated potassium channels in cell-attached and outside-out patches from large layer 5 cortical pyramidal neurons of the rat. J Physiol 525(Pt 3):611–620
36. Andreasen M, Lambert JD (1995) Regenerative properties of pyramidal cell dendrites in area CA1 of the rat hippocampus. J Physiol 483(Pt 2):421–441
37. Hoffman DA, Magee JC, Colbert CM, Johnston D (1997) K+ channel regulation of signal propagation in dendrites of hippocampal pyramidal neurons. Nature 387(6636): 869–875
38. Johnston D, Hoffman DA, Magee JC, Poolos NP, Watanabe S, Colbert CM, Migliore M (2000) Dendritic potassium channels in hippocampal pyramidal neurons. J Physiol 525(Pt 1): 75–81
39. Korngreen A, Sakmann B (2000) Voltage-gated K+ channels in layer 5 neocortical pyramidal neurones from young rats: subtypes and gradients. J Physiol 525(Pt 3):621–639

40. Magee JC (1998) Dendritic hyperpolarization-activated currents modify the integrative properties of hippocampal CA1 pyramidal neurons. J Neurosci 18(19):7613–7624
41. Williams SR, Stuart GJ (2000) Site independence of EPSP time course is mediated by dendritic I(h) in neocortical pyramidal neurons. J Neurophysiol 83(5):3177–3182
42. Bliss TV, Lomo T (1973) Long-lasting potentiation of synaptic transmission in the dentate area of the anaesthetized rabbit following stimulation of the perforant path. J Physiol 232(2):331–356
43. Daoudal G, Hanada Y, Debanne D (2002) Bidirectional plasticity of excitatory postsynaptic potential (EPSP)-spike coupling in CA1 hippocampal pyramidal neurons. Proc Natl Acad Sci U S A 99(22):14512–14517
44. Ishikawa M, Mu P, Moyer JT, Wolf JA, Quock RM, Davies NM, Hu XT, Schluter OM, Dong Y (2009) Homeostatic synapse-driven membrane plasticity in nucleus accumbens neurons. J Neurosci 29(18):5820–5831
45. Howell D (2010) Fundamental statistics for the behavioral sciences. Cengage Learning, Belmont, CA

Chapter 16

Run-Up and Run-Down

Nicholas Graziane and Yan Dong

Abstract

A burden associated with electrophysiology is potential artifacts that look like physiological events, but are instead mistakenly incorporated into the experimental recordings. Two such artifacts discussed in this chapter are current run-up and current run-down. Simply put, baseline responses increase (run-up) or decrease (run-down) over time. Membrane channels recorded using cell-attached, whole-cell, and excised-patch configurations are all susceptible to run-up/run-down. The purpose of this chapter is to discuss empirically derived causes of run-up and run-down so that both errors can be either identified or avoided before or during current measurements. Unfortunately, data showing causes and solutions to run-up and run-down in all membrane channels does not exist. However, we discuss receptors and cell types that have been studied for the purpose of keeping the experimenter aware of problem solving strategies during electrophysiological measurements.

Key words GTP, ATP, Protein kinase, Protein phosphatase, pH, Washout

1 Introduction

A burden associated with electrophysiology is potential artifacts that look like physiological events, but are instead mistakenly incorporated into the experimental recordings. Two such artifacts discussed in this chapter are current run-up and current run-down. Simply put, baseline responses increase (run-up) or decrease (run-down) over time. Membrane channels recorded using cell-attached, whole-cell, and excised-patch configurations are all susceptible to run-up/run-down. The purpose of this chapter is to discuss empirically derived causes of run-up and run-down so that both errors can be either identified or avoided before or during current measurements. Unfortunately, data showing causes and solutions to run-up and run-down in all membrane channels does not exist. However, we discuss receptors and cell types that have been studied for the purpose of keeping the experimenter aware of problem solving strategies during electrophysiological measurements.

Nicholas Graziane and Yan Dong, *Electrophysiological Analysis of Synaptic Transmission*, Neuromethods, vol. 112,
DOI 10.1007/978-1-4939-3274-0_16, © Springer Science+Business Media New York 2016

2 Run-Up

As mentioned in the introduction, run-up (see Fig. 12.3c) is defined by channels whose conductance is increased during baseline responses. Run-up is hard to appreciate especially when testing synaptic plasticity. This is because current run-up can mask true biological effects. However, before the frustrated experimenter decides to fling something expensive across the lab when confronted with run-up, they should realize that the problem might be worth investigating or even better, already investigated with the potential remedy. Below we list bioactive molecules critically involved in run-up:

1. GTP: Guanosine-5′-triphosphate (GTP) is an energy source for a cell, a substrate for metabolic reactions, and is essential for many signal transduction processes. Its role in run-up is most likely through the activation of G-proteins or through effects on the cytoskeleton [1].

   How GTP affects run-up has been directly tested in Ca^{2+} channels. Intracellular GTP (0.5 mM) can have dual effects on the speed of Ca^{2+} current run-up depending on the extracellular charge carrier (Ba^{2+} or Ca^{2+}) present in the bath. In the presence of Ca^{2+}, GTP slows the run-up of Ca^{2+} currents, whereas in the presence of Ba^{2+}, GTP enhances run-up [2].

   GTP is required for run-up of rapidly adapting mechanically activated channels (RA). For those unfamiliar, RA channels are expressed on rat dorsal root ganglion neurons (DRG), are nonselective cation channels, and are activated by mechanical stimulation (typically a glass micropipette moved across the cellular membrane). In the whole-cell patch configuration RA current run-up occurs when GTP (0.33 mM) or GTPγS (a non-hydrolyzable GTP analog) is added to the internal solution. In the absence of GTP or in the presence of GDP, RA current run-up can be prevented [3].

2. ATP: Adenosine triphosphate (ATP) is a cellular energy source, a substrate for signal transduction pathways, involved in transcription, and also acts as a neurotransmitter recognized by purinergic receptors. Diminished ATP in the recording solution can cause the ATP-sensitive K^{+} channel conductance or the calcium-activated K^{+} channel conductance to run up. Diminished ATP can be caused by "washout" during whole-cell recordings [4, 5] or purposefully as Hall and Armstrong left ATP out of solution during excised-patch measurements [6].

   ATP can also have dual roles on the speed of Ca^{2+} current run-up. In the absence of ATP run up occurs within ~3 min, while in the presence of ATP run-up is slowed taking ~18 min to reach an equivalent magnitude [7].

From what we know so far about GTP and ATP playing a role in run-up, why would we include ATP and GTP in our internal solutions? First, ATP and GTP can affect channel types differently, and second, ATP and GTP are critically important in the prevention of run-down (see dephosphorylation below).

3. Protein kinases: Protein kinase A (PKA) activation causes the run-up of glycine currents, as in trigeminal neurons [8]. However, this occurs when cesium methanesulfonate is prominent in the internal solution. If potassium fluoride is the major component, PKA has little effect on glycine-current run-up [9]. These results demonstrate how internal solution components can affect channel properties and remind us that internal solution should be considered carefully before experimentation.

   PKC activation increases run-up of $GABA_c$ receptor mediated currents, while inhibiting PKC decreases run-up. PKA activation has opposing effects as activating PKA suppresses run-up, while inhibiting PKA enhances $GABA_c$ receptor-mediated current run-up [10]. Together with Gu and Huang [8], we see that PKA activation can cause run-up of glycine currents, but has opposing effects on $GABA_c$ currents. These data demonstrate the varying effects bioactive molecules have on different membrane channels. Calmodulin-dependent protein kinase II (CaMKII) mediates run-up of connexin 36 channels (connexin 36 is a gap junction protein that exhibits run-up 5–10 min following cell break-in) [11]. Run-up of connexin 36-mediated currents can be prevented when the intracellular compartment is perfused with CaMKII inhibitors. These results demonstrate that phosphorylation of specific subunits of membrane channels may be required for run-up.

4. Protein phosphatases: Protein phosphatases can reverse the phosphorylation of channels, thus preventing run-up. Empirically, inhibiting protein phosphatases can shorten the time it takes for Ca^{2+} current run-up to the peak amplitude [12]. In addition, protein phosphatase activation can block run-up of connexin 36-mediated currents [11].

5. pH: pH-sensitive channels are susceptible to run-up due to changes in proton concentrations. For example, increases in intracellular pH (alkaline solution) can enhance TRAAK K^+ channels currents [13], while increases in extracellular pH enhance NMDAR function [14]. As discussed in Chap. 4, testing the internal and external solution's pH before experimentation may potentially limit day-to-day variations in pH-sensitive channel currents that are susceptible to run-up.

3 Run-Down

Run-down (see Fig. 12.3c) refers to the decrease in baseline channel conductance over time. Many channels are susceptible to run-down in whole-cell, and excised-patch configurations due to "washout" of cytosolic components necessary for channel function. In addition to "washout," there are other factors contributing to run-down and are listed below.

1. Washout: Molecules weighing 100–500 MW diffuse from the cytoplasm to the micropipette [15]. This elicits the loss of ATP and phosphorylating molecules, resulting in dephosphorylation of channels, and thus the run-down of channel currents [16–21]. For example, voltage-gated calcium channels (VGCCs) require channel phosphorylation to sustain functional currents. Following whole-cell patch clamp, VGCC current runs down because substrates necessary for channel phosphorylation are washed out [18, 22]. Luckily, the perforated-patch clamp technique can be implemented when washout is regularly experienced (see Chap. 7).
2. Dephosphorylation: As mentioned above, sustained channel conductance is dependent upon phosphorylation. Run-down has been known to occur in VGCC, but can be prevented when ATP or creatine phosphokinase is added to the internal solution [18, 23]. Channel dephosphorylation can also cause run-down by destabilizing synaptic-channel locations. Synaptic AMPA receptor-mediated currents run down when PKA activity is inhibited from binding to A-kinase anchoring proteins (AKAPs) [24, 25].

   Ca^{2+}-activated K^+ channel-mediated currents run-down following channel dephosphorylation. Run-down is caused by activation of GTP-mediated protein phosphatase1. To minimize run-down of Ca^{2+}-activated K^+ channels, ATP can be added to the internal solution [26].

   Thus far we have seen that GTP can cause run-up as well as run-down of currents. Why then is GTP included in internal solutions? As stated in Chap. 4, GTP is often necessary to maintain G protein and GTP-binding protein activation in order to prevent current run-down. For example, adenosine-activated K^+ current run-down can be prevented by including GTP in the internal micropipette solution during whole-cell patch clamp recordings [27].
3. Proteolytic degradation of channel proteins: Proteases can elicit current run-down in L-type Ca^{2+} channels and this proteolytic-induced run-down can be prevented by incorporating protease inhibitors such as leupeptin in the internal solution [28]. A possible mechanism mediating protease current

run-down is through depolymerization of filamentous actin. Preventing protease-induced depolymerization can inhibit run-down of cardiac ATP-sensitive K^+ currents [29] and *N*-methyl-D-aspartate (NMDA) receptor currents [30].

4. Mechanical strain: Mechanical strain refers to any manipulation that can alter the cellular membrane configuration deforming the cytoskeleton (e.g., pressure application, probes, invasive intracellular recording configurations). A channel's conductance can be regulated by membrane conformational changes. For example, TRAAK K^+ channel currents run-down as the negative pressure applied decreases as shown in inside-out patches [13]. In addition, Ca^{2+} channels expressed on myocytes from the rat basilar artery show decreased inward current following negative pressure application [31]. It should be noted that altered membrane configurations can be regulated by osmosis (i.e., a cell's shrinking and swelling depending upon hyperosmotic or hypoosmotic cytosolic solutions). Appropriate osmolality ranges ensure that the experimenter has controlled osmotic-induced changes in cell size, thus preventing potential alterations in baseline conductance measurements during electrophysiological recordings.

4 Summary

Run-up and run-down are common in intracellular electrophysiological recordings. Bioactive substrates regulating phosphorylation, dephosphorylation, proteolysis, and membrane conformations are critically involved in both processes. By understanding channel properties before experimentation, necessary components can be added to the external and/or internal solutions, thus decreasing the probability of run-up and/or run-down from occurring.

References

1. Mitchison T, Kirschner M (1984) Dynamic instability of microtubule growth. Nature 312(5991):237–242
2. Wagner JJ, Alger BE (1994) GTP modulates run-up of whole-cell Ca2+ channel current in a Ca(2+)-dependent manner. J Neurophysiol 71:814
3. Jia Z, Ikeda R, Ling J, Gu JG (2013) GTP-dependent run-up of Piezo2-type mechanically activated currents in rat dorsal root ganglion neurons. Mol Brain 6:57
4. Rorsman P, Trube G (1985) Glucose dependent K+-channels in pancreatic beta-cells are regulated by intracellular ATP. Pflugers Arch 405(4):305–309
5. Trube G, Rorsman P, Ohno-Shosaku T (1986) Opposite effects of tolbutamide and diazoxide on the ATP-dependent K+ channel in mouse pancreatic beta-cells. Pflugers Arch 407(5):493–499
6. Hall SK, Armstrong DL (2000) Conditional and Unconditional Inhibition of Calcium-activated Potassium Channels by Reversible Protein Phosphorylation. J Biol Chem 275(6): 3749–3754
7. Elhamdani A, Bossu J-L, Feltz A (1995) ATP and G proteins affect the runup of the Ca2+ current in bovine chromaffin cells. Pflugers Arch 430(3):410–419
8. Gu Y, Huang LY (1998) Cross-modulation of glycine-activated Cl- channels by protein kinase

C and cAMP-dependent protein kinase in the rat. J Physiol 506(Pt 2):331–339
9. Song YM, Huang LY (1990) Modulation of glycine receptor chloride channels by cAMP-dependent protein kinase in spinal trigeminal neurons. Nature 348(6298):242–245
10. Jung CS, Lee SJ, Paik SS, Bai SH (2000) Run-up of gamma-aminobutyric acid(C) responses in catfish retinal cone-horizontal cell axon-terminals is modulated by protein kinase A and C. Neurosci Lett 282(1-2):53–56
11. del Corsso C, Iglesias R, Zoidl G, Dermietzel R, Spray DC (2012) Calmodulin dependent protein kinase increases conductance at gap junctions formed by the neuronal gap junction protein connexin36. Brain Res 1487:69–77
12. Mironov SL, Lux HD (1991) Calmodulin antagonists and protein phosphatase inhibitor okadaic acid fasten the 'run-up' of high-voltage activated calcium current in rat hippocampal neurones. Neurosci Lett 133(2):175–178
13. Kim Y, Bang H, Gnatenco C, Kim D (2001) Synergistic interaction and the role of C-terminus in the activation of TRAAK K+ channels by pressure, free fatty acids and alkali. Pflugers Arch 442(1):64–72
14. Gray AT, Buck LT, Feiner JR, Bickler PE (1997) Interactive effects of pH and temperature on N-methyl-D-aspartate receptor activity in rat cortical brain slices. J Neurosurg Anesthesiol 9(2):180–187
15. Horn R, Marty A (1988) Muscarinic activation of ionic currents measured by a new whole-cell recording method. J Gen Physiol 92(2):145–159
16. Armstrong D, Eckert R (1987) Voltage-activated calcium channels that must be phosphorylated to respond to membrane depolarization. Proc Natl Acad Sci U S A 84(8):2518–2522
17. Becq F (1996) Ionic channel rundown in excised membrane patches. Biochim Biophys Acta 1286(1):53–63
18. Chad J, Kalman D, Armstrong D (1987) The role of cyclic AMP-dependent phosphorylation in the maintenance and modulation of voltage-activated calcium channels. Soc Gen Physiol Ser 42:167–186
19. Horn R, Korn SJ (1992) Prevention of rundown in electrophysiological recording. Methods Enzymol 207:149–155
20. Hoshi T (1995) Regulation of voltage dependence of the KAT1 channel by intracellular factors. J Gen Physiol 105(3):309–328
21. Tang XD, Hoshi T (1999) Rundown of the hyperpolarization-activated KAT1 channel involves slowing of the opening transitions regulated by phosphorylation. Biophys J 76(6):3089–3098
22. Korn SJ, Horn R (1989) Influence of sodium-calcium exchange on calcium current rundown and the duration of calcium-dependent chloride currents in pituitary cells, studied with whole cell and perforated patch recording. J Gen Physiol 94(5):789–812
23. Forscher P, Oxford GS (1985) Modulation of calcium channels by norepinephrine in internally dialyzed avian sensory neurons. J Gen Physiol 85(5):743–763
24. Rosenmund C, Carr DW, Bergeson SE, Nilaver G, Scott JD, Westbrook GL (1994) Anchoring of protein kinase A is required for modulation of AMPA/kainate receptors on hippocampal neurons. Nature 368(6474):853–856
25. Tavalin SJ, Colledge M, Hell JW, Langeberg LK, Huganir RL, Scott JD (2002) Regulation of GluR1 by the A-kinase anchoring protein 79 (AKAP79) signaling complex shares properties with long-term depression. J Neurosci 22(8): 3044–3051
26. Bielefeldt K, Jackson MB (1994) Phosphorylation and dephosphorylation modulate a Ca(2+)-activated K+ channel in rat peptidergic nerve terminals. J Physiol 475(2):241–254
27. Trussell LO, Jackson MB (1987) Dependence of an adenosine-activated potassium current on a GTP-binding protein in mammalian central neurons. J Neurosci 7(10):3306–3316
28. Belles B, Hescheler J, Trautwein W, Blomgren K, Karlsson JO (1988) A possible physiological role of the Ca-dependent protease calpain and its inhibitor calpastatin on the Ca current in guinea pig myocytes. Pflugers Arch 412(5): 554–556
29. Furukawa T, Yamane T-i, Terai T, Katayama Y, Hiraoka M (1996) Functional linkage of the cardiac ATP-sensitive K+ channel to the actin cytoskeleton. Pflugers Arch 431(4):504–512
30. Rosenmund C, Westbrook GL (1993) Rundown of N-methyl-D-aspartate channels during whole-cell recording in rat hippocampal neurons: role of Ca2+ and ATP. J Physiol 470:705–729
31. Langton PD (1993) Calcium channel currents recorded from isolated myocytes of rat basilar artery are stretch sensitive. J Physiol 471:1–11

Chapter 17

Kinetics of Synaptic Current

Nicholas Graziane and Yan Dong

Abstract

Current kinetics takes into account three key components of receptor function. They include an open or activated state, a deactivated state, and a desensitized state (Fig. 1) As the ligand binds to the receptor, a conformational change takes place allowing pore formation and ion permeability, defining an activated state. The deactivated state refers to a receptor transitioning from a bound to an unbound agonist state with decreasing ionic permeability as the channel closes. This process occurs as the agonist concentration becomes zero. Finally, a desensitized state refers to a reduced response to an agonist often due to prolonged agonist exposure (i.e., the receptor is in a non-conducting state despite agonist being bound to the receptor). Desensitization can be altered by neurotransmitter clearance from the synaptic cleft via diffusion, degradation, or reuptake through transporters expressed on neuronal or glial cells. Prolonged exposure to neurotransmission may induce desensitization of receptors, while rapid removal of the neurotransmitter from the synaptic cleft may reduce desensitization.

Key words Non-NMDARs, NMDARs, Nicotinic acetylcholine receptors, 5-HT3Rs, Purinergic P2X receptors, GABAA receptors, Glycine receptors

1 Introduction

Current kinetics takes into account three key components of receptor function. They include an open or activated state, a deactivated state, and a desensitized state (Fig. 1) As the ligand binds to the receptor, a conformational change takes place allowing pore formation and ion permeability, defining an activated state. The deactivated state refers to a receptor transitioning from a bound to an unbound agonist state with decreasing ionic permeability as the channel closes. This process occurs as the agonist concentration becomes zero. Finally, a desensitized state refers to a reduced response to an agonist often due to prolonged agonist exposure (i.e., the receptor is in a non-conducting state despite agonist being bound to the receptor). Desensitization can be altered by neurotransmitter clearance from the synaptic cleft via diffusion, degradation, or reuptake through transporters expressed on neuronal or

Nicholas Graziane and Yan Dong, *Electrophysiological Analysis of Synaptic Transmission*, Neuromethods, vol. 112,
DOI 10.1007/978-1-4939-3274-0_17,

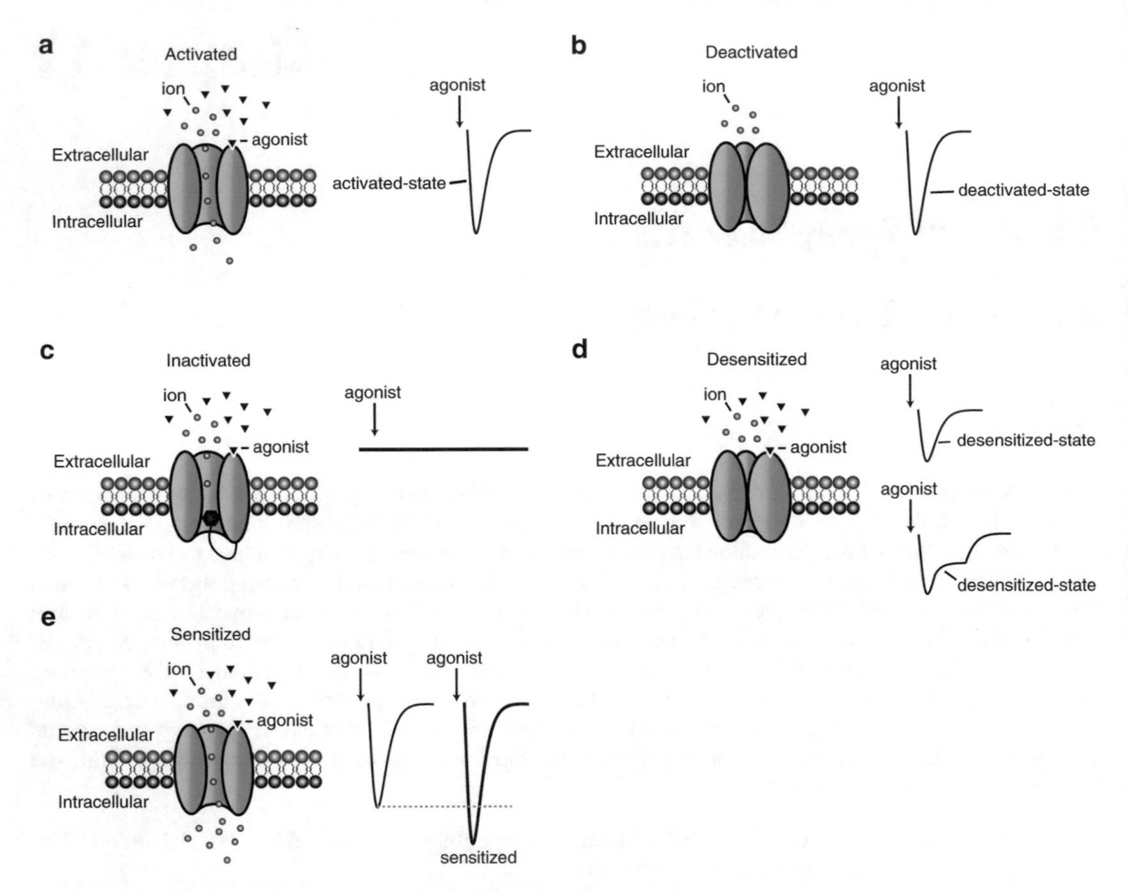

Fig. 1 Receptor/channel states. (**a**) Agonist binds to the channel inducing pore formation and ion + influx. The result is an inward current whose activation-state is illustrated as the rise time to peak current. (**b**) Agonist is cleared from the extracellular space, resulting in the channel closing, thus preventing ion influx. The deactivation phase of the current is illustrated as the decay phase. (**c**) In the presence of an agonist, the channel is inactive as ions are unable to pass through the pore due to peptide block (*ball* and *chain* model). (**d**) In the presence of an agonist, the channel pore closes preventing ion permeability. The current traces illustrate time points whereby the receptor becomes desensitized either synaptically (*top*) or during local agonist perfusion (*bottom*). (**e**) An agonist binds to a sensitized receptor, resulting in an amplified response demonstrated by the current traces

glial cells. Prolonged exposure to neurotransmission may induce desensitization of receptors, while rapid removal of the neurotransmitter from the synaptic cleft may reduce desensitization[1].

[1] There are two other key components to receptor function, which include inactivation and sensitization. When a channel/receptor loses permeability to an ion or is unable to activate downstream signaling processes in the presence of an agonist, it is considered inactive. An example is the "ball and chain" model of inactivation in which inactivation is caused by peptide (the ball) insertion into the open channel pore, thus blocking ion flow through the channel [1]. Sensitization, on the other hand, refers to a potentiated response to a given stimulus. For example, administration of an agonist at the same concentration previously used can trigger greater postsynaptic currents in sensitized receptors.

A receptor's activated, deactivated, and desensitized states can be studied using microscopic kinetic measurements. These measurements are performed using single-channel recordings in the cell-attached patch configuration. The receptor agonist is perfused from the recording electrode onto the receptor inducing open- and closed-channel states. The limitation of microscopic measurements is that they do not directly identify how a channel functions at a synapse. This limitation is caused by incomplete analysis of synaptic geometry and transmitter release. To overcome this limitation, macroscopic current kinetics using the whole-cell patch configuration are often measured to identify synaptic receptor states.

Below we have discussed microscopic and macroscopic current kinetics as they pertain to common synaptically expressed receptor subtypes in the CNS. The aspiring electrophysiologist should pay close attention to receptor kinetics during current measurements as they can provide insight into subunit expression or functional state of the receptor (e.g., desensitized). If there are noticeable irregularities in the receptor kinetics being measured in control conditions, it is important to check agonist/ neurotransmitter concentrations, exposure times to agonists/ neurotransmitters, and pH levels of the solutions as all can alter receptor kinetics [2].

2 Excitatory Ionotropic Receptors

2.1 Non-NMDA (N-Methyl-d-Aspartate) Receptors

Non-NMDA receptors constitute AMPA (α-amino-3-hydroxy-5-methyl-4-isoxazolepropionic acid) receptors (AMPARs) and kainate receptors. They form tetrameric channels (four subunits) assembled from a combination of GluA1-4 for AMPARs and GluA5-7, KA-1, and KA-2 for kainate receptors [3–12]. Commonly, AMPARs are composed of GluA1/GluA2 or GluA2/GluA3, although GluA2-lacking AMPARs can form functional channels with high Ca^{2+} permeability [13, 14]. All four subunits of AMPARs can exist in two alternatively spliced versions, flip and flop. These alternatively spliced versions can regulate channel desensitization and steady-state currents [15]. AMPAR desensitization can be interfered with by cyclothiazide (a positive allosteric modulator of AMPARs eliminating rapid desensitization), while kainate desensitization can be slowed by concanavalin A [16].

Non-NMDA receptors mediate fast excitatory transmission when activated by presynaptically released glutamate, typically with rise times that are <1 ms and decay time constants of 1–8 ms (see Fig. 9.1) [17–22]. The deactivation rate at room temperature is fast (0.5–3 ms), with varying desensitization times possibly related to the unitary conductance [23–28].

2.2 N-Methyl-d-Aspartate Receptors (NMDARs)

NMDARs are heteromeric complexes formed as a tetramer. There are three subunit subtypes known currently as GluN1, GluN2, and GluN3, which are recently changed from the previous NR1, NR2, and NR3 terminology. GluN1 subunits can be alternatively spliced from a single gene generating up to eight different GluN1 subunits. Encoding GluN2 and GluN3 subunits are six separate genes. There are four different GluN2 subunits (A, B, C, and D) and two GluN3 subunits (A and B). Functional assembly of these subunits is commonly GluN1/GluN2 with GluN2A being enriched in mature synapses and GluN2B being enriched in developing synapses. However, ternary complexes GluN1/GluN2/GluN3 are also thought to exist [29]. Activating NMDARs requires binding of both glycine and glutamate [30]. Once the glycine and glutamate are bound to the receptor, the channel opens allowing permeability of both Na^+ and Ca^{2+} ions with a ratio of ~9:1 Na^+ to Ca^{2+} and a conductance of 40–50 pS [16, 31]. Of course, ionic permeability can only occur once the Mg^{2+} block is removed from the channel pore [32]; at −90 mV the Mg^{2+} blockade is maximal, at −55 mV the Mg^{2+} block is incomplete, and at +50 mV the Mg^{2+} block is minimal [33].

NMDAR kinetics are slower than that of AMPA or kainate receptors with rise times that are ~20 ms and bi-exponential decay time constants of 40 and 100–200 ms (See Fig. 9.1) [32, 34–36]. The slow onset of NMDAR current most likely stems from the latency between glutamate binding and the first opening of the channel [35]. This latency, very likely, prevents synaptic NMDAR-mediated current during a postsynaptic potential as Mg^{2+} blocks the channel as soon as the spine repolarizes. However, if a second postsynaptic potential follows while the agonist is still bound, the depolarized dendrite unblocks NMDARs.

2.3 Neuronal Nicotinic Acetylcholine Receptors (nAChRs)

nAChRs belong to the superfamily of ligand-gated ion channels which includes glycine, $GABA_A$, and serotonin ($5\text{-}HT_3$) receptors. nAChRs are pentameric channels, which are permeable to Na^+, K^+, and Ca^{2+} (α7 containing nAChRs are Ca^{2+} permeable). The pentameric channels are formed in αβ combinations with the standard neuronal subunits α2–α6 and β2–β4. In addition to the standard neuronal subunits, α7–α9 subunits can form homomeric nAChRs that are inhibited by α-bungarotoxin. The predominant nAChR subunit composition in the mammalian brain is α4β2 or α7 [37–41]. However, this is not always true for all brain regions as the medial habenula and locus coeruleus express mainly α3 and β4 subunits [42].

nAChRs are activated by endogenous acetylcholine or by exogenous nicotine. Upon agonist binding, nAChRs are stabilized in an open conformation for several milliseconds. After activation, the channel can enter into either a closed resting state or a closed desensitized state that is unresponsive to agonist activation for several milliseconds. The rate at which the channel proceeds through the open, resting, or desensitized conformational states depends upon the subunit composition, peptide transmitters, protein kinases, cytoskeleton, calcium signals, and other signaling controls. However, data show that the burst lengths of nAChRs is fast, arranging from 5 to 10 ms [43–47].

At the synapse, acetylcholine (ACh) is rapidly released into the cleft at about 1 mmol/L. Nicotinic receptors are then immediately activated before ACh can be hydrolyzed by acetylcholinesterase. This rapid ACh release and hydrolysis is thought to prevent nAChR desensitization. However, nAChR can enter a highly complex desensitization state dependent upon the rate of synaptic firing, the application of exogenous agonists as seen with smokers and nicotine, or the application of acetylcholinesterase inhibitors, which are used in Alzheimer's treatments [48, 49]. There are three types of currents produced by nAChR activation, which is dependent upon the subunit composition. Type IA currents (α7-containing nAChRs) have short-lived open times, fast inactivation kinetics and are fast-desensitizing [50]. Also, Type IA currents characteristically run down, which is associated with high-energy phosphate compounds and an intracellular Mg^{2+}-dependent inward rectification [51–53]. In contrast to Type IA currents, Type II currents (α4β2-containing nAChRs) desensitize slowly, with minimal rundown, and show Mg^{2+}-independent inward rectification. Type III currents (α3β4-containing nAChRs), like Type II currents, desensitize slowly. Some neurons express both α7 and α4β2 subunits producing a Type IB current, which contains a fast component similar to Type IA currents and a slow component similar to type II currents [51, 52].

The type of current generated (type IA, IB, II, or III) can functionally affect the signal processing and temporal summation. Fast currents allow for rapid processing, whereas slow, prolonged currents may provide effective temporal integration of signals or provide extra time for an organized cellular response [54].

2.4 5-Hydroxytryptamine-3 Receptors (5-HT3Rs)

5-HT3Rs are closely related to nAChRs both in molecular structure (pentameric assembly of subunits) [55] and in agonist-induced activation and desensitization of ionic currents [56–59]. 5-HT3Rs are expressed both presynaptically and postsynaptically in many areas of the brain with the highest expression in the brainstem. Presynaptically, they regulate neurotransmitter/neuromodulator release such as the release of dopamine, GABA, substance P, and acetylcholine [60–63], while postsynaptically they pass cationic current capable of producing membrane depolarizations. In addition, 5-HT3Rs modulate behavioral functions including anxiety, cognition, alcohol addiction, depression, and inflammatory pain [64, 65].

5-HT3Rs are composed of five distinct subunits (A–E) typically expressed as heteromeric receptors, although the A subunit can functionally exist as a homomer [55]. These subunits function by producing rapidly activating and desensitizing inward currents (fast and slow mean-open channel time constants of ~0.38 and ~5.0, respectively) (see Fig. 9.2) permeable to Na^+, K^+, Ca^{2+}, and small organic cations [55, 66, 67].

3 Purinergic P2X Receptors

P2X receptors are activated by ATP and are classified as cation channels with substantial Ca^{2+} permeability [68]. There are seven P2x receptor subunits ($P2X_{1-7}$) expressed in the CNS. These subunits are posited to form trimers with multiple possible receptor subunit combinations [69–71]. P2X receptors are distributed widely throughout the CNS and expressed on both neurons ($P2X_2$, $P2X_4$, and $P2X_6$ are most abundant in neurons) and glia ($P2X_7$) [72–77]. Postsynaptically, P2X receptors elicit fast synaptic transmission [rise times are sometimes <1 ms, but generally <2 ms and a decay time constant are ~17 ms (see Fig. 9.3) [78]. Presynaptically, P2X receptors elicit glutamate release via Ca^{2+} entry into the presynaptic terminal [79].

The currents generated on the postsynaptic neuron are smaller in comparison to EPSCs mediated by glutamate (20–50 pA versus >1 nA) [78]. However, this may be attributed to the brain slice preparation used to measure the currents. As cells are damaged during the brain slicing process, they may release high levels of ATP and thus desensitize or cause the internalization of P2X receptors [80], thus producing the lower EPSC amplitudes. On the other hand, these small amplitude currents may match physiological function and potentially contribute to synaptic remodeling [80]. Still another explanation is that P2X receptors being situated at the peripheral postsynaptic density (where AMPAR density is low) may be activated less efficiently by the spillover of extracellular ATP [76, 77, 81].

Interestingly, ATP is co-released with other neurotransmitters such as ACh [82], GABA [83], or glutamate [84]. Co-release is not uncommon in the CNS. Other examples are the co-releases of dopamine with glutamate [85] or dopamine with GABA [86]. The problem is that co-release of neurotransmitters can make it difficult to isolate P2X receptor-mediated currents. If isolating P2X receptor-mediated currents is necessary, a pharmacological cocktail consisting of ionotropic and metabotropic antagonists to block AChRs, GABARs, and glutamate receptors can be applied.

P2X receptor-mediated currents have an importance role in pain sensation; P2X receptors are involved in chronic or persistent nociceptive behaviors following nerve injury or inflammation [76]. In addition, P2X receptors mediate taste sensations. As taste buds release ATP, P2X receptors are activated ($P2X_{2/3}$ heteromers) on gustatory nerves [87].

4 Inhibitory Ionotropic Receptors

4.1 γ-Aminobutyric acid$_A$ Receptors (GABA$_A$Rs)[1]

GABA$_A$Rs are activated by the neurotransmitter GABA. Once activated, a net inward flow of anions (chloride and bicarbonate) passes through the channel, hyperpolarizing the membrane in mature neurons[2]. There are 19 different subunits (α1–6, β1–3, γ1–3, δ, ε, θ, π, and ρ1–3) that can be combined in a pentameric assembly to form a GABA$_A$ channel. The most common subunit combinations of GABA$_A$Rs are α1/β2/γ2 or α2/β3/γ2, making up ~60 % and ~20 % of all GABA$_A$Rs, respectively [89]. These subunits are typically assembled as two α subunits, two β subunits, and one γ subunit [90]. Subunit assembly along with receptor phosphorylation, scaffolding proteins, and experimentally introduced regulators can all differentially affect GABA$_A$R kinetics.

The intracellular loops of the β and γ GABA$_A$R subunits contain phosphorylation sites. As GABA$_A$Rs become phosphorylated, the IPSC decay time is shortened [91]. These changes in decay kinetics attributed to channel phosphorylation are likely due to altered GABA unbinding to the channel as well as changes in gating kinetics. The functional relevance of fast desensitization is that it prolongs GABAergic responses by keeping receptors in a bound conformation allowing channels to open [92, 93].

Scaffolding proteins such as gephyrin and GABA$_A$R-associated protein (GABARAP) are critically involved in clustering and stabilizing GABA$_A$Rs at the synapse, affecting GABA$_A$R kinetics. Clustered GABA$_A$Rs deactivate faster and desensitize slower than GABA$_A$Rs that are diffusely distributed at the synapse [94]. Stabilizing GABA$_A$Rs allows protein kinases to be targeted to GABA$_A$R phosphorylation sites, causing kinetic properties to be altered.

Differences in experimental protocols can drastically affect GABA$_A$R kinetics. Low GABA concentrations or extended exposure times can impact GABA$_A$R kinetics by promoting desensitization [95] (see Table 1). Low recording temperatures (24 °C) generate slower GABA$_A$R activation, desensitization, and deactivation when compared to physiological recording temperatures (34 °C). The means by which GABA$_A$Rs are activated (local GABA perfusion, electrically evoked, etc.) produces incomparable

[1] In addition to GABA$_A$Rs there are GABA$_C$Rs, which are mainly located in the retina and are composed of rho subunits. Since 2008, it is recommended that GABA$_C$Rs become known as GABA$_A$-ρ [88].

[2] During embryonic development and at birth GABA$_A$Rs and GlyRs play different roles in regulating membrane potentials. Their activation elicits depolarization of membrane voltage causing the cell to become more excitable. This effect is due to the positive chloride equilibrium potential, which results in the efflux of chloride from the postsynaptic cell upon receptor activation.

Table 1
$GABA_AR$ channel kinetics

Subunits	Application duration	τ_{fast} (ms)	τ_{slow} (ms)
α1β2γ2	1 ms 10–20 s	12.7 ± 3.0 2.1 ± 0.2	175.0 ± 36[a] 35.7 ± 4.8[b]
α6β2γ2	1 ms	51.0 ± 9.0	272.0 ± 36[a]
α1α6β2γ2	1 ms	14.4 ± 2.0	161.0 ± 18[a]
α1β2	1 ms 10–20 s	15.0 ± 1.0 1.3 ± 0.1	271 ± 158[a] 35.6 ± 4.4[b]
α6β2	1 ms	80.0 ± 16	477.0 ± 263[a]
α1α6γ2	1 ms	15.7 ± 2	159.0 ± 26[a]
α1γ2	10–20 s	3.6 ± 0.5	35.9 ± 6.6[b]

[a][96]
[b][97]

kinetic differences. Electrically evoked $GABA_AR$ currents typically show a total activation to deactivation time of ~60 ms. Therefore, if the experimenter is considering using a paired-pulse protocol, a duration of 100 ms may be necessary as the typical 50 ms duration used in for excitatory currents may not be long enough to allow for the $GABA_AR$ channel to completely deactivate.

Lastly, $GABA_AR$ subunit expression can modulate kinetic properties. For example, α2/3 subunit-enriched $GABA_ARs$ produce slower activation and inactivation of IPSCs in comparison to α1 subunit-enriched $GABA_ARs$. Below, Table 1 lists desensitization times of different $GABA_AR$ subtypes when GABA was locally applied for the durations specified. We encourage the reader to first notice the differences in desensitization within each reference [96, 97]. Then notice the differences in desensitization rates that occur as GABA application times vary (between references) [96, 97].

4.2 Glycine Receptors (GlyRs)

GlyRs are pentameric receptors composed of α and β subunits arranged 2α:3β [98], expressed in the spinal cord and brainstem at high levels but also exist throughout the CNS. They are activated by presynaptically released glycine and are sensitive to the antagonist strychnine. Upon activation, GlyRs increase chloride conductance in the postsynaptic cell, inducing hyperpolarizing membrane potentials[3]. The influx of chloride ions through GlyRs is characterized by a fast rising phase with a bi-exponential decay phase. The rapid rise in current is mediated by the synchronous activation of GlyRs, whereas the bi-exponential decay phase suggests that GlyRs undergo asynchronous closure [99, 100]. Like many receptors, subunit composition is critically involved in dictating receptor

Table 2
Glycinergic IPSC time constants

Rat spinal neuron developmental stage	Rise time (10–90 %) (ms)	Decay time constant (ms)	Mean time constant (ms)
Embryonic day 20	0.91 ± 0.16[a]	Fast 14.8, slow 48.1[a]	27.0 ± 9.1[a]
Postnatal day 16–22	0.76 ± 0.22[a]	Fast 4.73, slow 14.0[a]	10.2 ± 2.4[a]

[a][101]

kinetics. During early development glycinergic IPSCs in the spinal cord are relatively slow compared to later in development (rat spinal neurons). Glycinergic IPSC mean time constant at embryonic day 20 is ~27 ms versus ~10 ms at postnatal day 22 [101]. The shortening of the glycinergic IPSC time is likely due to the molecular switching from the α2 to the α1 subunit at the postsynaptic membrane [101] (Table 2).

5 Summary

In this chapter we discuss the kinetics of synaptic currents generated by excitatory and inhibitory ionotropic receptors. The slow excitatory postsynaptic currents (EPSCs) are primarily generated by NMDARs, while non-NMDARs, nAChRs, and 5-HT3Rs elicit relatively faster EPSCs. Inhibitory postsynaptic currents are generated by $GABA_A$Rs and GlyRs with relatively comparable expression levels throughout the CNS as well as similar receptor kinetics.

It is important to note that these receptors are made up of protein subunits that undergo conformational changes, which initiate activation, deactivation, and desensitization phases. Therefore, changes in subunit gating caused by subunit expression, temperature, or protein–protein interactions can affect the current kinetics within each ionotropic receptor group discussed.

References

1. Armstrong CM (1981) Sodium channels and gating currents. Physiol Rev 61(3):644–683
2. Barberis A, Mozrzymas JW, Ortinski PI, Vicini S (2007) Desensitization and binding properties determine distinct alpha1beta2gamma2 and alpha3beta2gamma2 GABA(A) receptor-channel kinetic behavior. Eur J Neurosci 25(9):2726–2740
3. Bettler B, Boulter J, Hermans-Borgmeyer I, O'Shea-Greenfield A, Deneris ES, Moll C, Borgmeyer U, Hollmann M, Heinemann S (1990) Cloning of a novel glutamate receptor subunit, GluR5: expression in the nervous system during development. Neuron 5(5):583–595
4. Bettler B, Egebjerg J, Sharma G, Pecht G, Hermans-Borgmeyer I, Moll C, Stevens CF, Heinemann S (1992) Cloning of a putative glutamate receptor: a low affinity kainate-binding subunit. Neuron 8(2):257–265
5. Boulter J, Hollmann M, O'Shea-Greenfield A, Hartley M, Deneris E, Maron C, Heinemann S (1990) Molecular cloning and functional expression of glutamate receptor

subunit genes. Science 249(4972): 1033–1037

6. Egebjerg J, Bettler B, Hermans-Borgmeyer I, Heinemann S (1991) Cloning of a cDNA for a glutamate receptor subunit activated by kainate but not AMPA. Nature 351(6329): 745–748
7. Herb A, Burnashev N, Werner P, Sakmann B, Wisden W, Seeburg PH (1992) The KA-2 subunit of excitatory amino acid receptors shows widespread expression in brain and forms ion channels with distantly related subunits. Neuron 8(4):775–785
8. Keinanen K, Wisden W, Sommer B, Werner P, Herb A, Verdoorn TA, Sakmann B, Seeburg PH (1990) A family of AMPA-selective glutamate receptors. Science 249(4968):556–560
9. Morita T, Sakimura K, Kushiya E, Yamazaki M, Meguro H, Araki K, Abe T, Mori KJ, Mishina M (1992) Cloning and functional expression of a cDNA encoding the mouse beta 2 subunit of the kainate-selective glutamate receptor channel. Brain Res Mol Brain Res 14(1–2):143–146
10. Sakimura K, Morita T, Kushiya E, Mishina M (1992) Primary structure and expression of the gamma 2 subunit of the glutamate receptor channel selective for kainate. Neuron 8(2):267–274
11. Sommer B, Burnashev N, Verdoorn TA, Keinanen K, Sakmann B, Seeburg PH (1992) A glutamate receptor channel with high affinity for domoate and kainate. EMBO J 11(4):1651–1656
12. Werner P, Voigt M, Keinanen K, Wisden W, Seeburg PH (1991) Cloning of a putative high-affinity kainate receptor expressed predominantly in hippocampal CA3 cells. Nature 351(6329):742–744
13. Hollmann M, Hartley M, Heinemann S (1991) Ca2+ permeability of KA-AMPA--gated glutamate receptor channels depends on subunit composition. Science 252(5007): 851–853
14. Hume RI, Dingledine R, Heinemann SF (1991) Identification of a site in glutamate receptor subunits that controls calcium permeability. Science 253(5023):1028–1031
15. Sommer B, Keinanen K, Verdoorn TA, Wisden W, Burnashev N, Herb A, Kohler M, Takagi T, Sakmann B, Seeburg PH (1990) Flip and flop: a cell-specific functional switch in glutamate-operated channels of the CNS. Science 249(4976):1580–1585
16. Edmonds B, Gibb AJ, Colquhoun D (1995) Mechanisms of activation of glutamate receptors and the time course of excitatory synaptic currents. Annu Rev Physiol 57:495–519
17. Kiskin NI, Krishtal OA, Tsyndrenko A (1986) Excitatory amino acid receptors in hippocampal neurons: kainate fails to desensitize them. Neurosci Lett 63(3):225–230
18. Llano I, Marty A, Armstrong CM, Konnerth A (1991) Synaptic- and agonist-induced excitatory currents of Purkinje cells in rat cerebellar slices. J Physiol 434:183–213
19. Nelson PG, Pun RY, Westbrook GL (1986) Synaptic excitation in cultures of mouse spinal cord neurones: receptor pharmacology and behaviour of synaptic currents. J Physiol 372:169–190
20. Patneau DK, Mayer ML (1990) Structure-activity relationships for amino acid transmitter candidates acting at N-methyl-D-aspartate and quisqualate receptors. J Neurosci 10(7):2385–2399
21. Silver RA, Traynelis SF, Cull-Candy SG (1992) Rapid-time-course miniature and evoked excitatory currents at cerebellar synapses in situ. Nature 355(6356):163–166
22. Trussell LO, Fischbach GD (1989) Glutamate receptor desensitization and its role in synaptic transmission. Neuron 3(2):209–218
23. Colquhoun D, Jonas P, Sakmann B (1992) Action of brief pulses of glutamate on AMPA/kainate receptors in patches from different neurones of rat hippocampal slices. J Physiol 458:261–287
24. Hestrin S (1992) Activation and desensitization of glutamate-activated channels mediating fast excitatory synaptic currents in the visual cortex. Neuron 9(5):991–999
25. Hestrin S (1993) Different glutamate receptor channels mediate fast excitatory synaptic currents in inhibitory and excitatory cortical neurons. Neuron 11(6):1083–1091
26. Raman IM, Trussell LO (1992) The kinetics of the response to glutamate and kainate in neurons of the avian cochlear nucleus. Neuron 9(1):173–186
27. Tang CM, Shi QY, Katchman A, Lynch G (1991) Modulation of the time course of fast EPSCs and glutamate channel kinetics by aniracetam. Science 254(5029):288–290
28. Veruki ML, Morkve SH, Hartveit E (2003) Functional properties of spontaneous EPSCs and non-NMDA receptors in rod amacrine (AII) cells in the rat retina. J Physiol 549(Pt 3):759–774
29. Sasaki YF, Rothe T, Premkumar LS, Das S, Cui J, Talantova MV, Wong HK, Gong X, Chan SF, Zhang D, Nakanishi N, Sucher NJ,

Lipton SA (2002) Characterization and comparison of the NR3A subunit of the NMDA receptor in recombinant systems and primary cortical neurons. J Neurophysiol 87(4):2052–2063

30. Kleckner NW, Dingledine R (1988) Requirement for glycine in activation of NMDA-receptors expressed in Xenopus oocytes. Science 241(4867):835–837
31. Schneggenburger R, Zhou Z, Konnerth A, Neher E (1993) Fractional contribution of calcium to the cation current through glutamate receptor channels. Neuron 11(1): 133–143
32. Sah P, Hestrin S, Nicoll RA (1990) Properties of excitatory postsynaptic currents recorded in vitro from rat hippocampal interneurones. J Physiol 430:605–616
33. Jahr CE, Stevens CF (1990) Voltage dependence of NMDA-activated macroscopic conductances predicted by single-channel kinetics. J Neurosci 10(9):3178–3182
34. Hestrin S, Sah P, Nicoll RA (1990) Mechanisms generating the time course of dual component excitatory synaptic currents recorded in hippocampal slices. Neuron 5(3):247–253
35. Jahr CE (1992) High probability opening of NMDA receptor channels by L-glutamate. Science 255(5043):470–472
36. Lester RA, Clements JD, Westbrook GL, Jahr CE (1990) Channel kinetics determine the time course of NMDA receptor-mediated synaptic currents. Nature 346(6284):565–567
37. Charpantier E, Barneoud P, Moser P, Besnard F, Sgard F (1998) Nicotinic acetylcholine subunit mRNA expression in dopaminergic neurons of the rat substantia nigra and ventral tegmental area. Neuroreport 9(13): 3097–3101
38. Cimino M, Marini P, Fornasari D, Cattabeni F, Clementi F (1992) Distribution of nicotinic receptors in cynomolgus monkey brain and ganglia: localization of alpha 3 subunit mRNA, alpha-bungarotoxin and nicotine binding sites. Neuroscience 51(1):77–86
39. Clarke PB, Schwartz RD, Paul SM, Pert CB, Pert A (1985) Nicotinic binding in rat brain: autoradiographic comparison of [3H]acetylcholine, [3H]nicotine, and [125I]-alpha-bungarotoxin. J Neurosci 5(5):1307–1315
40. Seguela P, Wadiche J, Dineley-Miller K, Dani JA, Patrick JW (1993) Molecular cloning, functional properties, and distribution of rat brain alpha 7: a nicotinic cation channel highly permeable to calcium. J Neurosci 13(2):596–604
41. Wada E, Wada K, Boulter J, Deneris E, Heinemann S, Patrick J, Swanson LW (1989) Distribution of alpha 2, alpha 3, alpha 4, and beta 2 neuronal nicotinic receptor subunit mRNAs in the central nervous system: a hybridization histochemical study in the rat. J Comp Neurol 284(2):314–335
42. Mulle C, Vidal C, Benoit P, Changeux JP (1991) Existence of different subtypes of nicotinic acetylcholine receptors in the rat habenulo-interpeduncular system. J Neurosci 11(8):2588–2597
43. Kuba K, Tanaka E, Kumamoto E, Minota S (1989) Patch clamp experiments on nicotinic acetylcholine receptor-ion channels in bullfrog sympathetic ganglion cells. Pflugers Arch 414(2):105–112
44. Mathie A, Cull-Candy SG, Colquhoun D (1987) Single-channel and whole-cell currents evoked by acetylcholine in dissociated sympathetic neurons of the rat. Proc R Soc Lond B Biol Sci 232(1267):239–248
45. Moss BL, Schuetze SM, Role LW (1989) Functional properties and developmental regulation of nicotinic acetylcholine receptors on embryonic chicken sympathetic neurons. Neuron 3(5):597–607
46. Sargent PB (1993) The diversity of neuronal nicotinic acetylcholine receptors. Annu Rev Neurosci 16:403–443
47. Schofield GG, Weight FF, Adler M (1985) Single acetylcholine channel currents in sympathetic neurons. Brain Res 342(1): 200–203
48. Dani JA (2001) Overview of nicotinic receptors and their roles in the central nervous system. Biol Psychiatry 49(3):166–174
49. Quick MW, Lester RA (2002) Desensitization of neuronal nicotinic receptors. J Neurobiol 53(4):457–478
50. Castro NG, Albuquerque EX (1993) Brief-lifetime, fast-inactivating ion channels account for the alpha-bungarotoxin-sensitive nicotinic response in hippocampal neurons. Neurosci Lett 164(1–2):137–140
51. Albuquerque EX, Alkondon M, Pereira EF, Castro NG, Schrattenholz A, Barbosa CT, Bonfante-Cabarcas R, Aracava Y, Eisenberg HM, Maelicke A (1997) Properties of neuronal nicotinic acetylcholine receptors: pharmacological characterization and modulation of synaptic function. J Pharmacol Exp Ther 280(3):1117–1136
52. Alkondon M, Albuquerque EX (1993) Diversity of nicotinic acetylcholine receptors in rat hippocampal neurons. I. Pharmacological and functional evidence for distinct structural subtypes. J Pharmacol Exp Ther 265(3): 1455–1473

53. Castro NG, Albuquerque EX (1995) alpha-Bungarotoxin-sensitive hippocampal nicotinic receptor channel has a high calcium permeability. Biophys J 68(2):516–524
54. Papke RL (1993) The kinetic properties of neuronal nicotinic receptors: genetic basis of functional diversity. Prog Neurobiol 41(4): 509–531
55. Maricq AV, Peterson AS, Brake AJ, Myers RM, Julius D (1991) Primary structure and functional expression of the 5HT3 receptor, a serotonin-gated ion channel. Science 254(5030):432–437
56. Eisele JL, Bertrand S, Galzi JL, Devillers-Thiery A, Changeux JP, Bertrand D (1993) Chimaeric nicotinic-serotonergic receptor combines distinct ligand binding and channel specificities. Nature 366(6454):479–483
57. Neijt HC, te Duits IJ, Vijverberg HP (1988) Pharmacological characterization of serotonin 5-HT3 receptor-mediated electrical response in cultured mouse neuroblastoma cells. Neuropharmacology 27(3):301–307
58. van Hooft JA, Vijverberg HP (1996) Selection of distinct conformational states of the 5-HT3 receptor by full and partial agonists. Br J Pharmacol 117(5):839–846
59. Yakel JL, Lagrutta A, Adelman JP, North RA (1993) Single amino acid substitution affects desensitization of the 5-hydroxytryptamine type 3 receptor expressed in Xenopus oocytes. Proc Natl Acad Sci U S A 90(11): 5030–5033
60. Blandina P, Goldfarb J, Craddock-Royal B, Green JP (1989) Release of endogenous dopamine by stimulation of 5-hydroxytryptamine3 receptors in rat striatum. J Pharmacol Exp Ther 251(3):803–809
61. Lummis SC (2012) 5-HT(3) receptors. J Biol Chem 287(48):40239–40245
62. Miquel MC, Emerit MB, Nosjean A, Simon A, Rumajogee P, Brisorgueil MJ, Doucet E, Hamon M, Verge D (2002) Differential subcellular localization of the 5-HT3-As receptor subunit in the rat central nervous system. Eur J Neurosci 15(3):449–457
63. Thompson AJ, Lummis SC (2006) 5-HT3 receptors. Curr Pharm Des 12(28):3615–3630
64. Chameau P, van Hooft JA (2006) Serotonin 5-HT(3) receptors in the central nervous system. Cell Tissue Res 326(2):573–581
65. Thompson AJ, Lummis SC (2007) The 5-HT3 receptor as a therapeutic target. Expert Opin Ther Targets 11(4):527–540
66. Derkach V, Surprenant A, North RA (1989) 5-HT3 receptors are membrane ion channels. Nature 339(6227):706–709
67. Yang J (1990) Ion permeation through 5-hydroxytryptamine-gated channels in neuroblastoma N18 cells. J Gen Physiol 96(6): 1177–1198
68. Virginio C, North RA, Surprenant A (1998) Calcium permeability and block at homomeric and heteromeric P2X2 and P2X3 receptors, and P2X receptors in rat nodose neurones. J Physiol 510(Pt 1):27–35
69. Nicke A, Baumert HG, Rettinger J, Eichele A, Lambrecht G, Mutschler E, Schmalzing G (1998) P2X1 and P2X3 receptors form stable trimers: a novel structural motif of ligand-gated ion channels. EMBO J 17(11): 3016–3028
70. Stoop R, Thomas S, Rassendren F, Kawashima E, Buell G, Surprenant A, North RA (1999) Contribution of individual subunits to the multimeric P2X(2) receptor: estimates based on methanethiosulfonate block at T336C. Mol Pharmacol 56(5):973–981
71. Torres GE, Egan TM, Voigt MM (1999) Hetero-oligomeric assembly of P2X receptor subunits. Specificities exist with regard to possible partners. J Biol Chem 274(10): 6653–6659
72. Burnstock G, Knight GE (2004) Cellular distribution and functions of P2 receptor subtypes in different systems. Int Rev Cytol 240:31–304
73. Collo G, Neidhart S, Kawashima E, Kosco-Vilbois M, North RA, Buell G (1997) Tissue distribution of the P2X7 receptor. Neuropharmacology 36(9):1277–1283
74. Collo G, North RA, Kawashima E, Merlo-Pich E, Neidhart S, Surprenant A, Buell G (1996) Cloning OF P2X5 and P2X6 receptors and the distribution and properties of an extended family of ATP-gated ion channels. J Neurosci 16(8):2495–2507
75. Khakh BS (2001) Molecular physiology of P2X receptors and ATP signalling at synapses. Nat Rev Neurosci 2(3):165–174
76. Khakh BS, North RA (2006) P2X receptors as cell-surface ATP sensors in health and disease. Nature 442(7102):527–532
77. Rubio ME, Soto F (2001) Distinct Localization of P2X receptors at excitatory postsynaptic specializations. J Neurosci 21(2):641–653
78. Edwards FA, Gibb AJ, Colquhoun D (1992) ATP receptor-mediated synaptic currents in the central nervous system. Nature 359(6391):144–147
79. Shigetomi E, Kato F (2004) Action potential-independent release of glutamate by Ca2+ entry through presynaptic P2X receptors elicits

postsynaptic firing in the brainstem autonomic network. J Neurosci 24(12):3125–3135
80. North RA (2002) Molecular physiology of P2X receptors. Physiol Rev 82(4):1013–1067
81. Masin M, Kerschensteiner D, Dumke K, Rubio ME, Soto F (2006) Fe65 interacts with P2X2 subunits at excitatory synapses and modulates receptor function. J Biol Chem 281(7):4100–4108
82. Zhang M, Zhong H, Vollmer C, Nurse CA (2000) Co-release of ATP and ACh mediates hypoxic signalling at rat carotid body chemoreceptors. J Physiol 525(Pt 1): 143–158
83. Jo YH, Role LW (2002) Coordinate release of ATP and GABA at in vitro synapses of lateral hypothalamic neurons. J Neurosci 22(12):4794–4804
84. Pankratov Y, Lalo U, Verkhratsky A, North RA (2007) Quantal release of ATP in mouse cortex. J Gen Physiol 129(3):257–265
85. Stuber GD, Hnasko TS, Britt JP, Edwards RH, Bonci A (2010) Dopaminergic terminals in the nucleus accumbens but not the dorsal striatum corelease glutamate. J Neurosci 30(24):8229–8233
86. Hirasawa H, Betensky RA, Raviola E (2012) Corelease of dopamine and GABA by a retinal dopaminergic neuron. J Neurosci 32(38):13281–13291
87. Finger TE, Danilova V, Barrows J, Bartel DL, Vigers AJ, Stone L, Hellekant G, Kinnamon SC (2005) ATP signaling is crucial for communication from taste buds to gustatory nerves. Science 310(5753):1495–1499
88. Olsen RW, Sieghart W (2008) International Union of Pharmacology. LXX. Subtypes of gamma-aminobutyric acid(A) receptors: classification on the basis of subunit composition, pharmacology, and function. Update. Pharmacol Rev 60(3):243–260
89. Gonzalez-Burgos G, Lewis DA (2008) GABA neurons and the mechanisms of network oscillations: implications for understanding cortical dysfunction in schizophrenia. Schizophr Bull 34(5):944–961
90. Farrant M, Nusser Z (2005) Variations on an inhibitory theme: phasic and tonic activation of GABA(A) receptors. Nat Rev Neurosci 6(3):215–229
91. Jones MV, Westbrook GL (1997) Shaping of IPSCs by endogenous calcineurin activity. J Neurosci 17(20):7626–7633
92. Jones MV, Westbrook GL (1995) Desensitized states prolong GABAA channel responses to brief agonist pulses. Neuron 15(1):181–191
93. Maconochie DJ, Zempel JM, Steinbach JH (1994) How quickly can GABAA receptors open? Neuron 12(1):61–71
94. Chen L, Wang H, Vicini S, Olsen RW (2000) The gamma-aminobutyric acid type A (GABAA) receptor-associated protein (GABARAP) promotes GABAA receptor clustering and modulates the channel kinetics. Proc Natl Acad Sci U S A 97(21):11557–11562
95. Overstreet LS, Jones MV, Westbrook GL (2000) Slow desensitization regulates the availability of synaptic GABA(A) receptors. J Neurosci 20(21):7914–7921
96. Tia S, Wang JF, Kotchabhakdi N, Vicini S (1996) Distinct deactivation and desensitization kinetics of recombinant GABAA receptors. Neuropharmacology 35(9–10):1375–1382
97. Verdoorn TA, Draguhn A, Ymer S, Seeburg PH, Sakmann B (1990) Functional properties of recombinant rat GABAA receptors depend upon subunit composition. Neuron 4(6):919–928
98. Grudzinska J, Schemm R, Haeger S, Nicke A, Schmalzing G, Betz H, Laube B (2005) The beta subunit determines the ligand binding properties of synaptic glycine receptors. Neuron 45(5):727–739
99. Katz B, Miledi R (1973) The binding of acetylcholine to receptors and its removal from the synaptic cleft. J Physiol 231(3):549–574
100. Laube B (2002) Potentiation of inhibitory glycinergic neurotransmission by Zn2+: a synergistic interplay between presynaptic P2X2 and postsynaptic glycine receptors. Eur J Neurosci 16(6):1025–1036
101. Takahashi T, Momiyama A, Hirai K, Hishinuma F, Akagi H (1992) Functional correlation of fetal and adult forms of glycine receptors with developmental changes in inhibitory synaptic receptor channels. Neuron 9(6):1155–1161

Part IV

Experimentations with Computational Components

Chapter 18

Measurement of a Single Synapse

Nicholas Graziane and Yan Dong

Abstract

Electrophysiological measurements of single synapses are challenging given the size of a single synapse relative to a patch pipette. In addition, one has to take into account the limitations of microscopes in that they need to provide acceptable visualization of a single synapse for patching. However, despite these limitations, researchers have successfully measured single synaptic function along dendrites. The purpose of this chapter is to introduce the techniques that can be implemented to measure single synaptic function. Included in this chapter are such techniques as localized perfusion, localized electrical stimulation, photostimulation, and imaging. These techniques are designed with the assumption that multiple excitatory synapses do not contact a single spine, but rather only one synapse per spine. Whereas this assumption is supported by some empirical data [1], other data suggest otherwise [2], meaning that a complete understanding of the anatomical region is necessary before beginning single synapse experiments.

Key words Local perfusion, Local stimulation, Photostimulation, Imaging, Presynaptic vesicle dyes

1 Introduction

Electrophysiological measurements of single synapses are challenging given the size of a single synapse relative to a patch pipette. In addition, one has to take into account the limitations of microscopes in that they need to provide acceptable visualization of a single synapse for patching. However, despite these limitations, researchers have successfully measured single synaptic function along dendrites. The purpose of this chapter is to introduce the techniques that can be implemented to measure single synaptic function. Included in this chapter are such techniques as localized perfusion, localized electrical stimulation, photostimulation, and imaging. These techniques are designed with the assumption that multiple excitatory synapses do not contact a single spine, but rather only one synapse per spine. Whereas this assumption is supported by some empirical data [1], other data suggest otherwise [2], meaning that a complete understanding of the anatomical region is necessary before beginning single synapse experiments.

Nicholas Graziane and Yan Dong, *Electrophysiological Analysis of Synaptic Transmission*, Neuromethods, vol. 112,
DOI 10.1007/978-1-4939-3274-0_18,

2 Approaches for Determining Currents Generated at a Single Synapse

2.1 Localized Perfusion Techniques

Local perfusion techniques are useful strategies that restrict solutions to perfuse to specific regions, which allows for the localized activation of single synapses. This technique is typically performed in cell cultures given the ability to control the cell density and synaptic connections or in regions where few synaptic contacts are active (e.g., motor neurons, dentate granule cells). The perfusate used can contain any factor that initiates neurotransmitter release [e.g., Ca^{2+}, K^{+} (high K^{+} elicits action potentials), sucrose (sucrose makes the solution hypertonic causing exocytosis of presynaptic vesicles)] or activates postsynaptic responses directly (e.g., receptor agonists). For example, bathing cells in low Ca^{2+} solution blocks transmission throughout the preparation. In this environment, single synapses can be excited by microperfusing the synapse with a Ca^{2+}-containing solution [3, 4]. To elegantly demonstrate that the microperfused solution is localized to ~30 μm across, Bekkers and Stevens used a double-barreled pipette each with an internal diameter of 1–2 μm [3]. One pipette was filled with the Ca^{2+}-containing solution (to activate the synapses), while the other pipette was filled a dye (to visualize the localization of perfusion). A Picospritzer (General Valve) was used to provide gated pressure to each barrel at 1–3 psi, thus expelling the solution in a narrow stream. Since the activation of a single synapse was elicited using this technique, the authors measured currents using the whole-cell patch configuration. Another example of local perfusion again uses the double-barrel technique. However, in this approach one pipette with positive pressure delivers the solution, while the second pipette with negative pressure removes the perfused solution (Fig. 1) [5]. The removal of solution restricts the perfusate to a localized region, enabling measurements of activity at single synapses.

2.2 Localized Stimulation Techniques

Localized stimulation techniques are similar to localized perfusion techniques in that they require low density synaptic contacts in order to reliably initiate excitation of one synapse (Fig. 2). The difference however is that localized stimulation of synapses is through electrical pulses. These electrical pulses are generated through a fine-tipped glass pipette (~2 μm tip diameter) positioned nearby the targeted synapse [6, 7]. To avoid transmitter release at non-stimulated boutons, tetrodotoxin (TTX) is included in the bathing solution. In order to target single synapses, visualization is necessary. Typically dyes are introduced to label the synaptic targets. One option is to introduce the dye into the recorded cell and allow it to perfuse the cell (10–15 min) until spines are visualized. A second option is to label presynaptic vesicles (e.g., RH414), thus identifying synaptic boutons that can be targeted for measurements [7]. With the synaptic boutons imaged, the localized stimulation of synaptic boutons can elicit currents that are measured simply by using the whole-cell patch configuration.

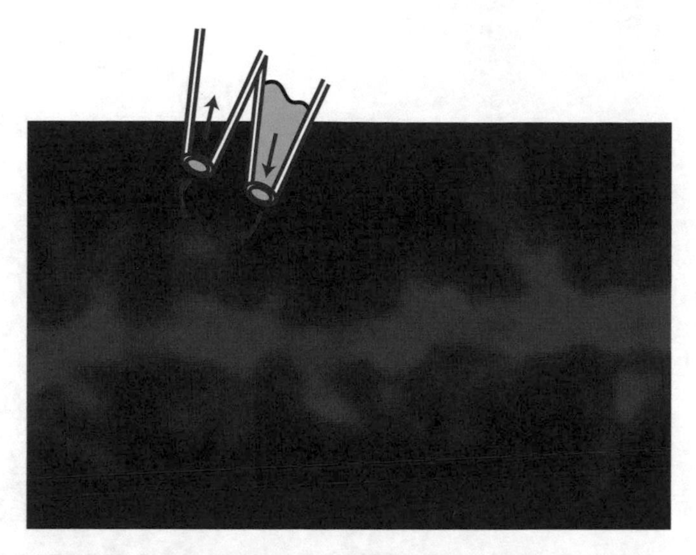

Fig. 1 Double barrel local perfusion targeting a single synapse. One micropipette is filled with a solution that initiates neurotransmitter release. Positive pressure expels the solution from one micropipette and the solutions is removed from a second micropipette via negative pressure

Fig. 2 Localized stimulation of a single synapse using electrical pulses generated by a fine-tipped glass pipette

2.3 Photostimulation

Photostimulation encompasses methods that use light to uncage a compound, rendering it bioactive (Fig. 3). The process works when a caged compound absorbs a photon. The absorbed photon breaks the covalent bond between the light sensitive caged group and the rest of the molecule. As the caged group breaks free, the uncaged compound transforms from an inactive to an active molecule. These caged compounds have minimal interaction with the biological system in their inactive state and they quickly enter the bioactive state when exposed to light. Photostimulation is a useful measure of single synapses activity because the light used to uncage

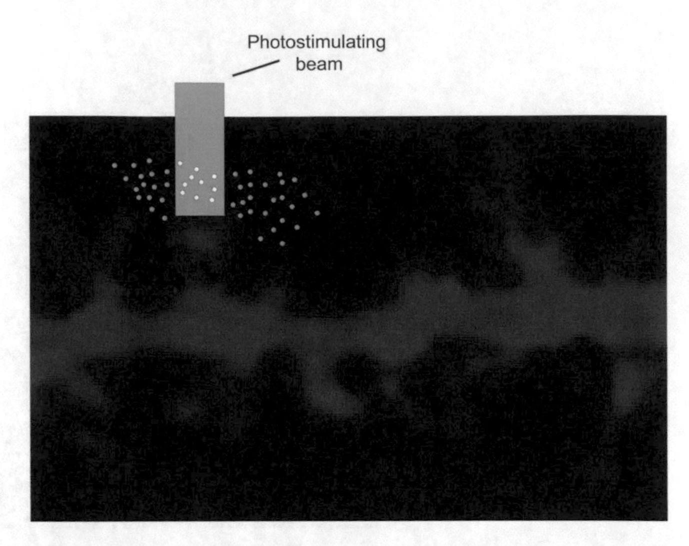

Fig. 3 Photostimulation of caged compounds (*green dots*) uncages the compounds (*white dots*) leading to their activation within a small region encompassing a synapse

the compound can be localized to a small region dictated by the diameter of the laser beam. This restricts the uncaging of the molecules to locations nearby the synapse (within ~1 μm of the laser). For example, Matsuzaki et al. used caged glutamate to test how LTP at hippocampal CA1 pyramidal neuron synapses alters synaptic morphology. Using two-photon photolysis of caged glutamate, the authors were able to uncage glutamate nearby a single synapse [8]. In doing this, they were able to detect morphological changes at the activated single synapse as well as the currents produced. Two-photon uncaging of glutamate has also been used by other investigators to determine changes in NMDAR subunit composition at single synapses [9] or the activation of intracellular signaling molecules associated with synaptic plasticity (e.g., CaMKII) [10].

2.4 Imaging

Quantal events are responses generated by the release of one vesicle from a synapse. The measurement tracks the currents generated by the spontaneous release of the quantal content in the presence of TTX or strontium. This approach is a straightforward and easy strategy to isolate single synaptic currents. However, the caveat is that specific synapses cannot be tracked since the recorded currents are produced from a random pool of synapses. In addition, action potential-dependent synchronous release of neurotransmitters cannot be measured at a specific synapse. To overcome these problems, more direct sampling approaches have been developed. For example, patching directly onto the synaptic bouton has been successfully accomplished [11]. First, boutons had to be visualized using the fluorescent dye, FM1-43. Once a bouton is isolated,

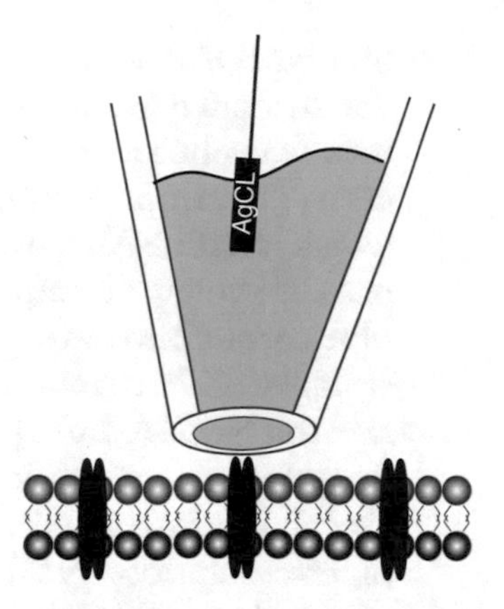

Fig. 4 A loose patch. A micropipette is positioned just above the cell membrane so that a gigaseal is not formed. Notice that the membrane is not deformed in a loose patch as is the case in a cell-attached patch in which negative pressure elicits a gigaseal (see Fig. 1.2)

a loose-patch[1] (Fig. 4) can be formed onto the dendritic membrane. Using this preparation, miniature excitatory postsynaptic currents (mEPSCs) are recorded without signal contamination from neighboring synapses as those signals are too small to be detected above the background noise. By isolating single synapses using this method, the sampled EPSC kinetics, amplitude, and frequency can be analyzed. In addition, this technique allows the experimenter to simultaneously record from the same cell in the whole cell patch configuration (with a second micropipette), making it possible to draw comparisons between functions of a single synapse and multiple synaptic connections innervating the cell's dendrites.

Another approach to measure quantal events at single synapses is to image NMDAR-mediated Ca^{2+} influx into individual spines [assuming that one synapse is present at an individual spine and

[1] A loose patch (Fig. 4) is formed in the absence of the conventional gigaseal, which is necessary for all other patch clamp techniques (e.g., cell-attached, inside out, outside out, whole cell). Instead, in a loose patch clamp, a loose seal is formed with a low electrical resistance (2–11 MΩ). Typically, the micropipettes used for patch-clamp experiments have resistances of 2–5 MΩ. Therefore, watching the resistance of the micropipette while approaching the membrane can indicate when a loose patch is forming (the resistance will increase). As the distance between the micropipette and the membrane decreases, the resistance increases. By observing the changes in resistance, the experimenter can avoid coming too close to the membrane and thus forming a seal. Instead, the micropipette should stay far enough away from the membrane so that a gigaseal is not formed leaving the membrane intact (no suction is used). Since the loose patch method is a noninvasive patch, the EPSCs are visualized as outward and not inward currents. This patch-clamp technique is useful in that the micropipette can be removed from the membrane sampling multiple locations without any damage to the cell.

that NMDAR activation is not saturated following an action potential mediated release of glutamate [12–14]] [15]. This method uses two-photon laser scanning microscopy[2] and implements necessary electrophysiological and pharmacological manipulations, which isolate NMDAR-mediated Ca^{2+} currents. For example, neurons are voltage clamped at +10 mV in order to remove the Mg^{2+} block from NMDARs, while maintaining inactivation of voltage-sensitive Ca^{2+} channels. AMPARs are blocked pharmacologically and the NMDAR antagonist APV is used to demonstrate the specificity of the Ca^{2+} transients produced [15]. Spines are imaged using Ca^{2+} insensitive fluorophores, which are loaded into the internal pipette solution. NMDAR-mediated Ca^{2+} transients are evoked using electrical stimulation and imaged using a Ca^{2+} indicator, which is also loaded into the internal pipette solution. The Ca^{2+} indicator fluoresces when bound to Ca^{2+}. Increases in fluorescence intensity detected at a single synapse following electrically evoked stimuli are then mediated by NMDAR-mediated Ca^{2+} currents.

2.5 Imaging Presynaptic Vesicles

Monitoring presynaptic vesicles can be performed by using dyes. A commonly used lipophilic dye is FM 1-43, which is introduced into vesicles by exposing the slice or culture preparation to FM 1-43 in the bath solution (e.g., 16 μM). Once FM 1-43 is introduced into the bath solution, an evoked stimulus is used (e.g., 5 or 10 stimuli delivered at 10 Hz) to cause action potential induced release of neurotransmitters. Following vesicle fusion and neurotransmitter release, the presynaptic vesicle undergoes endocytosis during which time the FM 1-43 dye is taken up into the vesicle. After full staining of the vesicles, the FM 1-43 dye is washed out of solution (e.g., 15 min with dye-free Tyrode solution containing zero Ca^{2+} to minimize spontaneous exocytosis). The FM 1-43 loaded vesicle is then reloaded with neurotransmitter and the release can be tracked using fluorescence detection [17]. This approach has provided significant information explaining action potential-induced neurotransmitter release at a synaptic bouton. For example, data suggest that multiple quanta are released from a single bouton following action potential-induced release of neurotransmitters [18], and that this multivesicular release is consistent between trials [18, 19].

---

[2] Two-photon laser scanning microscopy is a fluorescent imaging technique that enables deep tissue penetration, reduced light diffraction in the tissue and limits phototoxicity [16]. It works by exciting a fluorophore (a fluorescent chemical compound that reemits light when excited) using two photons of low energy (i.e., two photons with half of the wavelength needed to excite the fluorophore). The simultaneous absorption of the two photons by the fluorophore results in fluorescence emission. What gives this imaging technique the advantages listed above is that the probability of two photons simultaneously exciting a fluorophore is very low. Therefore, the two photons must be concentrated spatially and temporally at the focal plane. Only then at the focal plane are fluorophores excited, thus drastically reducing background noise.

3 Summary

A great challenge of electrophysiological measurements is isolating a single synapse and quantifying its current input. With techniques such as localized perfusion, localized electrical stimulation, photostimulation, and imaging, measurements at a single synapse is possible. Each technique has been implemented successfully and has contributed necessary information to our understanding of synaptic function.

References

1. Schikorski T, Stevens CF (1997) Quantitative ultrastructural analysis of hippocampal excitatory synapses. J Neurosci 17(15):5858–5867
2. Toni N, Buchs PA, Nikonenko I, Bron CR, Muller D (1999) LTP promotes formation of multiple spine synapses between a single axon terminal and a dendrite. Nature 402(6760): 421–425
3. Bekkers JM, Stevens CF (1995) Quantal analysis of EPSCs recorded from small numbers of synapses in hippocampal cultures. J Neurophysiol 73(3):1145–1156
4. Kraszewski K, Grantyn R (1992) Unitary, quantal and miniature GABA-activated synaptic chloride currents in cultured neurons from the rat superior colliculus. Neuroscience 47(3):555–570
5. Veselovsky NS, Engert F, Lux HD (1996) Fast local superfusion technique. Pflugers Arch 432(2):351–354
6. Chen G, Harata NC, Tsien RW (2004) Paired-pulse depression of unitary quantal amplitude at single hippocampal synapses. Proc Natl Acad Sci U S A 101(4):1063–1068
7. Kirischuk S, Veselovsky N, Grantyn R (1999) Relationship between presynaptic calcium transients and postsynaptic currents at single gamma-aminobutyric acid (GABA)ergic boutons. Proc Natl Acad Sci U S A 96(13): 7520–7525
8. Matsuzaki M, Honkura N, Ellis-Davies GC, Kasai H (2004) Structural basis of long-term potentiation in single dendritic spines. Nature 429(6993):761–766
9. Lee MC, Yasuda R, Ehlers MD (2010) Metaplasticity at single glutamatergic synapses. Neuron 66(6):859–870
10. Lee SJ, Escobedo-Lozoya Y, Szatmari EM, Yasuda R (2009) Activation of CaMKII in single dendritic spines during long-term potentiation. Nature 458(7236):299–304
11. Forti L, Bossi M, Bergamaschi A, Villa A, Malgaroli A (1997) Loose-patch recordings of single quanta at individual hippocampal synapses. Nature 388(6645):874–878
12. Mainen ZF, Malinow R, Svoboda K (1999) Synaptic calcium transients in single spines indicate that NMDA receptors are not saturated. Nature 399(6732):151–155
13. McAllister AK, Stevens CF (2000) Nonsaturation of AMPA and NMDA receptors at hippocampal synapses. Proc Natl Acad Sci U S A 97(11):6173–6178
14. Umemiya M, Senda M, Murphy TH (1999) Behaviour of NMDA and AMPA receptor-mediated miniature EPSCs at rat cortical neuron synapses identified by calcium imaging. J Physiol 521(Pt 1):113–122
15. Oertner TG, Sabatini BL, Nimchinsky EA, Svoboda K (2002) Facilitation at single synapses probed with optical quantal analysis. Nat Neurosci 5(7):657–664
16. Denk W, Strickler JH, Webb WW (1990) Two-photon laser scanning fluorescence microscopy. Science 248(4951):73–76
17. Aravanis AM, Pyle JL, Tsien RW (2003) Single synaptic vesicles fusing transiently and successively without loss of identity. Nature 423(6940):643–647
18. Prange O, Murphy TH (1999) Analysis of multiquantal transmitter release from single cultured cortical neuron terminals. J Neurophysiol 81(4):1810–1817
19. Nauen DW (2011) Methods of measuring activity at individual synapses: a review of techniques and the findings they have made possible. J Neurosci Methods 194(2): 195–205

Chapter 19

Measurement of Silent Synapses

Nicholas Graziane and Yan Dong

Abstract

In this chapter we describe the approaches that can be implemented to estimate the level of excitatory silent synapses. For each approach there are technical considerations, which we describe at the end of the chapter. However, before laying out the experimental design, a brief description of silent synapses as well as their functional role in physiology will precede.

Key words Coefficient of variation, Minimal stimulation

1 Introduction

In this chapter we describe the approaches that can be implemented to estimate the level of excitatory silent synapses. For each approach there are technical considerations, which we describe at the end of the chapter. However, before laying out the experimental design, a brief description of silent synapses as well as their functional role in physiology will precede.

2 Silent Synapses

A silent synapse is a term used to describe an excitatory synapse that is non-conducting at resting membrane potentials. This essentially means that AMPARs are either nonfunctional or absent from the postsynaptic membrane leaving only functional NMDARs. Since NMDARs are non-conducting at resting membrane potentials, these synapses are considered silent and can only be detected at depolarizing potentials when NMDARs are activated (i.e., AMPAR-silent, NMDAR-only synapses). As for silent synapse expression, they are highly enriched in the neonatal brain, but decrease to lower levels in adulthood (the levels of silent synapses are dependent upon the brain region selected) [1–3].

Nicholas Graziane and Yan Dong, *Electrophysiological Analysis of Synaptic Transmission*, Neuromethods, vol. 112,
DOI 10.1007/978-1-4939-3274-0_19, © Springer Science+Business Media New York 2016

It is posited that silent synapses are destined for two outcomes; either synaptic stabilization or synaptic elimination. Synapses can be stabilized by correlated presynaptic and postsynaptic activity. This activity-dependent plasticity can either stabilize or traffic AMPARs into the postsynaptic membrane [4, 5]. Whatever the mechanism, silent synapses appear to be efficient plasticity substrates that enable fast presynaptic-to-postsynaptic connections to occur during activity-dependent plasticity [1, 6–13]. Since activity-dependent plasticity, such as LTP (see Chap. 12), is believed to be required for learning and memory, silent synapse are thought to be key substrates for encoding new learned behaviors [4].

Silent synapses can potentially act as substrates for synaptic elimination/pruning as synapses that were once functional lose AMPARs leaving only NMDARs. The belief is that glutamatergic synapses are disassembled following the sequential removal of AMPARs and NMDARs. Support of this comes from studies in Alzheimer's disease models in which synaptic elimination is preceded by AMPAR and NMDAR synaptic removal [5, 14, 15]. Therefore, the detection of silent synapses can be a critical time window that follows AMPAR synaptic removal and precedes NMDAR synaptic elimination.

Aberrant regulation of synaptic stabilization and/or elimination is posited to underlie pathological conditions such as addiction and neurodegenerative disorders. For example, in addiction research, cocaine administration triggers silent synapse formation, which is detectable up to 5 days post-treatment, but disappears by day 7 [16, 17]. The generation of these silent synapses is predicted to precede synaptic stabilization leading to LTP-like mechanisms (thus the disappearance of silent synapses by day 7 post-treatment) [18, 19]. Reversing the formation of these cocaine-generated synapses (using LTD) has been shown to block cocaine-seeking behaviors in drug abstinent animals [17]. These results suggest that newly formed synapses contribute to addiction-like behavior and that reversing the formation of these synapses can prevent drug-seeking. If this is true, preventing silent synapse formation could be a potential therapeutic target for addiction. In addition, aberrant synaptic elimination may be the cause of neurodegenerative diseases such as Alzheimer's disease. It is posited that amyloid-β oligomers trigger the synaptic removal of AMPARs followed by the removal of NMDARs [5, 14, 15, 20, 21]. These results suggest that silent synapse formation is a predecessor to downstream synaptic reorganization that when gone awry, can lead to pathological conditions.

3 Measuring Silent Synapses

3.1 Coefficient of Variance Analysis (CV Analysis)

The CV of EPSCs measures how much the sample EPSC amplitudes vary from the mean EPSC amplitude. Therefore, a low CV represents small deviations between sampled EPSCs and the mean

EPSC current, whereas a high CV reflects large deviations between sampled EPSCs and the mean EPSC. Silent synapses can be calculated using the CV analysis approach [22–24] by taking the CV-NMDAR/CV-AMPAR. The following describes how to isolate both AMPAR and NMDAR currents. In the whole-cell patch configuration, stable (i.e., no current run-up/down is present) excitatory postsynaptic currents (EPSCs) can be measured by voltage-clamping the cell at a hyperpolarizing potential (e.g., –70 mV), thus isolating AMPAR-mediated currents. After 50 consecutive currents are recorded, the cell is then voltage-clamped at a depolarizing potential (+50 mV) and 50 consecutive stable currents are then measured, thus isolating NMDAR+AMPAR-mediated currents. Once the measurement is complete, the CV-NMDAR/CV-AMPAR can be calculated using the following formula at both hyperpolarizing (CV-AMPAR) and depolarizing potentials (CV-NMDAR):

$$CV = \sqrt{SV(EPSC) - SV(noise)} \,/\, mean(EPSC) \tag{1}$$

where SV(EPSC) is the sample variation of the peak current amplitude, SV(noise) is the sample variation of the root mean square noise and mean (EPSC) is the average of the EPSC peak amplitude. This calculation is easily done at hyperpolarizing potentials because only AMPAR-mediated currents are visible. However, at depolarizing potentials a dual component (AMPAR current + NMDAR current) is visible. In order to isolate the NMDAR currents at the depolarizing potential, the peak EPSC amplitude can be measured 35 ms after the peak amplitude from the EPSC recorded at –70 mV (see Fig. 9.1). At this time point, the dual EPSC is attributable to mainly NMDARs [16]. After the NMDAR component is properly isolated, the SV(EPSC) can be calculated followed by the CV using Eq. (1). Once the CV-NMDAR and the CV-AMPAR is calculated, the ratio CV-NMDAR to CV-AMPAR can then be calculated.

How should the experimenter interpret the CV-NMDAR/CV-AMPAR? A significant decrease in CV-NMDAR/CV-AMPAR between control and experimental groups suggests the presence of silent synapses. A decrease in the ratio can be caused by either a decrease in CV-NMDAR or an increase in CV-AMPAR. A decrease in CV-NMDAR reflects an increase in the number of release sites (N), which is expected when new synapses are generated (Fig. 1). Increases in N will decrease CV, which can be shown mathematically by substituting for mean EPSC in Eq. (1) with N, Pr, *and* Q. Remember from Chap. 11 that

$$meanEPSC(I) = N \Pr Q \tag{2}$$

where N refers to the number of release sites, Pr refers to the probability of release, and Q refers to the quantal size. Therefore, Eq. (1) becomes

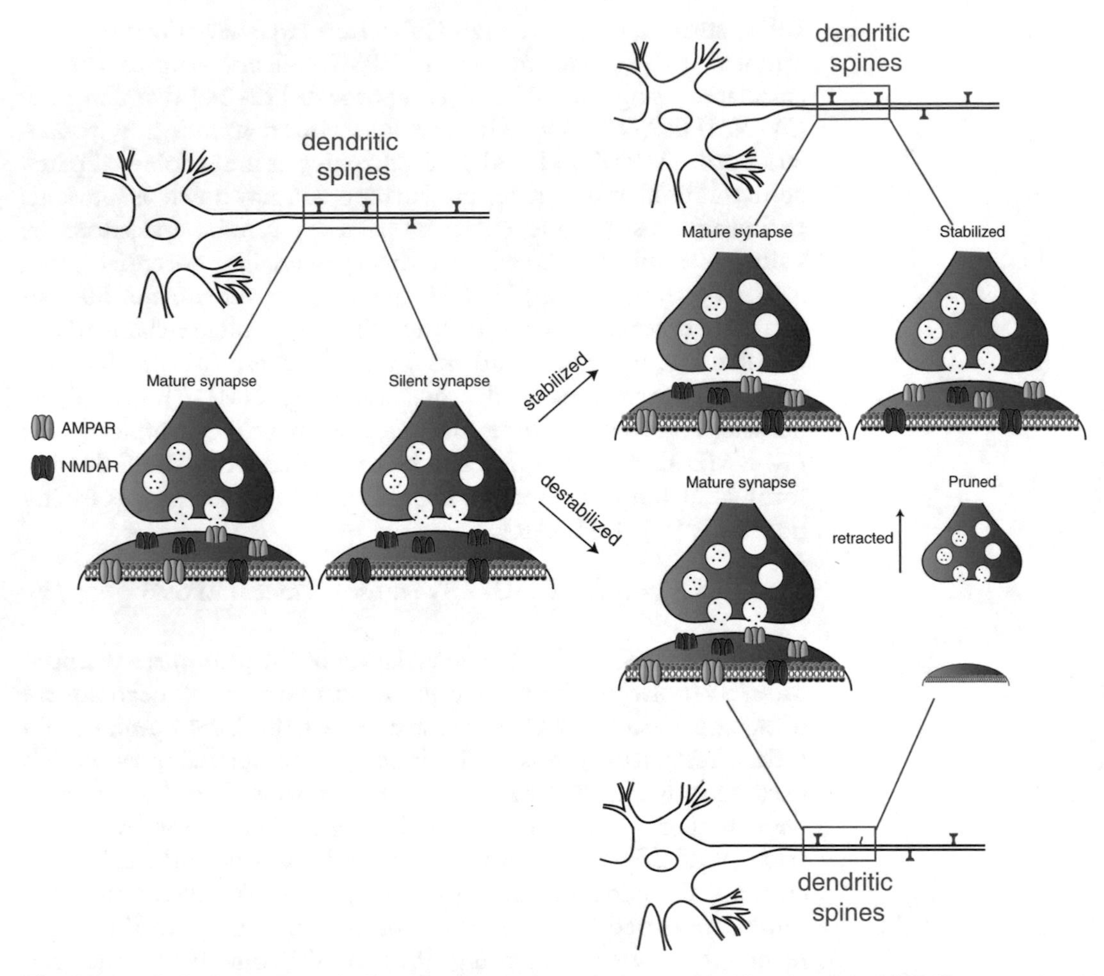

Fig. 1 Fate of silent synapses. A silent synapse is formed along a dendrite shown by NMDAR expression in the absence of AMPARs. This NMDAR-containing synapse is capable of recruiting AMPARs, thus forming a stabilized, functional synapse (*top*) or AMPARs are not recruited causing synaptic weakening and eventual elimination (*bottom*)

$$\mathrm{CV} = \sqrt{\mathrm{SV}(\mathrm{EPSC}) - \mathrm{SV}(\mathrm{noise})} \, / \, N \, \mathrm{Pr} \, Q \tag{3}$$

Assuming that Pr and *Q* remain constant during the measurement, an increase in *N* reduces the calculated CV. In this case, *N* increases as newly generated synapses are formed expressing only NMDARs. This causes the CV-NMDAR/CV-AMPAR ratio to decrease as CV-NMDAR decreases and CV-AMPAR remains the same.

Another possibility is that silent synapses are generated via silencing of AMPARs at once functional synapses. This could occur when AMPARs are removed from the postsynaptic membrane. If this is the case, the changes will be reflected in *Q*. So as functional AMPARs decrease, *Q* becomes smaller causing CV to increase (see Eq. 3). Therefore, CV-NMDAR is unchanged, while CV-AMPAR increases. This causes the CV-NMDAR/CV-AMPAR to decrease suggesting silent synapses are present and potentially are generated by AMPAR endocytosis. An approach for measuring AMPAR endocytosis is to bath-apply AMPA and measure the change in the holding current [25]. A smaller change in the holding current before and after AMPA application in experiment versus control groups suggests AMPAR endocytosis; however, this approach includes both synaptic and extrasynaptic AMPARs in the assay. Another approach to test whether silent synapses are generated by AMPAR endocytosis is to block activity-dependent AMPAR endocytosis in vivo [26–29] and then measure silent synapses in vitro. By blocking activity-dependent endocytosis of AMPARs, silent synapse generation can be blocked.

3.2 Minimal Stimulation Assay

Silent synapses can also be measured using the minimal stimulation approach. In whole-cell patch configuration, the cell is voltage clamped at a hyperpolarizing potential (e.g., –70 mV). Once small currents (<40 pA) are obtained at the hyperpolarizing potential, the stimulus intensity is reduced so that currents are evoked 50 % of the time. Once a 50/50 ratio of failures to successes is achieved, the stimulus intensity and frequency remain fixed for the duration of the experiment. Fifty current traces at –70 mV are measured followed by 50 current traces at a positive holding potential (e.g., +50 mV) such that the Mg^{2+} block can be removed from NMDARs (Fig. 2). The percent of silent synapses can then be calculated using the equation

$$= 1 - Ln\left(F_{-70\text{mV}}\right) / Ln\left(F_{+50\text{mV}}\right) \quad (4)$$

in which $F_{-70\text{ mV}}$ is the failure rate at –70 mV and $F_{+50\text{ mV}}$ is the failure rate at +50 mV [1, 7, 10].

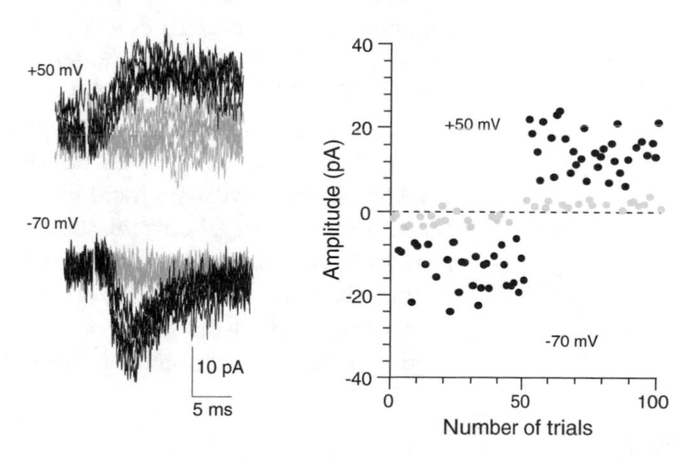

Fig. 2 Example traces and time course from a minimal stimulation assay in which current successes and failures (50/50) are elicited at –70 mV followed by depolarizing the neuron to +50 mV and recording the currents using the same minimal stimulation

4 Technical Considerations

When measuring silent synapses either by using the CV assay or the minimal stimulation assay, it is important to be sure that the access resistance remains consistent. The access resistance can be monitored throughout the recording and should not deviate by more than 20 % between conditions. A second technical consideration to consider is that since the neuron is held at both hyperpolarizing and depolarizing potentials, the reversal potential needs to be consistent between cells (see Chap. 13). For example, if the EPSC reversal potential is more depolarized in cell 2 versus cell 1, the driving force exerted on the ion at depolarized potentials will be smaller in cell 2 versus cell 1. This can cause an underestimation of the EPSC amplitude in the CV assay or it can cause an underestimation of the detectable successes in the minimal stimulation assay. Incorporating measurements with varying reversal potentials can skew the averaged data from the data set. Therefore, reversal potentials should be monitored and only recordings with similar reversal potentials (±1–3 mV) should be used for further analysis.

5 Summary

Silent synapses are posited to be key substrates for synaptic reorganization capable of affecting neonatal brain development as well as brain plasticity in adulthood. Aberrant synaptic stabilization or elimination is hypothesized to engender pathological conditions such as facilitating drug addiction or contributing to neurodegenerative diseases. With this ever-growing interest in silent synapses and their role in pathological conditions, electrophysiological approaches are being implemented in order to detect silent synapse formation in in vitro brain slices. One approach includes the CV assay whereby the ratio of the CV-NMDAR to CV-AMPAR is calculated. Decreases in this ratio between control and experimental groups suggest that an increase in the number of silent synapses has occurred. A second approach to measuring silent synapses includes the minimal stimulation assay in which the number of current successes detected at hyperpolarizing potentials is compared to the number of successes at depolarizing potentials using Eq. (4). Errors in attempting to measure silent synapses can arise from changes in access resistance as well as irregularities in the reversal potentials between cells.

References

1. Durand GM, Kovalchuk Y, Konnerth A (1996) Long-term potentiation and functional synapse induction in developing hippocampus. Nature 381(6577):71–75
2. Groc L, Gustafsson B, Hanse E (2006) AMPA signalling in nascent glutamatergic synapses: there and not there! Trends Neurosci 29(3): 132–139

3. Petralia RS, Esteban JA, Wang YX, Partridge JG, Zhao HM, Wenthold RJ, Malinow R (1999) Selective acquisition of AMPA receptors over postnatal development suggests a molecular basis for silent synapses. Nat Neurosci 2(1):31–36
4. Dong Y, Nestler EJ (2014) The neural rejuvenation hypothesis of cocaine addiction. Trends Pharmacol Sci 35(8):374–383
5. Hanse E, Seth H, Riebe I (2013) AMPA-silent synapses in brain development and pathology. Nat Rev Neurosci 14(12):839–850
6. Hill TC, Zito K (2013) LTP-induced long-term stabilization of individual nascent dendritic spines. J Neurosci 33(2):678–686
7. Isaac JT, Nicoll RA, Malenka RC (1995) Evidence for silent synapses: implications for the expression of LTP. Neuron 15(2):427–434
8. Katz LC, Shatz CJ (1996) Synaptic activity and the construction of cortical circuits. Science 274(5290):1133–1138
9. Kerchner GA, Nicoll RA (2008) Silent synapses and the emergence of a postsynaptic mechanism for LTP. Nat Rev Neurosci 9(11):813–825
10. Liao D, Hessler NA, Malinow R (1995) Activation of postsynaptically silent synapses during pairing-induced LTP in CA1 region of hippocampal slice. Nature 375(6530):400–404
11. Marie H, Morishita W, Yu X, Calakos N, Malenka RC (2005) Generation of silent synapses by acute in vivo expression of CaMKIV and CREB. Neuron 45(5):741–752
12. Montgomery JM, Pavlidis P, Madison DV (2001) Pair recordings reveal all-silent synaptic connections and the postsynaptic expression of long-term potentiation. Neuron 29(3):691–701
13. Rumpel S, Hatt H, Gottmann K (1998) Silent synapses in the developing rat visual cortex: evidence for postsynaptic expression of synaptic plasticity. J Neurosci 18(21):8863–8874
14. Hsieh H, Boehm J, Sato C, Iwatsubo T, Tomita T, Sisodia S, Malinow R (2006) AMPAR removal underlies Abeta-induced synaptic depression and dendritic spine loss. Neuron 52(5):831–843
15. Ting JT, Kelley BG, Lambert TJ, Cook DG, Sullivan JM (2007) Amyloid precursor protein overexpression depresses excitatory transmission through both presynaptic and postsynaptic mechanisms. Proc Natl Acad Sci U S A 104(1):353–358
16. Huang YH, Lin Y, Mu P, Lee BR, Brown TE, Wayman G, Marie H, Liu W, Yan Z, Sorg BA, Schluter OM, Zukin RS, Dong Y (2009) In vivo cocaine experience generates silent synapses. Neuron 63(1):40–47
17. Lee BR, Ma YY, Huang YH, Wang X, Otaka M, Ishikawa M, Neumann PA, Graziane NM, Brown TE, Suska A, Guo C, Lobo MK, Sesack SR, Wolf ME, Nestler EJ, Shaham Y, Schluter OM, Dong Y (2013) Maturation of silent synapses in amygdala-accumbens projection contributes to incubation of cocaine craving. Nat Neurosci 16(11):1644–1651
18. Pascoli V, Turiault M, Luscher C (2012) Reversal of cocaine-evoked synaptic potentiation resets drug-induced adaptive behaviour. Nature 481(7379):71–75
19. Robinson TE, Kolb B (1999) Alterations in the morphology of dendrites and dendritic spines in the nucleus accumbens and prefrontal cortex following repeated treatment with amphetamine or cocaine. Eur J Neurosci 11(5):1598–1604
20. Kessels HW, Nabavi S, Malinow R (2013) Metabotropic NMDA receptor function is required for beta-amyloid-induced synaptic depression. Proc Natl Acad Sci U S A 110(10):4033–4038
21. Snyder EM, Nong Y, Almeida CG, Paul S, Moran T, Choi EY, Nairn AC, Salter MW, Lombroso PJ, Gouras GK, Greengard P (2005) Regulation of NMDA receptor trafficking by amyloid-beta. Nat Neurosci 8(8):1051–1058
22. Faber DS, Korn H (1991) Applicability of the coefficient of variation method for analyzing synaptic plasticity. Biophys J 60(5):1288–1294
23. Kullmann DM (1994) Amplitude fluctuations of dual-component EPSCs in hippocampal pyramidal cells: implications for long-term potentiation. Neuron 12(5):1111–1120
24. Manabe T, Wyllie DJ, Perkel DJ, Nicoll RA (1993) Modulation of synaptic transmission and long-term potentiation: effects on paired pulse facilitation and EPSC variance in the CA1 region of the hippocampus. J Neurophysiol 70(4):1451–1459
25. Koya E, Cruz FC, Ator R, Golden SA, Hoffman AF, Lupica CR, Hope BT (2012) Silent synapses in selectively activated nucleus accumbens neurons following cocaine sensitization. Nat Neurosci 15(11):1556–1562
26. Ahmadian G, Ju W, Liu L, Wyszynski M, Lee SH, Dunah AW, Taghibiglou C, Wang Y, Lu J, Wong TP, Sheng M, Wang YT (2004) Tyrosine phosphorylation of GluR2 is required for insulin-stimulated AMPA receptor endocytosis and LTD. EMBO J 23(5):1040–1050
27. Wang Y, Ju W, Liu L, Fam S, D'Souza S, Taghibiglou C, Salter M, Wang YT (2004) alpha-Amino-3-hydroxy-5-methylisoxazole-4-propionic acid subtype glutamate receptor (AMPAR) endocytosis is essential for N-methyl-

D-aspartate-induced neuronal apoptosis. J Biol Chem 279(40):41267–41270

28. Wang YT (2008) Probing the role of AMPAR endocytosis and long-term depression in behavioural sensitization: relevance to treatment of brain disorders, including drug addiction. Br J Pharmacol 153(Suppl 1):S389–S395

29. Wong TP, Howland JG, Robillard JM, Ge Y, Yu W, Titterness AK, Brebner K, Liu L, Weinberg J, Christie BR, Phillips AG, Wang YT (2007) Hippocampal long-term depression mediates acute stress-induced spatial memory retrieval impairment. Proc Natl Acad Sci U S A 104(27):11471–11476

Chapter 20

Dendritic Patch

Nicholas Graziane and Yan Dong

Abstract

So far we have discussed patch clamp procedures that target the neuronal soma. However, electrical properties of neurons are critically influenced by dendrites, which not only make up a large portion of neuron's surface area and but also receive a large proportion of synaptic inputs. The electrical properties of dendritic membrane and somatic membrane can be vey different. Therefore, recording electrical properties directly from dendrites is desired in some experimental designs. Luckily, skilled electrophysiologists have optimized electrophysiological protocols that allow neuronal dendritic recording. Such recordings have provided important information regarding voltage-gated ion channel distribution and function [1–5], differences between somatic and dendritic membrane channel properties [6, 7], and synaptic integration of backpropagating action potentials combined with synaptic potentials [8–10]. The purpose of this chapter is to provide the beginning electrophysiologist with the necessary information required to maximize their success rate for dendritic recordings. While reading this chapter, we also would like to refer the reader to a superb detailed approach to dendritic patching documented by [11].

Key words Visualizing dendrites, Köhler illumination, Differential interference contrast, Technical considerations

1 Introduction

So far we have discussed patch clamp procedures that target the neuronal soma. However, electrical properties of neurons are critically influenced by dendrites, which not only make up a large portion of neuron's surface area and but also receive a large proportion of synaptic inputs. The electrical properties of dendritic membrane and somatic membrane can be vey different. Therefore, recording electrical properties directly from dendrites is desired in some experimental designs. Luckily, skilled electrophysiologists have optimized electrophysiological protocols that allow neuronal dendritic recording. Such recordings have provided important information regarding voltage-gated ion channel distribution and function [1–5], differences between somatic and dendritic membrane channel properties [6, 7], and synaptic integration of backpropagating action potentials combined with synaptic potentials [8–10].

Nicholas Graziane and Yan Dong, *Electrophysiological Analysis of Synaptic Transmission*, Neuromethods, vol. 112, DOI 10.1007/978-1-4939-3274-0_20,

Table 1
Normal aCSF ingredients [11]

Compound	Concentration (mM)
$NaCl^-$	125
KCl^-	2.5
$MgCl_2^-$	1
$CaCl_2^-$	2
$NaHCO_3^-$	25
$NaH_2PO_4^-$	1.25
Glucose	25
Osmolality (mmol/kg)	310

The purpose of this chapter is to provide the beginning electrophysiologist with the necessary information required to maximize their success rate for dendritic recordings. While reading this chapter, we also would like to refer the reader to a superb detailed approach to dendritic patching documented by [11].

2 External and Internal Solutions

External cutting solution: Normal aCSF cutting solution bubbled with carbogen (95 % O_2/5 % CO_2) can be used for brain slicing procedures. Normal aCSF ingredients taken from [11] are listed in Table 1 below. To minimize excitotoxicity, 50–100 % of $NaCl^-$ can be replaced with sucrose, the Ca^{2+}:Mg^{2+} ratio can be adjusted, or 2–10 mM kynurenic acid can be added to the slicing solution to block ionotropic glutamate transmission.

Internal solution for whole-cell recording: See Table 2 below.

3 Equipment

- Standard electrophysiological equipment can be used for brain slice preparation (e.g., vibratome for slice preparation, incubation chamber for cut brain slices) and for recordings (e.g., micromanipulators, patch-clamp amplifiers, and pipette pullers).
- Visualizing dendrites for patching: an upright microscope (placed on an air table to prevent vibrations) equipped with infrared-differential interference contrast (IR-DIC) or Dodt

Table 2
Standard intracellular solution for dendritic-patch clamping [11]

Compound	Concentration (mM)
K-Methanesulfonate	130
KCl	7
HEPES	10
EGTA (K)	0.05
Na_2ATP	2.0
MgATP	2.0
Na_2GTP	0.5

pH = 7.2–7.3 (KOH)
Osmolality = 290 mOsm
Aliquot and freeze (−20 °C for 1 month or −80 °C for 1 year)

contrast optics. Alternatively, a fluorescent microscope with a CCD camera [12] or a two-photon microscope equipped with scanning Dodt contrast optics can be used [13]. Between the camera and objective, a magnifier 2–4× should be available.

- Water-immersion objectives with 40–60× magnification, high numerical aperture (≥0.75), and a long working distance allowing electrodes to slide underneath for patching access.
- Video camera and monitor: CCD or Vidicon camera and controller (Hammamatsu C2400-07; Dage 1000; Optonis VX45). A black and white monitor is sufficient. Be sure that the camera can perform fast-image refresh rates and that the camera is sensitive to near infrared ranges. It is recommended to use a camera with a large detector and/or high spatial resolution.
- Manometer: necessary for monitoring pressure and suction changes (not typically used for somatic recordings).

4 Optimizing Optics

An extremely critical component to successful dendritic-patch clamping is being able to precisely locate and patch a dendrite. In order for this to occur, the optics must be optimized accurately. Here, we take you through the steps necessary to produce a clear image of a dendrite on your monitor.

Adjusting Köhler illumination: Köhler illumination is a common illumination procedure which is used for reflected light. For Köhler illumination to function properly, the light emitted from the lamp filament must be collected and focused at the place of the condenser aperture diaphragm. This can be easily realigned daily.

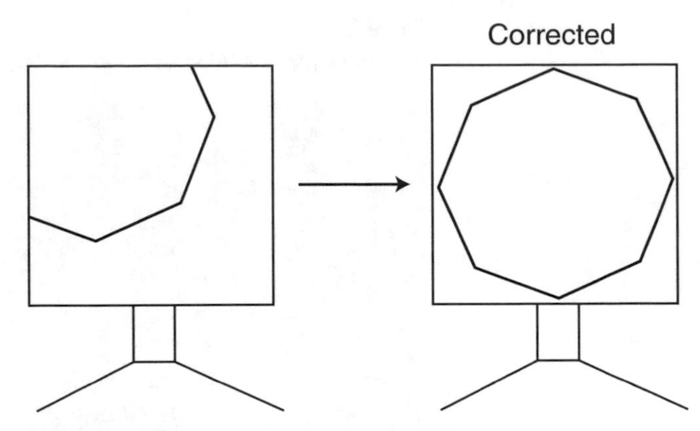

Fig. 1 Correcting Köhler illumination. The diagram shows an octagon poorly centered on the monitor receiving an output signal from a camera connected to the microscope (*left*). After the Köhler illumination is corrected, the octagon is centered within the monitor's screen

The Köhler illumination can be adjusted using the microscopic eyepieces, or, as many prefer, using the monitor, since that is what we use to visualize our neurons. Below details how to perform Köhler illumination readjustments:

1. Using the high magnification, focus on the brain slice and then move to an empty region in the recording chamber. Close the condenser aperture diaphragm until the light seen is 60–90 % visible (Fig. 1). If the octagon is not clear, move the condenser toward or away from the recording chamber until the octagon's edges are sharp. If the octagon is not centered, move the condenser from side-to-side. Once completed, move the infrared filter into the light path.

4.1 Adjusting DIC Optics (Conventional)

Differential interference contrast (DIC) is a beam-splitting interference system used to enhance the contrast of transparent samples. It works using two polarizers one located above the objective (analyzer) and the other below the condenser, and two DIC prisms with one located above the objective and the other below the specimen usually in the condenser (Fig. 2). Below the specimen, unpolarized light passes through the polarizer effectively polarized to 45°. This light then enters the Nomarski-modified Wollaston prism (DIC prism) and is separated into two rays polarized at 90° to each other (sampling and reference rays). The light then passes through the condenser, which focuses the light onto the sample. The sample is illuminated by two light sources, one polarized at 0° and the other polarized at 90°. These two rays travel through the slice experiencing different optical path lengths causing phase changes with one ray relative to the other. The separated rays can then travel through the objective lens where they are focused onto the

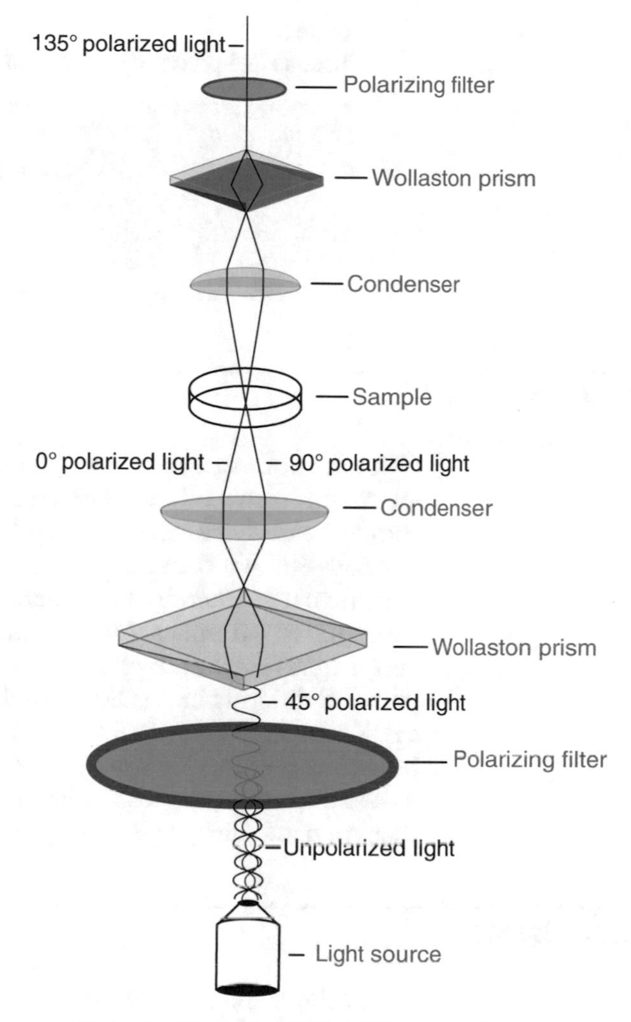

Fig. 2 Diagram illustrating the light path for DIC imaging. See text for description

second DIC prism. At the second DIC prism, the two rays are recombined and polarized at 135° causing interference, allowing for the image to be visualized (Fig. 2). To make sure that the DIC optics are properly adjusted, the microscope manufacturer can be contacted or these steps can be used:

1. Be sure that the polarizer and analyzer are 90° to one another. To do this, rotate the polarizer until the darkest image possible is visualized on the camera (be sure to turn off automatic gain controls on the camera controller).
2. Adjust the DIC prism above the objective to obtain a three-dimensional image. Typically one side of the image appears bright while the other side appears as the image is casting a shadow.
3. Adjust the camera so that the contrast is optimized using the light intensity, offset and gain controls.

Table 3
Tissue slice preparation for dendritic patch

Brain region	Anatomical plane
Hippocampus	Longitudinal
Neocortex	Sagittal
Substantia nigra	Coronal

5 Procedure

Once the optics are optimized, be sure that the dendrites are parallel to the plane of the slice (see Table 3). Patchable dendrites should be easily visualized at depths of about 50 μm in brain slices cut 150–300 μm thick. At this point the procedure is similar to any somatic patch whereby the experimenter lowers a patch electrode through the external solution until it rests on the dendrite. Applying negative pressure produces a gigaseal and depending on the protocol differing intracellular patch configurations can be achieved as discussed in Chap. 1 (for a detailed procedure list see [11]). For beginners looking for healthy, patchable dendrites, it is suggested to look for smooth, well-defined dendrites while avoiding those that are dark, faint, or swollen [11].

6 Troubleshooting

1. Losing a dendrite-of-interest in deeper layers of the slice. This problem can be resolved by lowing the patch electrode through the slice while holding positive pressure. The internal solution rushing out of the pipette clears the tissue debris allowing the dendrite to be seen.
2. Dendrites are hard to patch because they move away from the patch pipette. This is typically a problem when the slice quality is poor. However, if the slice quality is believed to be adequate, the experimenter can try to aim for dendrites close to branch points as this can help anchor the dendrite [11].

7 Limitations to Dendritic Patch

1. The size of dendrites that can be patched. Many patchable dendrites are located on proximal apical dendrites, while the finer branches remain unpatchable. Certain cell types such as stellate cells have very fine dendrites rendering dendritic recordings inaccessible to patch electrodes.

2. Patch electrodes for dendritic patch often have high resistance, making access resistance high (>20 MΩ). The high access resistance can distort current or potential measurements (see Chap. 3).
3. Using whole-cell patch configuration can cause "washout" of key bioactive substrates, disrupting cellular physiology. This of course can be overcome using the perforated patch technique.

8 Summary

Dendritic patch is a technically challenging technique that requires a high level of skill both during patching as well as during preparation. Keys to increase the success rate include (1) optimizing the optics so that dendrites can be clearly visualized, (2) preparing healthy brain slices on the appropriate anatomical plane and at the right thickness, and (3) targeting the most patchable dendrite based on morphology and appearance. With a successful recording, critically important details regarding dendritic function can be determined allowing the beginning electrophysiologist to contribute significantly to the field.

References

1. Bekkers JM (2000) Distribution and activation of voltage-gated potassium channels in cell-attached and outside-out patches from large layer 5 cortical pyramidal neurons of the rat. J Physiol 525(Pt 3):611–620
2. Hoffman DA, Magee JC, Colbert CM, Johnston D (1997) K+ channel regulation of signal propagation in dendrites of hippocampal pyramidal neurons. Nature 387(6636): 869–875
3. Korngreen A, Sakmann B (2000) Voltage-gated K+ channels in layer 5 neocortical pyramidal neurones from young rats: subtypes and gradients. J Physiol 525(Pt 3):621–639
4. Magee JC (1998) Dendritic hyperpolarization-activated currents modify the integrative properties of hippocampal CA1 pyramidal neurons. J Neurosci 18(19):7613–7624
5. Stuart G, Hausser M (1994) Initiation and spread of sodium action potentials in cerebellar Purkinje cells. Neuron 13(3):703–712
6. Magee JC, Cook EP (2000) Somatic EPSP amplitude is independent of synapse location in hippocampal pyramidal neurons. Nat Neurosci 3(9):895–903
7. Williams SR, Stuart GJ (2002) Dependence of EPSP efficacy on synapse location in neocortical pyramidal neurons. Science 295(5561): 1907–1910
8. Magee JC, Johnston D (1997) A synaptically controlled, associative signal for Hebbian plasticity in hippocampal neurons. Science 275 (5297):209–213
9. Stuart GJ, Hausser M (2001) Dendritic coincidence detection of EPSPs and action potentials. Nat Neurosci 4(1):63–71
10. Watanabe S, Hoffman DA, Migliore M, Johnston D (2002) Dendritic K+ channels contribute to spike-timing dependent long-term potentiation in hippocampal pyramidal neurons. Proc Natl Acad Sci U S A 99(12): 8366–8371
11. Davie JT, Kole MH, Letzkus JJ, Rancz EA, Spruston N, Stuart GJ, Hausser M (2006) Dendritic patch-clamp recording. Nat Protoc 1(3):1235–1247
12. Larkum ME, Senn W, Luscher HR (2004) Top-down dendritic input increases the gain of layer 5 pyramidal neurons. Cereb Cortex 14(10):1059–1070
13. Nevian T, Larkum ME, Polsky A, Schiller J (2007) Properties of basal dendrites of layer 5 pyramidal neurons: a direct patch-clamp recording study. Nat Neurosci 10(2):206–214

Part V

Experimentations with Molecular and Visual Components

Chapter 21

Electrophysiological and Visual Tags

Nicholas Graziane and Yan Dong

Abstract

Just like we can identify friends based on physical characteristics (voice, appearance, etc.), so too can we identify synaptic proteins or neurons based on their electrophysiological characteristics. This chapter discusses different approaches to electrophysiologically identify specific synaptic receptors or specific neuronal types including examples in which these approaches have been implemented experimentally.

Key words Calcium-permeable AMPARs, GluN2B, Dopamine neurons, Capacitance, Input resistance, Resting membrane potential, Fluorescent dyes, Fluorescent tags, pHluorins

1 Introduction

Just like we can identify friends based on physical characteristics (voice, appearance, etc.), so too can we identify synaptic proteins or neurons based on their electrophysiological characteristics. This chapter discusses different approaches to electrophysiologically identify specific synaptic receptors or specific neuronal types including examples in which these approaches have been implemented experimentally.

Additionally, we discuss visual tags that have been used to monitor both presynaptic and postsynaptic functions. We first describe the postsynaptic visual tags that have been used to support or add additional information to electrophysiological experiments followed by a description of presynaptic markers and their advantages/disadvantages.

The purpose of this chapter is to acquaint the beginning electrophysiologist with two concepts. First, electrophysiology can be used to identify receptors and/or neurons as well as measure the transfer of electrical information between cells or their intrinsic membrane excitability. Second, by including visual tagging experiments with electrophysiological research, powerful analysis of presynaptic and postsynaptic function is achievable.

Nicholas Graziane and Yan Dong, *Electrophysiological Analysis of Synaptic Transmission*, Neuromethods, vol. 112,
DOI 10.1007/978-1-4939-3274-0_21, © Springer Science+Business Media New York 2016

2 Tags Used in Electrophysiology Experiments

2.1 Electrophysiological Tags

Electrophysiological methods are used to identify specific receptors based on their electrophysiological characteristics. These characteristics are distinct for the given receptor rendering it easy to identify among multiple receptors expressed on the same cell. Because of these distinct characteristics, the receptor is essentially tagged for identification. For example, calcium-permeable AMPARs (CPAMPARs) (i.e., GluA2-lacking AMPARs) can be identified postsynaptically. This can be done by measuring the current–voltage relationship of AMPARs (*I–V* curve). Since CPAMPARs inwardly rectify at positive holding potentials, they can be easily identified from calcium-impermeable AMPARs, which do not show such inward rectification (i.e., calcium-impermeable AMPARs show linear *I–V* relationships) [1, 2]. The inward rectification of CPAMPARs is caused by polyamines, which block the channel at positive holding potentials (>0 mV). Therefore, to measure the CPAMPAR *I–V* relationship, spermine (0.1 mM) needs to be added to the internal solution so that CPAMPARs become blocked at positive holding potentials.

Demonstrating how using the CPAMPAR tag can be implemented experimentally, Hayashi et al. [2] investigated whether LTP induction or activation of CaMKII induced AMPAR synaptic insertion by overexpressing the GluA1 subunit. If LTP or CaMKII activation induced AMPAR synaptic insertion, the overexpressed GluA1 subunit would be preferentially inserted into the synapse leading to current rectification. The authors observed increases in current rectification following LTP or CaMKII activation suggesting AMPARs are inserted into the synapse [2].

Electrophysiological tags are also present on NMDARs. This is because GluN2A-containing NMDARs possess significantly faster decay kinetics than GluN2B-containing NMDARs [3]. Therefore, neurons can be voltage-clamped at, e.g., +40 mV, and NMDAR currents can be measured. The time it takes for the NMDAR current to decay to ½ its peak amplitude can be conveniently calculated and this decay time can manifest whether NMDAR synaptic expression is predominantly GluN2A- or GluN2B-containing NMDARs.

In addition, NMDARs are blocked by Mg^{2+} at resting membrane potentials. This characteristic can be taken advantage of in order to tag NMDARs. Remember that functional assembly of NMDARs is commonly GluN1/GluN2 with GluN2A being predominantly expressed in mature synapses and GluN2B being expressed in developing, newly generated synapses. In addition, ternary complexes GluN1/GluN2/GluN3 are also thought to exist [4]. The noticeable similarity between NMDAR functional assembly is that GluN1 is expressed in all subtypes. Taking advantage of this characteristic, researchers have mutated the GluN1 subunit (N598R mutation), causing decreased Mg^{2+}-binding affinity for functionally expressed NMDARs at hyperpolarizing potentials [5]. Introducing this mutant receptor into the preparation, NMDAR synaptic insertion can be investigated. For example, Huang et al. investigate whether cocaine induces synaptic insertion of NMDARs on medium spiny neurons (MSNs) in the nucleus accumbens (NAc). To test this, the authors evoke currents while voltage clamping the neuron at −90 mV (complete Mg^{2+} block of NMDARs). They then measure the current amplitude at 0 ms (AMPAR mediated) and at 35 ms after the peak current amplitude (NMDAR mediated). Once the current amplitudes are measured, they define a ratio of the current amplitude at 35 ms to the current amplitude at 0 ms (I_{35ms}/I_{0ms}) to indicate the number of synaptic NMDARs not blocked by Mg^{2+}. They find that exposure to cocaine increases I_{35ms}/I_{0ms} in NAc MSNs expressing the mutant GluN1 subunit, while saline controls show no increase in I_{35ms}/I_{0ms}. The authors conclude that mutated GluN1-containing NMDARs are delivered to the postsynaptic membrane in cocaine-treated, but not saline-treated animals [3].

Neurons, like receptors, can also be tagged as they too possess electrophysiologically defined characteristics. For example, dopamine (DA) neurons in the ventral tegmental area (VTA) are often identified using two electrophysiological measurements. First, DA neurons in the VTA, compared to GABAergic or glutamatergic neurons, can be identified by the presence of I_H currents with 95 % accuracy [6]. I_H currents can be activated by hyperpolarizing the neuron in voltage-clamp mode (e.g., dropping the voltage from −50 to −100 mV for 500 ms) and measuring the "sag" current produced (Fig. 1). Despite some groups reporting high success

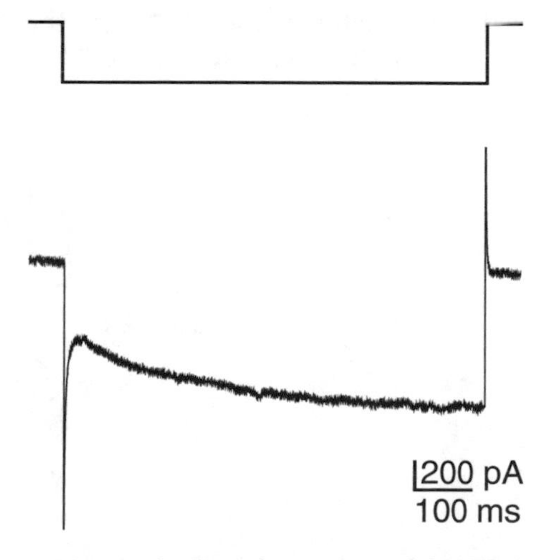

Fig. 1 Sag (IH) currents seen in dopamine neurons of the VTA. A voltage clamped cell is given a hyperpolarized from −50 to −100 mV inducing the sag current illustrated

rates in identifying DA neurons using the I_H assay, still other groups show no distinction [7]. Therefore, post hoc immunocytochemical assays are recommended. Second, DA neurons are pacemaker cells, firing action potentials (APs) at 1–5 Hz in in vitro brain slices. By simply using a cell-attached patch, the pacemaker activity of the DA neuron can be measured. Typically non DA neurons in the VTA fire significantly faster (e.g., 10 Hz), making it easy to distinguish between DA and non DA neurons [7].

Three other useful electrophysiological tags that can help identify cell types are the cell's resting membrane potential, capacitance, and input resistance. It is common that different cell types populating a specific brain region have incomparable resting membrane potentials. For example, in the NAc, MSNs have a resting membrane potential of ~–80 mV, while persistent and low threshold spike cells (PLTS cells) in the same brain region sit at ~–56 mV [8]. This difference can be easily identified electrophysiologically by current-clamping the cell in the whole-cell configuration and measuring the potential. The cell capacitance is proportionally related to the cell size. Therefore, measuring the cell capacitance can corroborate what is visually seen by the electrophysiologist. Calculating the cell capacitance is typically done by software. However, if the software used does not offer this feature, the capacitance can be calculated as follows: a small voltage step can be delivered that is long enough to cause the clamped current to come to a steady state. The integrated transient charge can then be divided by the voltage-clamp step size (Fig. 2) [9].

Lastly, the input resistance can be measured to determine electrophysiological characteristics of a cell. The input resistance is a measure of the number of conductive channels that are open on the cell's surface and can vary from cell type to cell type. For example, rat MSNs in the NAc have input resistances around ~200 ± 80 MΩ, while PLTS cells in the same region have input

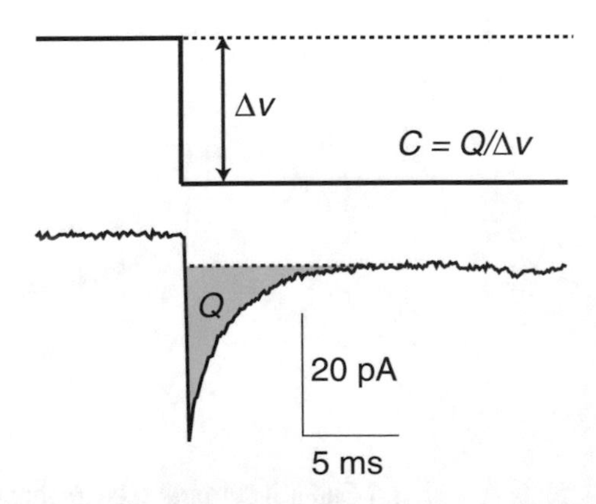

Fig. 2 Illustration of capacitance calculation. The transient charge (*Q*) produced after a hyperpolarizing step is divided by the voltage step size (Δ*v*) to give the cell capacitance

resistances of ~640 ± 245 MΩ (experiments performed with Ringer's solution and K^+-methanesulfonate internal solution) [8]. The input resistance can be calculated as discussed in Chap. 15.

2.2 Postsynaptic Visual Tags

Visual tags are often used by electrophysiologists to monitor receptor trafficking (e.g., exocytosis, endocytosis, and lateral movement) and are useful in that they enable detailed time lapse analysis of tagged molecules. Commonly used visual tags are fluorescent tags such as green fluorescent protein (GFP), which can be fused to a protein of interest using standard molecular biology techniques [10]. Fusing GFP to synaptic receptors has given electrophysiologists receptor tracking information during synaptic plasticity induction. For example, Shi et al. monitor changes in AMPAR distribution following LTP induction in the hippocampal CA1 region [11]. By fusing GFP to the GluA1 subunit, the authors are able to visualize the rapid delivery of tagged receptors into dendritic spines. These results (1) correlate nicely with the observed increases in EPSP amplitudes following LTP induction and (2) provide valuable support for a postsynaptic mechanism of LTP expression, which was controversial at the time. Importantly, before attempting the intended investigation, the authors empirically show that the fused GFP-GluA1 subunit can be trafficked to the cell surface and is functional (studies done in organotypic slice culture). Therefore, we should point out to the beginning electrophysiologist considering of using a visual tag that performing basic tests of biological function are a critically important first step to any investigation. This is because the chimeric molecule can sometimes behave differently than the native molecule [12].

Fluorescent tags are also used in photobleaching experiments. These experiments can monitor the diffusion properties of a fluorescently tagged protein throughout the cellular membrane by a process known as FRAP (fluorescence recovery after photobleaching). FRAP experiments excite fluorophores in a region of interest using high illumination intensity with the time duration dependent upon the region's size [e.g., ~2 s for spines and ~3–5 min for dendrites [13]]. Once the region is photobleached, the time it takes for the fluorescence to recover is measured. An example of using photobleaching in electrophysiologically relevant experiments is demonstrated by Makino and Malinow who investigated whether AMPARs moved into synapses from non-synaptic sites or from intracellular sites following LTP induction [13]. By using a pH-sensitive fluorescent tag, they are able to protect receptors from photobleaching that are trafficked in low pH intracellular compartments, but not receptors expressed at extrasynaptic sites. Therefore, if LTP drives receptors into the synapse that originate from extrasynaptic pools, following photobleaching the tagged receptors should produce small increases in fluorescence. However, if receptors protected from photobleaching are inserted at the synapse, a large increase in fluorescence should be detected. The authors find that following LTP induction in the hippocampal CA1 region, the fluorescent intensity observed after photobleaching correlates with AMPAR insertion from extrasynaptic pools. These data clearly demonstrate the use of visual tags to add valuable additional data to associated electrophysiological measurements.

Calcium imaging is another visual tag that is used in combination with electrophysiology. Calcium imaging is a technique whereby calcium-selective chelators (e.g., EGTA or BAPTA) are hybridized to fluorescent chromophores [14, 15]. Upon Ca^{2+} binding to the calcium chelators, intramolecular conformational changes occur altering the emitted fluorescence, which can be detected using video imaging[1] (e.g., CCD cameras, confocal microscopy). Often, these hybridized molecules (e.g., Fura-2, Indo-1, fluo-2, fluo-4, Oregon Green BAPTA-1AM) are introduced intracellularly via the internal solution in a whole-cell or sharp electrode configuration. Electroporation[2] is also an alternative to loading single cells [17–19]. As intracellular Ca^{2+} concentrations change via influx, efflux, or exchange with internal stores, the fluorescence can be detected providing a quantitative value of intracellular Ca^{2+} concentrations. Typical contributors to neuronal calcium signaling are voltage-gated calcium channels, NMDARs, calcium-permeable AMPARs (GluA2-lacking), metabotropic glutamate receptors, and calcium release from internal stores (most often the endoplasmic reticulum) [20].

Additionally, calcium imaging can be used to monitor the activity of neuronal circuits. To monitor circuits, the tissue is stained using an air pressure pulse [21–23], which applies the caged calcium chelators to the extracellular space. The caged chelators, in addition to being hybridized to the fluorophore, are also hybridized to an ester. This ester serves two purposes: one, it prevents the calcium chelators from binding to calcium, thus rendering the molecule caged. Two, the ester group is lipophilic enabling the molecule to penetrate cellular membrane. Once the dye has penetrated the membrane, the ester group is removed by intracellular esterases, leaving the carboxyl group free to bind to calcium [24].

---

[1] Charged-coupled device (CCD) camera or confocal microscopy are acquisition options for video imaging with distinct advantages and disadvantages. CCD cameras are advantageous over confocal microscopy in that they acquire images 10-100 times faster for a specified visual field and the epifluorescence is less damaging. On the other hand, confocal microscopy has superior spatial resolution along the XY-plane allowing for single-cell detection without interference from cells above or below [16].

[2] Electroporation is the use of electrical current, which disrupts the membrane integrity forming pores in addition to excreting the charged hybridized fluorescent molecules out of the pipette.

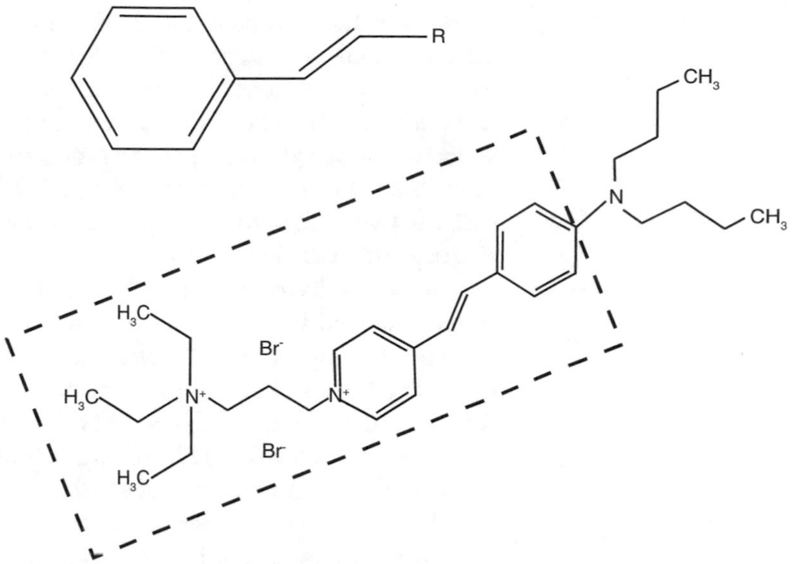

Fig. 3 An illustration of the structure of a styryl group (*top*) and its location (*dashed box*) in a popularly used fluorescent dye, FM1-43 dye (*bottom*)

2.3 Presynaptic Visual Tags

Presynaptic visual tags are also extensively used in electrophysiology dating back to the 1950s with the work of Bernard Katz [25, 26]. Presynaptic tags include fluorescent dyes and fluorescent tags.

Fluorescent dyes used for tagging presynaptic vesicles are composed of a styryl group[3] (Fig. 3). The purpose of the styryl group is to provide solubility in water, but also to bind to, but not cross, the lipid membrane (i.e., styryls are amphipathic). How do dyes get incorporated into the presynaptic vesicles? The dye is introduced into the medium and binds to the cellular membrane. Upon stimulation, presynaptic terminals release neurotransmitters via the fusion of neurotransmitter-containing vesicles to the cellular membrane. As exocytosis occurs, the vesicles fuse to the membrane where the dye is bound. As the subsequent endocytosis of the vesicle occurs, the dye becomes fully incorporated into the vesicle. The experimenter then washes the dye out of the medium to clear out the background signals. However, we know that not all vesicles fuse to the membrane the same way. For example, vesicles can fuse via kiss-and-run or full collapse (clathrin-dynamin dependent). Depending on how the vesicle fuses to the membrane can dictate how much dye is loaded into the vesicle. Since kiss-and-run vesicles may not be fully loaded with dye, their fluorescence intensity can

[3] A styryl group (Fig. 3) is a derivative of styrene with hydrogen removed from the omega carbon (the carbon + methyl group at the end of the carbon chain that is farthest away from the carboxyl group). This group is a univalent radical providing a charged head group preventing the molecule from traversing the membrane.

be weaker than a vesicle undergoing full collapse. This of course is advantageous for an experimenter distinguishing between two modes of release. The time between kiss-and-run and full collapse of vesicles can also be taken advantage of for research. If the dye is washed out at varying time intervals, rapid and slow endocytotic components can be differentiated [27–29]. For example, washing out the dye before full collapse takes place can leave only quickly endocytosed vesicles labeled.

So far we have discussed that endocytotic pools can be differentially labeled based on the vesicle fusion properties or based on the speed of dye washout following a stimulus. Additionally, the dye's structure can affect endocytotic labeling. For example, more lipophilic dyes (e.g., FM1-43, FM1-84) label vesicles and cisternae (larger surface membrane infoldings), while less hydrophobic dyes (FM2-10) label unitary vesicles [30].

An alternative approach visualizing presynaptic endocytosis is to monitor presynaptic exocytotic mechanisms. Following dye uptake into the synaptic vesicles, the dye is washed out of the preparation. With dye located exclusively intracellularly, the presynaptic terminal is stimulated inducing neurotransmitter and dye release into the extracellular compartment. The fluorescence intensity is then measured.

Quantifying endocytotic dye uptake and exocytotic dye excretion has contributed significant information to understanding of presynaptic function. For example, the amount of dye loaded in a single vesicle matched the dye excreted during exocytosis suggesting that synaptic vesicles recycle without combining with other intracellular compartments (if vesicles mixed with endosomal compartments the fluorescence intensity would decrease upon exocytosis) [31]. Other studies have shown that there are active and inactive synaptic vesicle pools as well as resting pools that remain inactive except during synaptic fatigue triggered by intense stimulations [30, 32–40]. Clearly then an advantage of using dyes is that they allow researchers to monitor vesicular cycling.

A disadvantage of dyes is that nonspecific background staining occurs during dye incubation. This effect is pronounced in sliced tissue as the unhealthy tissue at the surface of the slice causes excessive background staining [41, 42]. Therefore, dye rerelease is critically important for selecting a region of interest. However, dye rerelease presents its own challenges mainly due to the kinetics of destaining (fluorescent loss). Typically, dyes are inserted into membranes within a millisecond range and are excreted within a few milliseconds [43, 44]. However, slow destaining rates have been reported [45–48] possibly due the number of dye loaded vesicles releasing dye simultaneously (simultaneous release slows fluorescent-loss kinetics), the dye's lipophilic tail length (increases in the tail length can slow the departitioning rate), or the presence of a postsynaptic cell, which can lead to nonspecific binding of the dyes to extracellular membranes once excreted presynaptically [30].

Given the aforementioned shortcomings of fluorescent dyes in monitoring presynaptic vesicular function, fluorescent tags have been developed to overcome these limitations. A common fluorescent tag used is pHluorin, which is hybridized to a synaptic vesicle protein [49]. This pH-sensitive GFP is inactive while inside the acidic environment of a vesicle (pH ~5.5), but upon exposure to the extracellular pH, the fluorescence intensity is greatly increased. pHluorin can be conjugated to synaptobrevin, synaptophysin, vesicular glutamate transporter (Vglut1), or other v-SNARE proteins enabling this tag to specifically associate with synaptic vesicles. This is an advantage over dyes in that dyes adhere to lipids potentially leading to nonspecific binding. An additional advantage of pHluorin is that pHluorin tags can be conjugated to other fluorescence proteins allowing the experimenter to simultaneously monitor the trafficking of many vesicular molecules.

The disadvantage of pHluorins is that transfection methods can induce toxicity. In addition, coupling pHluorin with vesicular proteins does not directly reflect endocytosis of vesicles. For example, instead of endocytosis, lateral diffusion is often observed in pHluorin conjugated to synaptobrevin probes. The interpretation of these results is either due to vesicular movement or it can be due to re-clustering or re-internalization of synaptobrevin [30]. Lastly, fluorescent activity is dependent upon re-acidification of the presynaptic vesicle. Therefore, delays in internalization vs fluorescence intensity may be observed [50, 51].

3 Summary

In this chapter we discussed electrophysiological characteristics of CPAMPARs and NMDAR subunits. In addition, we described electrophysiological measurements that can help identify cell types such as input resistance, capacitance, and resting membrane potentials. Furthermore, we discussed postsynaptic visual tags such as GFP fusion or calcium imaging and presynaptic visual tags such as fluorescent dyes and tags.

These tagging strategies have been used extensively by neuroscientists to generate valuable information in understanding synaptic function. To the beginning electrophysiologist, it is paramount to understand how the proper use of these techniques, while keeping in mind their limitations, can lead to the accumulation of high quality data allowing the researcher to accurately interpret their results. In addition, with the continued development of technological advances, researchers should continue to peruse the literature in search of better techniques that can be used to complement their electrophysiological measurements.

References

1. Cull-Candy S, Kelly L, Farrant M (2006) Regulation of Ca2+-permeable AMPA receptors: synaptic plasticity and beyond. Curr Opin Neurobiol 16(3):288–297
2. Hayashi Y, Shi SH, Esteban JA, Piccini A, Poncer JC, Malinow R (2000) Driving AMPA receptors into synapses by LTP and CaMKII: requirement for GluR1 and PDZ domain interaction. Science 287(5461):2262–2267
3. Huang YH, Lin Y, Mu P, Lee BR, Brown TE, Wayman G, Marie H, Liu W, Yan Z, Sorg BA, Schluter OM, Zukin RS, Dong Y (2009) In vivo cocaine experience generates silent synapses. Neuron 63(1):40–47
4. Sasaki YF, Rothe T, Premkumar LS, Das S, Cui J, Talantova MV, Wong HK, Gong X, Chan SF, Zhang D, Nakanishi N, Sucher NJ, Lipton SA (2002) Characterization and comparison of the NR3A subunit of the NMDA receptor in recombinant systems and primary cortical neurons. J Neurophysiol 87(4):2052–2063
5. Barria A, Malinow R (2005) NMDA receptor subunit composition controls synaptic plasticity by regulating binding to CaMKII. Neuron 48(2):289–301
6. Chen BT, Bowers MS, Martin M, Hopf FW, Guillory AM, Carelli RM, Chou JK, Bonci A (2008) Cocaine but not natural reward self-administration nor passive cocaine infusion produces persistent LTP in the VTA. Neuron 59(2):288–297
7. Margolis EB, Lock H, Hjelmstad GO, Fields HL (2006) The ventral tegmental area revisited: is there an electrophysiological marker for dopaminergic neurons? J Physiol 577(Pt 3): 907–924
8. Kawaguchi Y (1993) Physiological, morphological, and histochemical characterization of three classes of interneurons in rat neostriatum. J Neurosci 13(11):4908–4923
9. Taylor AL (2012) What we talk about when we talk about capacitance measured with the voltage-clamp step method. J Comput Neurosci 32(1):167–175
10. Snapp E (2005) Design and use of fluorescent fusion proteins in cell biology. Curr Protoc Cell Biol Chapter 21: Unit 21.24
11. Shi SH, Hayashi Y, Petralia RS, Zaman SH, Wenthold RJ, Svoboda K, Malinow R (1999) Rapid spine delivery and redistribution of AMPA receptors after synaptic NMDA receptor activation. Science 284(5421):1811–1816
12. Hanson DA, Ziegler SF (2004) Fusion of green fluorescent protein to the C-terminus of granulysin alters its intracellular localization in comparison to the native molecule. J Negat Results Biomed 3:2
13. Makino H, Malinow R (2009) AMPA receptor incorporation into synapses during LTP: the role of lateral movement and exocytosis. Neuron 64(3):381–390
14. Tsien RY (1980) New calcium indicators and buffers with high selectivity against magnesium and protons: design, synthesis, and properties of prototype structures. Biochemistry 19(11): 2396–2404
15. Williams DA, Fogarty KE, Tsien RY, Fay FS (1985) Calcium gradients in single smooth muscle cells revealed by the digital imaging microscope using Fura-2. Nature 318(6046): 558–561
16. Lacar B, Young SZ, Platel JC, Bordey A (2010) Imaging and recording subventricular zone progenitor cells in live tissue of postnatal mice. Front Neurosci 4:43
17. Judkewitz B, Rizzi M, Kitamura K, Hausser M (2009) Targeted single-cell electroporation of mammalian neurons in vivo. Nat Protoc 4(6):862–869
18. Kitamura K, Judkewitz B, Kano M, Denk W, Hausser M (2008) Targeted patch-clamp recordings and single-cell electroporation of unlabeled neurons in vivo. Nat Methods 5(1):61–67
19. Nevian T, Helmchen F (2007) Calcium indicator loading of neurons using single-cell electroporation. Pflugers Arch 454(4):675–688
20. Grienberger C, Konnerth A (2012) Imaging Calcium in Neurons. Neuron 73(5):862–885
21. Connor JA, Razani-Boroujerdi S, Greenwood AC, Cormier RJ, Petrozzino JJ, Lin RC (1999) Reduced voltage-dependent Ca2+ signaling in CA1 neurons after brief ischemia in gerbils. J Neurophysiol 81(1):299–306
22. Garaschuk O, Milos RI, Konnerth A (2006) Targeted bulk-loading of fluorescent indicators for two-photon brain imaging in vivo. Nat Protoc 1(1):380–386
23. Stosiek C, Garaschuk O, Holthoff K, Konnerth A (2003) In vivo two-photon calcium imaging of neuronal networks. Proc Natl Acad Sci U S A 100(12):7319–7324
24. Tsien RY (1981) A non-disruptive technique for loading calcium buffers and indicators into cells. Nature 290(5806):527–528
25. Fatt P, Katz B (1950) Some observations on biological noise. Nature 166(4223):597–598
26. Katz B (2003) Neural transmitter release: from quantal secretion to exocytosis and beyond. J Neurocytol 32(5-8):437–446

27. Chung C, Barylko B, Leitz J, Liu X, Kavalali ET (2010) Acute dynamin inhibition dissects synaptic vesicle recycling pathways that drive spontaneous and evoked neurotransmission. J Neurosci 30(4):1363–1376
28. Groemer TW, Klingauf J (2007) Synaptic vesicles recycling spontaneously and during activity belong to the same vesicle pool. Nat Neurosci 10(2):145–147
29. Sara Y, Virmani T, Deak F, Liu X, Kavalali ET (2005) An isolated pool of vesicles recycles at rest and drives spontaneous neurotransmission. Neuron 45(4):563–573
30. Kavalali ET, Jorgensen EM (2014) Visualizing presynaptic function. Nat Neurosci 17(1): 10–16
31. Murthy VN, Stevens CF (1998) Synaptic vesicles retain their identity through the endocytic cycle. Nature 392(6675):497–501
32. Betz WJ, Bewick GS (1992) Optical analysis of synaptic vesicle recycling at the frog neuromuscular junction. Science 255(5041):200–203
33. Denker A, Bethani I, Krohnert K, Korber C, Horstmann H, Wilhelm BG, Barysch SV, Kuner T, Neher E, Rizzoli SO (2011) A small pool of vesicles maintains synaptic activity in vivo. Proc Natl Acad Sci U S A 108(41): 17177–17182
34. Harata N, Pyle JL, Aravanis AM, Mozhayeva M, Kavalali ET, Tsien RW (2001) Limited numbers of recycling vesicles in small CNS nerve terminals: implications for neural signaling and vesicular cycling. Trends Neurosci 24(11):637–643
35. Harata N, Ryan TA, Smith SJ, Buchanan J, Tsien RW (2001) Visualizing recycling synaptic vesicles in hippocampal neurons by FM 1-43 photoconversion. Proc Natl Acad Sci U S A 98(22):12748–12753
36. Henkel AW, Lubke J, Betz WJ (1996) FM1-43 dye ultrastructural localization in and release from frog motor nerve terminals. Proc Natl Acad Sci U S A 93(5):1918–1923
37. Marra V, Burden JJ, Thorpe JR, Smith IT, Smith SL, Hausser M, Branco T, Staras K (2012) A preferentially segregated recycling vesicle pool of limited size supports neurotransmission in native central synapses. Neuron 76(3):579–589
38. Pyle JL, Kavalali ET, Piedras-Renteria ES, Tsien RW (2000) Rapid reuse of readily releasable pool vesicles at hippocampal synapses. Neuron 28(1):221–231
39. Sara Y, Mozhayeva MG, Liu X, Kavalali ET (2002) Fast vesicle recycling supports neurotransmission during sustained stimulation at hippocampal synapses. J Neurosci 22(5): 1608–1617
40. Schikorski T, Stevens CF (2001) Morphological correlates of functionally defined synaptic vesicle populations. Nat Neurosci 4(4):391–395
41. Kay AR, Alfonso A, Alford S, Cline HT, Holgado AM, Sakmann B, Snitsarev VA, Stricker TP, Takahashi M, Wu LG (1999) Imaging synaptic activity in intact brain and slices with FM1-43 in C. elegans, lamprey, and rat. Neuron 24(4):809–817
42. Pyle JL, Kavalali ET, Choi S, Tsien RW (1999) Visualization of synaptic activity in hippocampal slices with FM1-43 enabled by fluorescence quenching. Neuron 24(4):803–808
43. Wu Y, Yeh FL, Mao F, Chapman ER (2009) Biophysical characterization of styryl dye-membrane interactions. Biophys J 97(1): 101–109
44. Zenisek D, Steyer JA, Feldman ME, Almers W (2002) A membrane marker leaves synaptic vesicles in milliseconds after exocytosis in retinal bipolar cells. Neuron 35(6):1085–1097
45. Kavalali ET, Klingauf J, Tsien RW (1999) Properties of fast endocytosis at hippocampal synapses. Philos Trans R Soc Lond B Biol Sci 354(1381):337–346
46. Klingauf J, Kavalali ET, Tsien RW (1998) Kinetics and regulation of fast endocytosis at hippocampal synapses. Nature 394(6693): 581–585
47. Richards DA, Guatimosim C, Betz WJ (2000) Two endocytic recycling routes selectively fill two vesicle pools in frog motor nerve terminals. Neuron 27(3):551–559
48. Richards DA, Guatimosim C, Rizzoli SO, Betz WJ (2003) Synaptic vesicle pools at the frog neuromuscular junction. Neuron 39(3): 529–541
49. Miesenbock G (2012) Synapto-pHluorins: genetically encoded reporters of synaptic transmission. Cold Spring Harb Protoc 2012(2): 213–217
50. Atluri PP, Ryan TA (2006) The kinetics of synaptic vesicle reacidification at hippocampal nerve terminals. J Neurosci 26(8):2313–2320
51. Granseth B, Odermatt B, Royle SJ, Lagnado L (2006) Clathrin-mediated endocytosis is the dominant mechanism of vesicle retrieval at hippocampal synapses. Neuron 51(6):773–786

Part VI

In Vivo Recordings

Chapter 22

Extracellular Recordings

Nicholas Graziane and Yan Dong

Abstract

The common approach to scientific research is to follow the theory of reductionism, which dissects complex scientific questions into basic components, thus limiting experimental variables and potential confounds in the experimental results. In electrophysiology, the reductionist's approach has successfully provided a plethora of groundbreaking findings especially through the use of ex vivo measurements whereby the experimenter can more easily manipulate the circuit and/or neuron. However, by taking this ex vivo approach, many additional findings may go undiscovered and the physiological relevance may be questioned as neurons or circuits are removed from their in vivo milieu. Therefore, in vivo electrophysiology is a useful tool for the beginning electrophysiologist to become acquainted with as it extends ex vivo findings into physiologically relevant discoveries.

Keywords Action potentials, Synaptic currents, Intrinsic currents, Calcium spikes, Afterhyperpolarization currents, Gap junctions, Glia, Ephaptic conduction, Electroencephalogram, Electrocorticogram, Local field potentials

1 Introduction

The common approach to scientific research is to follow the theory of reductionism, which dissects complex scientific questions into basic components, thus limiting experimental variables and potential confounds in the experimental results. In electrophysiology, the reductionist's approach has successfully provided a plethora of groundbreaking findings especially through the use of ex vivo measurements whereby the experimenter can more easily manipulate the circuit and/or neuron. However, by taking this ex vivo approach, many additional findings may go undiscovered and the physiological relevance may be questioned as neurons or circuits are removed from their in vivo milieu. Therefore, in vivo electrophysiology is a useful tool for the beginning electrophysiologist to become acquainted with as it extends ex vivo findings into physiologically relevant discoveries.

In this chapter, we answer key questions that are relevant to in vivo electrophysiological measurement such as what generates

Nicholas Graziane and Yan Dong, *Electrophysiological Analysis of Synaptic Transmission*, Neuromethods, vol. 112,
DOI 10.1007/978-1-4939-3274-0_22,

in vivo electrophysiological signals and how can these signals be measured? We also include a practical approach for in vivo electrophysiological measurements in anesthetized and freely moving animals. At the conclusion of this chapter, we expect the beginning electrophysiologist to be able to identify potential sources contributing to the electrical signals recorded as well as their possible location within the neuronal population. In addition, we anticipate that the practical guide will assist the beginning electrophysiologist as they begin setting up for in vivo measurements.

2 Electrical Theory Revisited

Extracellular fields recorded from neuronal tissue are generated by the flow of ions across a membrane occurring at spines, dendrites, somas, axons, axon terminals, and glia. There are multiple sources that generate extracellular fields including synaptic currents, intrinsic currents, calcium spikes, action potentials, afterhyperpolarization currents, gap junctions, glia, and ephaptic conduction.

2.1 Synaptic

Although a synaptic event would be difficult to detect as a local field potential, the overlap of the relatively slow synaptic potentials at multiple synaptic sites can lead to a measurable signal. These measurable signals are referred to as sinks or sources depending upon the flow of ions across the lipid membrane (see Chap. 2) and the measured signal size is influenced by the distance between the recording and the location of the sink or source. For example, when receptors at excitatory synapses are activated, positively charged ions flow down the electrical gradient toward the cytosol leaving negatively charged ions, which accumulate along the lipid membrane (an active current sink). The flow of positive ions intracellularly causes an opposing ionic flux of positive ions from the intracellular compartment to the extracellular compartment at adjacent regions (passive current source). This flow of ions in a pyramidal neuron creates a dipole, which contributes to a current decay of $1/r^2$. Inhibitory currents mediated by $GABA_A$ receptors can also contribute to the extracellular field in depolarized membranes. Normally, when a cell is at its resting membrane potential, which is near the Cl^- equilibrium potential, inhibitory fields are minimal. However, the hyperpolarizing effects of Cl^- in a depolarized membrane can generate substantial extracellular field currents [1–3].

2.2 Intrinsic Currents

Intrinsic currents contribute to the oscillation of membrane potentials and are generated by voltage-dependent hyperpolarization-activated cyclic nucleotide (HCN)-gated channels (I_h channels) or by T-type calcium channels (I_T channels). Both types of channels contribute to pacemaker activity in designated cell types via activation by synaptically induced voltage changes occurring at a particular frequency range [4]. Therefore, an extracellular field triggered by intrinsic currents is both frequency- and voltage-dependent. The contribution of intrinsic currents to the field potential comes from synchronous membrane potential fluctuations from a group of neurons, typically inhibitory interneurons.

2.3 Calcium Spikes

Dendritic calcium spikes are large (10–50 mV) and long lasting (10–100 ms), thus contributing significantly to the extracellular field [5, 6]. It would be assumed then that calcium spikes are easily distinguished in an in vivo measurement. However, calcium spikes are often generated by NMDAR-dependent EPSPs, which can interfere with the recorded calcium spike signal [7–9]. To differentiate between calcium spikes and EPSPs, calcium spikes can generate fields across the laminar boundaries of afferent inputs due to their active propagation within a cell [1]. Making calcium spikes easier to detect is when they occur independently of NMDAR-dependent EPSPs and are instead triggered by back propagating somatic action potentials [10].

2.4 Action Potentials

Action potentials are detected in extracellular recordings as "spikes" due to their short duration of activity (<2 ms) typically in layers with a high density of cell bodies and axons. With the synchronous firing of many neurons, these spikes can contribute to the high frequency components of the field potential, which can provide valuable information. For example, increases in spiking activity can be identified by the broadening of the frequency spectrum. Additionally, specific cell types can be distinguished based on the power distribution produced from the activated neurons [11–13]. As the power of the high frequency bands increase, the spike synchrony produced by a population of neurons increases, which causes concomitant increases in the lower-frequency range due to AHPs [14, 15].

2.5 After-hyperpolarizations

Calcium spikes and action potentials activate Ca^{2+} -activated K^{+} channels, eliciting afterhyperpolarization (AHP) potentials. AHPs contribute to the extracellular field due to their size and duration, which compare to the size and duration of synaptic events [16]. It is posited that synchronized AHPs contribute to long-lasting field potentials (0.5–2 s) that occur during unexpected stimuli or the initiation of movement [17, 18].

2.6 Gap Junctions

Gap junctions contribute indirectly to the extracellular field by coordinating the activity of a group of neurons via changes in membrane excitability. These changes in membrane excitability can potentially shift neurons to an excitable state inducing the activation of voltage-gated channels, which can be detected extracellularly.

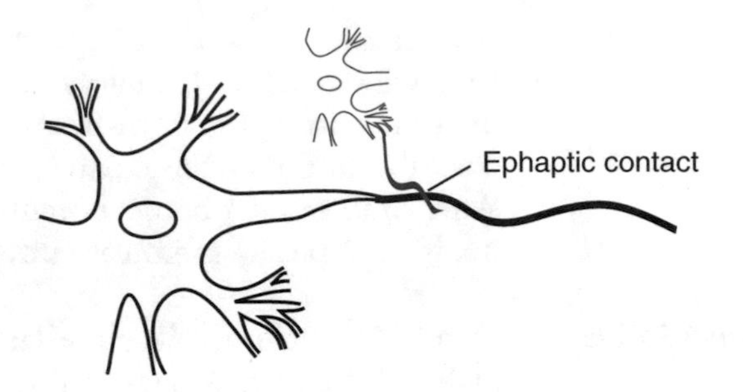

Fig. 1 Ephaptic contact between two dendrites on adjacent neurons

2.7 Glia

There is evidence suggesting that glia contribute to the extracellular field at low frequencies (<0.1 Hz) arising from glia, vascular events, or glia–neuron interactions [1, 19–22].

2.8 Ephaptic Conduction

A location where axons or dendrites from different neurons touch, but synaptic connections are not formed is called an ephapse (Fig. 1). Ephaptic connections have been known to impart ephaptic transmission or cross talk between neurons especially those with limited myelination [23]. This ephaptically mediated transfer of electrical potentials from one neuron to another occurs because of the conducting properties of the extracellular medium. However, due to the high resistivity of the conducting medium combined with the transient nature of generated spikes, detecting ephaptic effects are unlikely. This applies to single neurons, but given the simultaneous activation of a population of neurons, strong gradients in extracellular voltage are generated, which can generate detectable fields triggered by ephaptic coupling [24, 25].

3 Types of Recordings

3.1 Electro-encephalogram (EEG)

EEGs record general brain activity via electrodes that are placed on the scalp. In order for signals to be detected, neurons with similar spatial orientations must be simultaneously activated producing potential gradients. The scalp electrodes then detect these potential gradients over time and the difference between the potentials measured from the scalp electrodes and the reference electrode is calculated. Most EEG signals are derived from cortical neurons since they are aligned and simultaneously fire, creating large enough electrical gradients to pass from the cortex through the cerebrospinal fluid, pia mater, arachnoid mater, and skull. The low conductivity of bone attenuates the electrical signal making the spatial resolution of EEGs a limitation.

3.2 Electrocorticogram (ECoG)

ECoGs are an invasive recording procedure whereby the electrode grid used to measure neuronal activity is placed on the subdural layer on the cortical surface. Field potentials, typically generated by cortical pyramidal cells, conduct from the cerebral cortex through the cerebrospinal fluid, pia mater, and arachnoid mater before reaching the subdural recording electrodes. With the electrode grid placed under the skull, the temporal resolution is ~5 ms and the spatial resolution is 1 cm [26, 27].

3.3 Local Field Potential (LFP)

LFPs are recorded from a microelectrode placed in the brain such that electrical activity from a population of neurons can be sampled. This approach is similar to in vitro extracellular recordings previously discussed (see Chap. 2).

3.4 Whole Cell

In vivo whole cell recordings are often accomplished using a blind patch clamp approach. Using this approach, the neurons of interest cannot be seen. However, the pipette resistance can indicate whether the pipette is nearby or touching a cell. Once in contact with a cell, negative pressure can be applied to rupture the membrane allowing for whole-cell access.

In order to perform a whole cell patch blindly, the experimenter needs to apply a square voltage step while the micropipette is lowered into the brain. Much like in vitro approaches, positive pressure is applied to the micropipette to prevent debris from clogging the pipette tip as it is lowered through the tissue. As the micropipette approaches a cell, the square wave current is reduced indicating that the positive pressure should be replaced with suction. Blind patching is a technique with high success in regions of high cell density. However, in regions with low cell density such as the sensory cortex, successful whole-cell configurations are achieved <5 % of the time [28].

4 Recording in Anesthetized Animals

Anesthetized animal preparations are often used for invasive in vivo electrophysiological measurements allowing for target specificity of the recording electrode in the immobile animal. In using this preparation, electrophysiologists can expect to record from single neurons for up to 3 h and from a healthy anesthetized animal for up to 12 h [28]. The anesthesia agents typically used for this procedure include urethane, chloral hydrate, isofluorane, barbiturates, and ketamine.

The selection of anesthesia should be considered carefully depending upon the surgical procedures used as well as the electrophysiological measurements that need to take place. Adequate pain relief should be considered as well as the duration of animal immobility that is required. For example, urethane or chloral hydrate-induced anesthesia can last for more than 8 h, while ketamine lasts less than 1 h [28–30]. Additionally, the anesthetic can alter the electrophysiological measurement. If the experimenter wants to preserve slow oscillations (up/down states), urethane or chloral hydrate can be used. However, if slow oscillations or bursting activity are not a concern, isofluorane can be used with the advantage of quick induction and recovery as well as fine control of the anesthesia depth [31, 32].

Another important component to consider while performing electrophysiological experiments in an anesthetized animal is monitoring the vital signs and artificial ventilation throughout the recording. The blood pressure of the animal (80–140 mmHg) cannot fall below 80 mmHg as this can alter neuronal function. The blood pressure can be monitored using a heparin-saline filled catheter placed in the right carotid artery lying just inside the aortic arch on one end and connected to a transducer on the other end [28]. The pCO_2 levels are monitored throughout the recording and maintained at 3.8 % using artificial ventilation with oxygen-enriched air. This ensures that the pO_2, pCO_2, and the acid–base balance of the blood remain intact. Additionally, controlling the ventilation at a small tidal volume and high respiration rate reduces respiratory-induced movements and allows the use of muscle relaxants to reduce motor activity, thus increasing recording stability [28]. Lastly, the body temperature is monitored using a rectal thermometer and maintained at 37–38 °C using a heating pad beneath the abdomen [28].

In order to measure brain activity, the anesthetized animal is placed in a stereotaxic apparatus and an incision is made between the ears exposing the bregma and lambda (Fig. 2). The head is positioned so that bregma and lambda are on the same horizontal plane and into the brain region of interest is identified using ste-

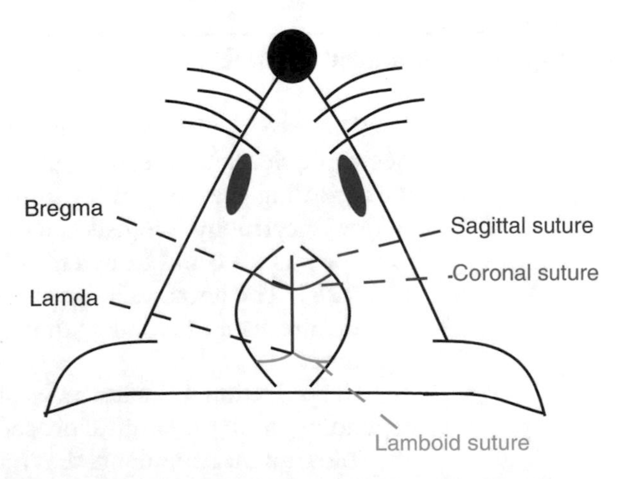

Fig. 2 Illustration of a rodent skull labeled with important landmarks useful for stereotaxic injections

reotaxic coordinates. Once located, a drill is used to drill a hole in the skull so that the micropipette can pass through until it reaches the intended target for measurements.

The micropipettes used for in vivo electrophysiology in anesthetized animals are fabricated from thin-walled borosilicate glass with resistances of 8–15 MΩ. They are pulled such that they have a long taper in order to reduce dimpling to the brain surface [28]. The micropipettes are filled with internal solution containing salts, pH buffers, Ca^{2+} chelators, and ATP, which is similar to what is used for in vitro preparations. The external solution perfusing the recorded area is artificial cerebral spinal fluid (aCSF) used in in vitro preparations.

Once the micropipette reaches the targeted brain region, field potentials as well as whole cell measurements can be obtained.

5 Recording in Freely Moving Animals

A limitation to recording from anesthetized animals is the concern of anesthetic effects on neuronal activity. In addition, the neuronal activity under anesthesia may mimic brain wave patterns during sleep and not necessarily reflect true activity that occurs in an awake and freely moving animal. To overcome these limitations, recording in freely moving animals has been made possible using dental acrylic to anchor the micropipette in place. This method, covered in detail by [33], is similar to the preparation used to record neurons in an anesthetized animal. However, an additional step is added in which the micropipette is anchored using dental acrylic. Once the acrylic hardens, the anesthesia is antagonized, allowing for the animal to quickly wake up and begin its behavioral task. With the micropipette firmly in place, electrophysiological recordings of the targeted region can be obtained in freely moving animals [33].

A second option is firmly fix the skull of an animal in place, while the rest of the animal is free to move. This is typically done using a spherical treadmill (a light ball floating on a thin layer of air). With the skull and micropipette firmly in place, the animal is free to run on the spherical treadmill. Using virtual reality, a visual scene that moves in response to the ball's rotation is presented to the animal. This allows animals to navigate through virtual mazes while the experimenter can record neuronal activity [34, 35].

6 Summary

This chapter discusses the multiple sources contributing to extracellular field potentials including synaptic currents, intrinsic currents, calcium spikes, action potentials, afterhyperpolarization currents, gap junctions, glia, and ephaptic conduction. Furthermore, we have discussed the strategies that are used to record extracellular field potentials such as EEG, EcOG, LFPs, and in vivo whole-cell recordings. Lastly, we have concluded with practical approaches describing extracellular in vivo electrophysiological recordings from anesthetized animals or from freely moving animals. By describing the concepts pertaining to in vivo electrophysiology accompanied with practical approaches, we hope to provide the beginning electrophysiologist with the necessary tools for proper data collection and interpretation of their in vivo recordings.

References

1. Buzsaki G, Anastassiou CA, Koch C (2012) The origin of extracellular fields and currents--EEG, ECoG, LFP and spikes. Nat Rev Neurosci 13(6):407–420
2. Glickfeld LL, Roberts JD, Somogyi P, Scanziani M (2009) Interneurons hyperpolarize pyramidal cells along their entire somatodendritic axis. Nat Neurosci 12(1):21–23
3. Trevelyan AJ (2009) The direct relationship between inhibitory currents and local field potentials. J Neurosci 29(48):15299–15307
4. Llinas RR (1988) The intrinsic electrophysiological properties of mammalian neurons: insights into central nervous system function. Science 242(4886):1654–1664
5. Schiller J, Schiller Y, Stuart G, Sakmann B (1997) Calcium action potentials restricted to distal apical dendrites of rat neocortical pyramidal neurons. J Physiol 505(Pt 3):605–616
6. Wong RK, Prince DA, Basbaum AI (1979) Intradendritic recordings from hippocampal neurons. Proc Natl Acad Sci U S A 76(2): 986–990
7. Hirsch JA, Alonso JM, Reid RC (1995) Visually evoked calcium action potentials in cat striate cortex. Nature 378(6557):612–616
8. Larkum ME, Nevian T, Sandler M, Polsky A, Schiller J (2009) Synaptic integration in tuft dendrites of layer 5 pyramidal neurons: a new unifying principle. Science 325(5941): 756–760
9. Schiller J, Major G, Koester HJ, Schiller Y (2000) NMDA spikes in basal dendrites of cortical pyramidal neurons. Nature 404(6775): 285–289
10. Stuart G, Spruston N, Häusser M (2007) Dendrites. Oxford University Press, Oxford
11. Belluscio MA, Mizuseki K, Schmidt R, Kempter R, Buzsaki G (2012) Cross-frequency phase-phase coupling between theta and gamma oscillations in the hippocampus. J Neurosci 32(2):423–435
12. Manning JR, Jacobs J, Fried I, Kahana MJ (2009) Broadband shifts in local field potential power spectra are correlated with single-neuron spiking in humans. J Neurosci 29(43):13613–13620
13. Miller KJ, Sorensen LB, Ojemann JG, den Nijs M (2009) Power-law scaling in the brain surface electric potential. PLoS Comput Biol 5(12):e1000609
14. Einevoll GT, Pettersen KH, Devor A, Ulbert I, Halgren E, Dale AM (2007) Laminar population analysis: estimating firing rates and evoked synaptic activity from multielectrode recordings in rat barrel cortex. J Neurophysiol 97(3):2174–2190
15. Zanos TP, Mineault PJ, Pack CC (2011) Removal of spurious correlations between spikes and local field potentials. J Neurophysiol 105(1):474–486
16. Buzsaki G, Bickford RG, Ponomareff G, Thal LJ, Mandel R, Gage FH (1988) Nucleus basalis and thalamic control of neocortical activity in the freely moving rat. J Neurosci 8(11):4007–4026
17. Kornhuber HH, Becker W, Taumer R, Hoehne O, Iwase K (1969) Cerebral potentials accompanying voluntary movements in man: readiness potential and reafferent potentials. Electroencephalogr Clin Neurophysiol 26(4):439
18. Walter WG, Cooper R, Aldridge VJ, McCallum WC, Winter AL (1964) Contingent negative variation: an electric sign of sensorimotor association and expectancy in the human brain. Nature 203:380–384
19. Hughes SW, Lorincz ML, Parri HR, Crunelli V (2011) Infraslow (<0.1 Hz) oscillations in

thalamic relay nuclei basic mechanisms and significance to health and disease states. Prog Brain Res 193:145–162

20. Kang J, Jiang L, Goldman SA, Nedergaard M (1998) Astrocyte-mediated potentiation of inhibitory synaptic transmission. Nat Neurosci 1(8):683–692
21. Poskanzer KE, Yuste R (2011) Astrocytic regulation of cortical UP states. Proc Natl Acad Sci U S A 108(45):18453–18458
22. Vanhatalo S, Palva JM, Holmes MD, Miller JW, Voipio J, Kaila K (2004) Infraslow oscillations modulate excitability and interictal epileptic activity in the human cortex during sleep. Proc Natl Acad Sci U S A 101(14):5053–5057
23. Rasminsky M (1980) Ephaptic transmission between single nerve fibres in the spinal nerve roots of dystrophic mice. J Physiol 305: 151–169
24. Jefferys JG (1995) Nonsynaptic modulation of neuronal activity in the brain: electric currents and extracellular ions. Physiol Rev 75(4): 689–723
25. McCormick DA, Contreras D (2001) On the cellular and network bases of epileptic seizures. Annu Rev Physiol 63:815–846
26. Asano E, Juhasz C, Shah A, Muzik O, Chugani DC, Shah J, Sood S, Chugani HT (2005) Origin and propagation of epileptic spasms delineated on electrocorticography. Epilepsia 46(7):1086–1097
27. Hashiguchi K, Morioka T, Yoshida F, Miyagi Y, Nagata S, Sakata A, Sasaki T (2007) Correlation between scalp-recorded electroencephalographic and electrocorticographic activities during ictal period. Seizure 16(3): 238–247
28. Furue H, Katafuchi T, Yoshimura M (2007) In vivo patch-clamp technique. In: Walz W (ed) Patch-clamp analysis, vol 38. Humana Press, Totowa, NJ, pp 229–251
29. Brunson DB (2008) Chapter 3 - Pharmacology of inhalation anesthetics. In: Fish RE, Brown MJ, Danneman PJ, Karas AZ (eds) Anesthesia and analgesia in laboratory animals, 2nd edn. Academic, San Diego, CA, pp 83–95
30. Meyer RE, Fish RE (2008) Chapter 2 - Pharmacology of injectable anesthetics, sedatives, and tranquilizers. In: Fish RE, Brown MJ, Danneman PJ, Karas AZ (eds) Anesthesia and analgesia in laboratory animals, 2nd edn. Academic, San Diego, CA, pp 27–82
31. Ferron JF, Kroeger D, Chever O, Amzica F (2009) Cortical inhibition during burst suppression induced with isoflurane anesthesia. J Neurosci 29(31):9850–9860
32. Kroeger D, Amzica F (2007) Hypersensitivity of the anesthesia-induced comatose brain. J Neurosci 27(39):10597–10607
33. Lee AK, Epsztein J, Brecht M (2009) Head-anchored whole-cell recordings in freely moving rats. Nat Protoc 4(3):385–392
34. Harvey CD, Collman F, Dombeck DA, Tank DW (2009) Intracellular dynamics of hippocampal place cells during virtual navigation. Nature 461(7266):941–946
35. Holscher C, Schnee A, Dahmen H, Setia L, Mallot HA (2005) Rats are able to navigate in virtual environments. J Exp Biol 208(Pt 3):561–569

ERRATUM

Electrophysiological Analysis of Synaptic Transmission

Nicholas Graziane and Yan Dong

Nicholas Graziane and Yan Dong, *Electrophysiological Analysis of Synaptic Transmission*, Neuromethods, vol. 112, DOI 10.1007/978-1-4939-3274-0, © Springer Science+Business Media New York 2016

DOI 10.1007/978-1-4939-3274-0_23

The sequence of the author names was incorrect on the cover.

The correct sequence is:
Nicholas Graziane
Yan Dong

The online version of the updated original book can be found at
http://dx.doi.org/10.1007/978-1-4939-3274-0

Nicholas Graziane and Yan Dong, *Electrophysiological Analysis of Synaptic Transmission*, Neuromethods, vol. 112, DOI 10.1007/978-1-4939-3274-0_23, © Springer Science+Business Media New York 2016

Index

A

B

C

D

E

Nicholas Graziane and Yan Dong, *Electrophysiological Analysis of Synaptic Transmission*, Neuromethods, vol. 112, DOI 10.1007/978-1-4939-3274-0,

G

H

I

T

V

Printed in the United States
By Bookmasters